International School of Physical Acoustics

Third Course on

ULTRASONIC SIGNAL PROCESSING

Oct 20 — 30, 1988
Erice, Italy

Editor
A. Alippi

Coordination by
G. Scarano

World Scientific
Singapore • New Jersey • London • Hong Kong

Published by

World Scientific Publishing Co. Pte. Ltd.,
P O Box 128, Farrer Road, Singapore 9128
USA office: 687 Hartwell Street, Teaneck, NJ 07666
UK office: 73 Lynton Mead, Totteridge, London N20 8DH

Library of Congress Cataloging-in-Publication data is available.

THIRD COURSE ON ULTRASONIC SIGNAL PROCESSING

ISBN 9971-50-864-8

Printed in Singapore by JBW Printers & Binders Pte. Ltd.

The organization of the Course has been made possible through the contribution of

Consiglio Nazionale delle Ricerche

Elettronica San Giorgio-Elsag

Ente Nazionale per l'Energia Elettrica

and the administrative support of

Istituto di Acustica "O.M. Corbino"

which all we wish to acknowledge.

International School of Physical Acoustics
Third course on
ULTRASONIC SIGNAL PROCESSING
Erice, Italy, October 20-30, 1988

DIRECTORS

Prof. A. Alippi
Institute of Acoustics, CNR
Rome, ITALY

Prof. W.G. Mayer
Dept of Physics
Georgetown University
Washington DC, USA

LECTURERS

L. Adler
Ohio State University
Columbus OH, USA

B.A. Auld
Stanford University
Stanford CA, USA

J. Attal
USTL, Lab. Microacoustique
Montpellier, FRANCE

D. Fitting
Nat. Inst. Standards and Technology
Bolder CO, USA

J.J. Gagnepain
CNRS
Besançon, FRANCE

Y.V. Gulyaev
Institute of Radio Engin. and Electronics
Moscow, USSR

A. Korpel
University of Iowa
Iowa City IA, USA

G. Quentin
University Paris VII
Paris, FRANCE

R. Reibold
Physikalish-Technische Bundesanstalt
Braunschweig, FRG

E. Verona
Institute of Acoustics, CNR
Rome, ITALY

FOREWORD

The International School of Physical Acoustics, aims at bringing to the attention of the attendants and of all those that may be interested in going through its Proceedings the problems and progresses of different aspects of elastic wave principles and applications. This third course is devoted to Ultrasonic Signal Processing.

The topic faces many aspects of Physical acoustics, and thus allows several experiences from different neighbouring fields to be presented within one common frame. That really falls into the scope of the School: exchanging experiences and comparing results through the search of a common language.

Ultrasounds is a powerful tool for physics and technology: it allows one to test matter and sound structures not invasively. That is why different fields, as microscopy, medical diagnostics, underground prospecting, transducer technology and material sciences in general, they all reserve a large area to ultrasonic technology.

The central problem then is the analysis and the interpretation of ultrasonic signals. This is not simply signal processing technique, but ultrasonic signal processing: one needs to know what the ultrasonic wave behaviour is, which is the physics of the generation and propagation: the interaction of elastic waves with matter, dispersion, scattering, absorption effects.

We hope that the effort of bringing together outstanding scholars and attendants to present their results and to participate in debates will contribute in the process of defining a common language and approach to the problems.

The course, as the previous ones, is divided in tutorial sections and sections devoted to advanced research papers. In the Proceedings, the tutorial papers have been collected in Section 1, Invited papers, while the advanced research papers in Section 2, Contributed papers. This latter is divided in two parts: Field reconstruction and imaging, and Transducers and materials. Obviously enough, every classification is a partial truth and one will probably find that some papers participate of both natures.

To conclude, I wish to extend my deepest thanks to all those who have contributed to make this course possible and successful: the people of the Ettore Majorana Centre for Scientific Culture at Erice for their warm support and hospitality; the staff of the Corbino Institute of Acoustics in Rome for the organization, particularly, Lucia Covi for her ability to make things going easy and Annamaria Moretti for making numbers satisfy arithmetics.

Adriano Alippi

TABLE OF CONTENTS

I. INVITED PAPERS

GENERAL PRINCIPLES AND TECHNIQUES OF ACOUSTIC SIGNAL PROCESSING

Bert A. Auld

Laboratoire de Physique et Métrologie des Oscillateurs
associé à l'Université de Franche-Comté-Besançon
32, avenue de l'Observatoire - 25000 Besançon - France

ABSTRACT

This paper is intended as a general introduction to the methods of ultrasonic signal processing. It will give general overview of concepts and their physical implemention, as an introduction to the more detailed papers that follow. Attention will be focused on the tapped delay line as a basic element in the signal processing chain. Both linear and nonlinear taps will be considered, and a variety of applications (analog Fourier transformers, variable bandwidth filters, variable delay lines, convolvers and correlators, etc.) will be discussed. Implementation of this principles will be illustrated by surface acoustic wave delay lines, acousto-optic modulators and hybrid semiconductor-piezoelectric structures.

I - INTRODUCTION

The first signal processing applications of ultrasonic waves were as simple frequency filters and short term memories. An electroded piezoelectric plate has an electrical impedance, measured between the two electrodes, that exhibits resonant behavior at the various mechanical resonances of the plate. Because of the low velocity of ultrasonic waves these resonators are much smaller than microwave electromagnetic resonators, and even most lumped circuit resonators. Such resonators were first used as frequency control elements in radio transmitter ocillators. In these applications, the most commonly used resonance is a half-wavelength standing wave between the parallel faces of the plate. These resonators were later combined as resonant circuit elements in various filter networks for communications and other applications.

The slow velocity of ultrasonic waves also makes them highly suitable for use in short term memory applications. In this case, the basic element is an ultrasonic delay line. This consists of a propagation medium in the form of a bar or rod with an electroded piezoelectric plate bonded to each end. The piezoelectric resonators in this application function as electromechanical transducers, used to convert an electrical signal into an ultrasonic wave and vice versa. In a rod of length in the order of centimeters the ultrasonic wave experiences a delay in the order of microseconds. An electrical signal applied to the input terminal can therefore be temporarily stored in the device for this period of time before exiting from the output terminal. This signal processing element is often used to compare an incoming signal with another signal arriving at a later time. The delay (or storage) time need not be limited to a single transit time of the line. In a recirculating delay line the output signal is returned to the input, with an amplification to compensate for conversion and propagation losses, and propagated a number of times through the delay medium.

In the devices described above the ultrasonic wave is a plane acoustic wave (of either longitudinal or shear type). These so-called bulk wave signal processing components have, by their very nature, limitations that make them unsuitable for many of the sophisticated signal processing functions required in modern radar and other electronic systems. Basically, the problem is that a separate bulk wave device is required for each processing operation. For example, bulk wave communications filters require a separate piezoelectric plate for each resonant element in the equivalent circuit of the filter. A second example is the tapped delay line, in which an input signal is sampled at a sequence of later times. As will be seen in Section II, such processing elements have many applications in radar and communications. Clearly, a single bulk wave delay line is unsuitable here because of the virtual impossiblility of effective piezoelectric coupling to the bulk wave within the volume of the delay line. To obtain samples of the input signal at various time delays, a separate bulk wave delay line would be required for each time-delayed sample. It was the need to overcome this problem that led to the development of Rayleigh wave, or surface acoustic wave (SAW), ultrasonic signal processing technology.

The tapped delay line is probably the most fundamental and most widely used element in modern signal processing systems. It need not be a SAW delay line, although the low velocity of ultrasonic waves makes such devices very compact. Alternatives are charge-coupled semiconductor devices, or even optical fibers. Even when an ultrasonic wave is used to provide the delay, the taps need not be piezoelectric. Because the deformation in an ultrasonic wave modulates the optical index of the medium, it is possible to sample the information by scattering a light beam from the ultrasonic wave. By using a series of individual light beams spaced along the path of the ultrasonic wave, tapped delay line readouts can be realized. This can be done even with bulk waves because the light beam can easily traverse the interior of a (transparent) ultrasonic delay medium. But, as will seen later, the SAW technology still offers advantages. Another important variant of the basic tapped delay line structure is the use of nonlinear, rather than linear, taps. Such components, which use a variety of combinations of inputs, have been developed with both discrete and continuous distributions of taps.

In a series of lectures prepared for a school, it is fitting that the material be selected from one or more textbooks. Although a number of books now avai able treat various aspects of ultrasonic signal processing, two were judged to have the most appropriate selection of topics for this course. They are :

G. S. Kino, *Acoustic Waves*, Prentice-Hall, Englewood Cliffs, New Jersey, 1987.

David P. Morgan, *Surface-Wave Devices for Signal Processing*, Elsevier, Amsterdam, 1985.

The material presented here has been selected from these two texts, each of which has an extensive list of references to the current literature.

II. GENERAL PRINCIPLES

It was noted in the introduction that the tapped delay line, in its various forms, is the most basic element of a modern signal processing system. This section will develop the elementary properties of a linearly tapped delay line and illustrate some of its properties.

(1) Transversal Filter

Figure 1 illustrates the circuit connections for a tapped delay line used in signal processing. The input to the line is at the upper left and the combined outputs of the taps, summed electrically, appear at the lower right. In the frequency domain the transfer function of this device consists of a combination of the parallel paths through the different taps. At each tap the phase of the transmitted signal is the ultrasonic phase delay from the input to that tap, and is equal to the angular frequency multiplied by the propagation time delay over the same path. The frequency response is therefore

$$H_s(\omega) = \sum_{n=1}^{N} v_n \, exp(-jn\omega\tau) \tag{1}$$

where τ_s is the time delay between adjacent taps. It is clear from this description that the frequency response can be controlled by selecting the strengths of the interfering terms in Equation (1). The structure shown in the figure, and called a **transversal filter,** can therefore be used to design and fabricate a variety of bandpass filters.

From the form of Equation (1) it can be seen that the phase shifts at the different taps repeat periodically in frequency, so that the transfer function is a periodic function of frequency. In the time domain, the impulse response of the transversal filter has the form

$$h_s(t) = \sum_{n=1}^{N} v_n \delta(t - n\tau_s) \tag{2}$$

where the contribution from each tap appears shifted in time by the total delay from the input to that tap. This feature is illustrated in Figure 1, where a narrow pulse at the input produces a series of delayed pulses at the output, each with an amplitude corresponding to the tap weighting v_n. In this case the weights are all of unit amplitude, but differing signs. The contribution of the first tap exits first, and so on.

The form of Equation (2) and the illustration in Figure 1 show that the impulse response of the structure samples at discrete times, **and over a finite interval**, any time waveform introduced at the input terminal. That is, the output signal has the standard form of the convolution theorem

$$y(t) = \int_{-\infty}^{+\infty} u(t-\tau) h_s(\tau) d\tau \tag{3}$$

except that the integration range is restricted to the total time delay of the line. Design of a frequency bandpass filter in the form of a transversal filter must therefore take into account both the laws of discrete sampling and the finite sampling interval. Because of these restrictions, the design of a transversal frequency filter will be described below in terms of its time domain response.

Another important application of transversal filters is for generating coded wave forms such as the one illustrated in Figure 1. These waveforms have importance in radar or communications, where they will be seen to provide reduced peak power requirements or improved signal-to-noise ratio and secrecy. Figure 2 illustrates another type of transversal filter, where the spacing of the taps is nonuniform, and the tap weights **alternate in sign**. In this case the impulse response consists of closely-spaced impulses arriving first, grading slowly into more widely-spaced impulses arriving last. If the laws of sampling theory are respected in the design of the filter, the impulse response can be approximated by a so-called chirp waveform. For a transversal filter with linearly graded tap spacings and a weighting amplitude function W_0, this has the form

$$h_s(t) = a\, W_0(t)\, exp\, j\, (\omega_0 t - \mu t^2/2) \quad (4)$$

The second factor is a sinusoid whose frequency decreases linearly with time,

$$\omega(t) = (\omega_0 - \mu t/2) \quad (5)$$

This form of the impulse response arises from the fact that the high frequency components are transferred by the closely-spaced taps at the beginning of the line in Figure 2, and therefore arrive first. The low frequency components are transferred by the other end of the line, and arrive last.

(2) Matched Filter

In the coded waveform applications noted above an essential element in the system is a transversal filter that is "matched" to the coded waveform. For both the digital coded (Fig. 1) and chirp coded (Fig. 2) waveforms the matched filter is a similar transversal filter with the tap weighting pattern reversed along the length of the line. In Figure 1, this means that the matched filter has the two minus taps on the left, and the single minus tap on the right. Consequently, the single minus pulse which arrives first in the waveform is delayed the most by the matched filter. And the double minus pulses which arrive last are delayed least. The result of a detailed analysis shows that all of the pulses arrive at the matched filter output at the same time, and produce a large narrow "spike." An input wavefunction with a different code does not experience this pulse compression and gives a weak time-dispersed signal.

A similar result is obtained from a chirp coded wave form. If the transversal filter shown in Figure 2 is to be used as a matched filter, the corresponding coded waveform must have the high frequencies at the back of the waveform and the low frequencies at the front.

$$u(t) = exp\, j\,(\omega_o t + \mu t^2/2) \tag{6}$$

Here, the instantaneous frequency increases with time. The coded waveform is called an up-chirp in contrast with the down-chirp of the matched filter impulse response (Equation (4). The response of the matched filter to be coded waveform can be calculated by applying the convolution integral in Equation (3). When the weighting amplitude function is a constant, this gives

$$\begin{aligned} y(t) &= \alpha \int_{-T/2}^{+T/2} exp\, j\,(\omega_o(t-\tau) + \mu_o(t-\tau)^2/2)\, exp\, j\,(\omega_o\tau - \mu\tau^2/2)\, d\tau \\ &= \alpha\, exp\, j((\omega_o t + \mu t^2/2) \int_{-T/2}^{+T/2} exp\, j\,(-\mu t \tau)\, d\tau \\ &= \alpha T \frac{sin\,(\mu T t/2)}{\mu T t/2}\, exp\, j(w_o t + \mu t^2/2) \end{aligned} \tag{7}$$

In Equation (7) the envelope of the compressed pulse has the form of a sinc function. In this function T is the length of the chirp and one-half μT is the bandwidth of the chirp. Consequently, the width of the compressed pulse is determined by the time-bandwidth product of the matched filter.

$$N = BT \tag{8}$$

(3) Analog Fourier Transform

Figure 3 shows in block diagram form how a chirp filter (Fig. 2) can be used to generate the Fourier transform of an arbitrary input signal. In this case, the filter, unlike Figure 2, has the form of an up-chirp. The incoming signal is pre-multiplied by a down-chirp in a mixer, then passed through the filter, and finally postmultiplied by the same down-chirp. The result, as shown in the figure, is the Fourier transform of the input signal (as a function of the argument μt) imposed on a carrier of frequency $2\,\omega_0$. This is easily proved by following the same steps as in Equation (7).

A chirp transform processor (Fig. 3) delivers at its output (**as a function of time**) the Fourier transform of the input signal, provided the input signal is shorter than the overall delay time of the filter. Using this element as a base, a wide variety of signal processing functions can be realised: Variable-frequency filtering, variable time delay, spectrum analysis, convolution and correlation, etc. All of these operations involve first converting an input signal into its Fourier transform. Since this transform is presented as a function of time, rather than frequency, all operations that are normally carried out in the frequency domain are now carried out in the time domain. For example, Figure 4 shows the block diagram of a variable-frequency filter realised by using chirp transform filters. The first filter transforms to the Fourier transform domain, where the "frequency" filtering is realised by modulating as a function of time. For example, a bandpass filter can be obtained by gating a portion of the Fourier transform signal and stopping the rest. The second chirp transform filter transforms back to the time domain, where the operation appears as frequency filtering. Obviously, the time modulation of the Fourier transformed signal can be easily changed electronically, giving a variable-frequency filter.

Other variants of the chirp transform filter in Figure 3 exist and many applications are described in the two texts cited in the Introduction.

III. SURFACE ACOUSTIC WAVE PROCESSORS

Figure 5 illustrates a typical SAW delay line structure. By contrast with the bulk acoustic wave delay lines described in the introduction, this structure is planar. The transducers at the left and the right are arrays of interdigital fingers that generate a periodic spatial variation of electric field in a piezoelectric substrate. At a frequency where the wavelength of the Rayleigh wave equals the period of the transducer, the transducer is resonant and excites the Rayleigh wave with maximum efficiency.

Because of its planar nature a SAW delay line is perfectly adapted to the realization of compact transversal filters. A tap is easily implemented by placing a pair of electrode fingers in the region between the input and output transducers of Figure 5 and connecting the electrodes to an electrical detector. In a transversal filter the output transducer is eliminated from Figure 5 and an array of two-finger taps is placed in the path of the Rayleigh wave. The electrical outputs of the taps are then connected together electrically, as in the summation elements of Figures 1 and 2. As an illustration, Figure 6 shows a SAW version of the chirp filter of Figure 2. Since the frequency bandwidth of an interdigital transducer varies inversely with the number of finger pairs, the input transducer of a chirp filter is made with a small number of fingers, as shown, to increase the time-bandwidth product. The structure shown in the figure is a down-chirp filter, which is a matched filter for an up-chirp waveform. Another type of transversal filter commonly used in SAW processors is illustrated in Figure 7. Here, the taps are realised as reflectors (etched grooves) rather than as transducers. The reflections from different parts of the array are superposed and picked up by the receiving transducer at the lower left of the figure. High frequency components of the input signal create constructively interfering reflections where the elements of the array are closely spaced. High frequency components are therefore reflected near the front end of the array, while low frequency components are reflected near the back. The reflective array structure is therefore a down -chirp filter, as in Figure 2.

Figure 8 illustrates a SAW transversal filter used for bandpass filtering. In this case the spacing of the taps is uniform, but the weighting is nonuniform. The figure illustrates one method of weighting, by varying the overlap of adjacent interdigital fingers. In this way, different taps intercept different fractions of the wide surface wave beam excited by the input transducer on the left. Design of the weighting pattern is best carried out in the time domain. As was noted in the introduction, the tap distribution of a transversal filter respresents a sampled version of the impulse response and can be directly designed to give a required frequency response. The problem is that the transversal filter can realize the required impulse response only over a finite period of time. An evaluation of the effect of this truncation is given in the textbooks cited.

The advantage of chirp transform filters in signal processing is that they permit direct analog processing of signals in the Fourier transform plane. A block diagram is given in Figure 9 for a SAW variable delay line. Its operation is based on the Fourier transform relations

$$f(t) \leftrightarrow F(\omega)$$

$$f(t+T) \leftrightarrow e^{(j\omega T)} F(\omega) \tag{9}$$

which show that multiplication of the Fourier transform by a phase factor produces a time shift in the time domain. This operation is carried out at the right of the figure, between the two chirp filters. At the output of the left-hand chirp filter appears the Fourier transform of the input signal at the left. This transform waveform appears as a function of time, as in Figure 3. At this point ω on the right side of Equation (9) is in reality a time variable, so that the quantity T corresponds to the **frequency** of the voltage controlled oscillator in Figure 9. Consequently, after a retransformation by the second chirp filter, the output is a time shifted version of the input signal, with the time shift directly proportional to the frequency of the voltage controlled oscillator.

Figure 10 illustrates another application of SAW technology to frequency filtering. This is a planar version of the simple bulk wave standing wave resonance in the thickness of a piezoelectric plate. In the SAW version the reflections occur at arrays of reflecting grooves, which are efficient reflectors in a band about the frequency where the groove spacing is one-half wavelength. This technology permits easier fabrication at high-frequencies and realisation of coupled resonator filters on the same crystal substrate.

IV. ACOUSTO-OPTIC PROCESSING

As this topic will be treated in detail in the paper by Korpel, only two simple examples will be given.

The acousto-optic interaction was described in the introduction as a method for continously tapping an ultrasonic beam. Figure 11 shows how this interaction can be realised in a planar geometry, between a surface acoustic wave and a guided optical beam. The optical beam samples the acoustic beam continuously, and displays the tap outputs as a function of the spatial corrdinate. These can then be combined as in the summation element of Figure 2 by, for example, using a lens to focus the acousto-optically scattered light beam onto a photodetector.

Another illustration of acousto-optic processing is shown in Figure 12, where the object is to perform the correlation of two signals 1 and 2. At the left, signal 1 modulates a light source, whose output is collimated and passed through an ultrsonic delay line excited at the lower end by signal 2. Using the Schlieren technique the image of the output of this acousto-optic modulator is projected onto an array of integrating photodetectors. Because the intensity modulation due to signal 2 is produced by an ultrasonic delay line, the modulation function is shifted as a function of x. In the figure V is the propagation velocity of the acoustic wave. When the product of the two intensities is time-integrated in the detector array, one obtains a spatial display of the correlation function.

V. OTHER PROCESSING TECHNIQUES

In many signal processing systems the operation of convolution is required. One approach to this problem is based on the chirp transform filter of Figure 3. Since a convolution in the time domain becomes a multiplication in the frequency domain , the procedure is straightforward. Chirp transform filters are used to obtain the Fourier transforms of the two functions in question. These transforms are multiplied in a mixer and the product is then converted to a convolution in the time domain by means of another chirp transform filter.

A more elegant and flexible approach is to use a tapped delay line with nonlinear taps and to input the two signals to be convolved into opposite ends of this line. Figure 13 compares this device with the linear transversal filter structure of Figures 1 and 2. In (a) the output is a sampled version of the convolution of the input signal with the tap weighting function. The nonlinear device in (b), by contrast, gives the output shown in the figure when the taps are all equally weighted. This can be converted to the form of a sampled convolution by defining the variable of the function f in the figure as τ_n.

SAW devices similar to Figure 13 (b) have been developed for special applications, but a more common approach is to use a continuously distributed nonlinearity. One example is the so-called acoustic convolver illustrated in Figure 14. In this device the nonlinearity (a quadratic dependence of the electric polarization, in a piezoelectric crystal, on the elastic strain) will be discussed in the paper by Gagnepain, together with the physical effect used for acousto-optic sampling. The two counter-propagating acoustic waves generate a polarization that is a continuous analog of the function in Figure 13 (b). The polarization is integrated by detecting the current flowing from the strip electrode shown in the figure, giving the convolved output. Since the two input frequencies are the same in this case, the output is at twice that frequency. The figure also shows a track-changer structure for reducing the width of the input surface wave beams. This increases the intensity of the two beams, and thereby the efficiency of the nonlinear mixing.

Other nonlinear convolvers use nonlinear mixing in a semiconductor or an array of semiconductor diodes placed in the piezoelectric fields of the surface waves, above the substrate surface. Variants of this device also have the capability of storing the reference signal against which the incoming signal is to be convolved. Special procedures are also used to obtain a correlation operation, rather than a convolution. These and other details are given in the two textbooks.

VI. SUMMARY

The aim of this paper has been to present some of the general features of analog signal processing using ultrasonic waves, with emphasis on linear surface acoustic wave techniques. A brief introduction to acousto-optic processing has been given, as an introduction to other papers in this volume concerning applications of this technique to the processing of both time waveforms and acoustic field distributions. The physical basis for the nonlinear processors discussed here will also be elaborated in one of the following papers.

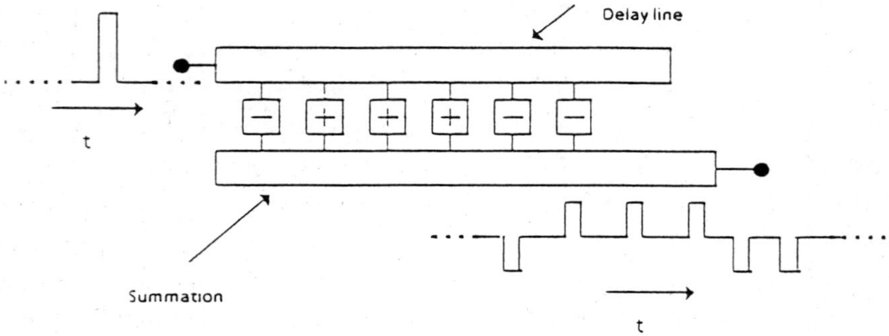

Figure 1 Transversal for Filter Digital Code Generation

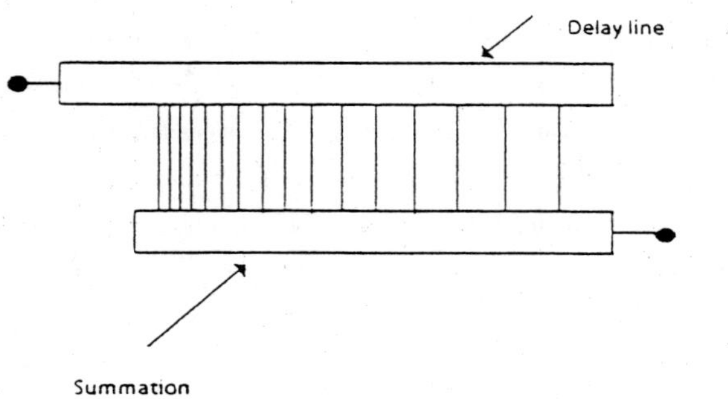

Figure 2 Transversal Filter for Chirp Waveform Generation

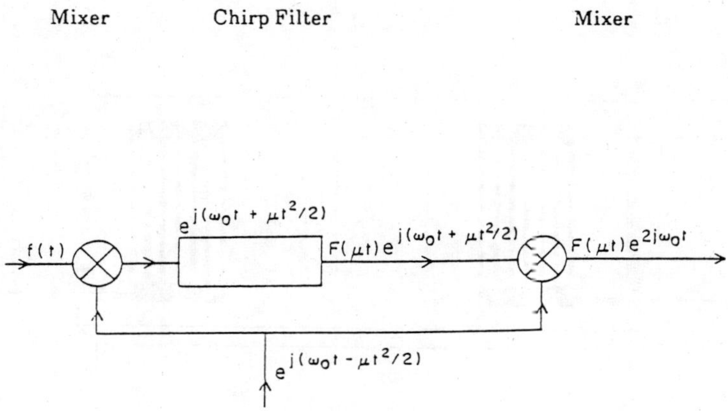

Figure 3 Basic Fourier Transform Operation with a Chirp Filter (After Kino)

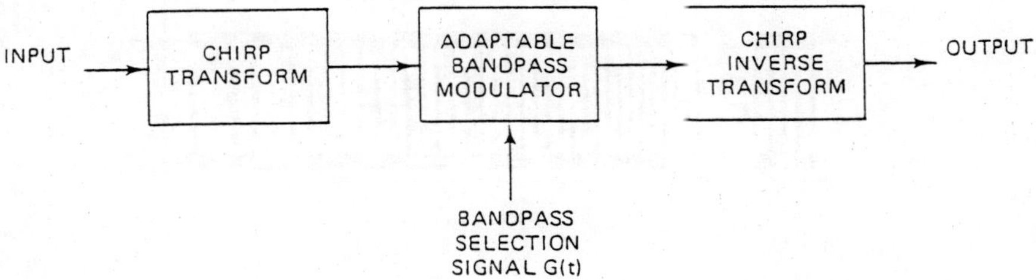

Figure 4 Variable Frequency Filter Based on Chirp Transform Operations (After Kino)

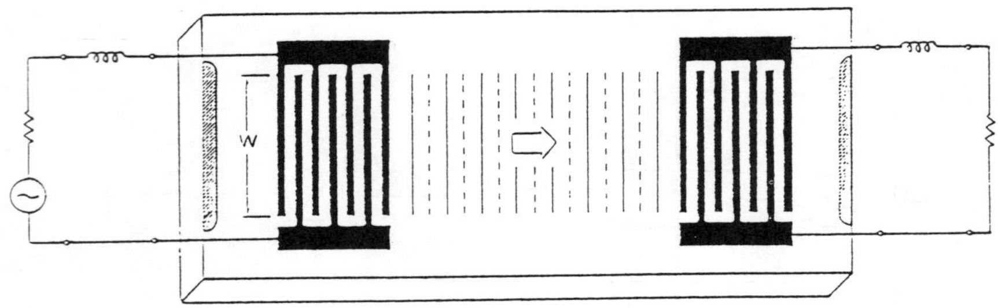

Figure 5 SAW Delay Line (After Morgan)

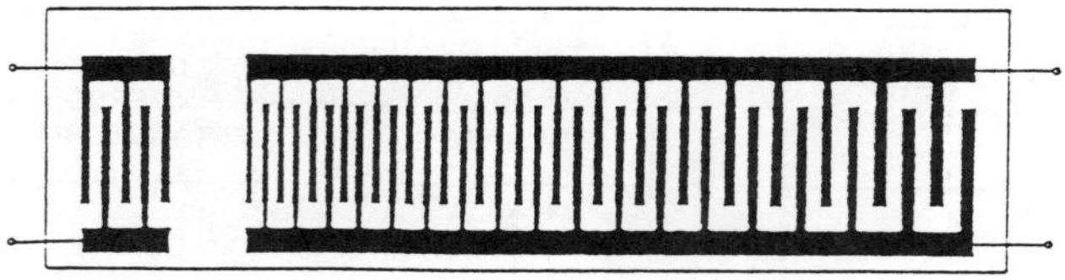

Figure 6 SAW Tapped Delay Line Corresponding to Figure 2 (After Morgan)

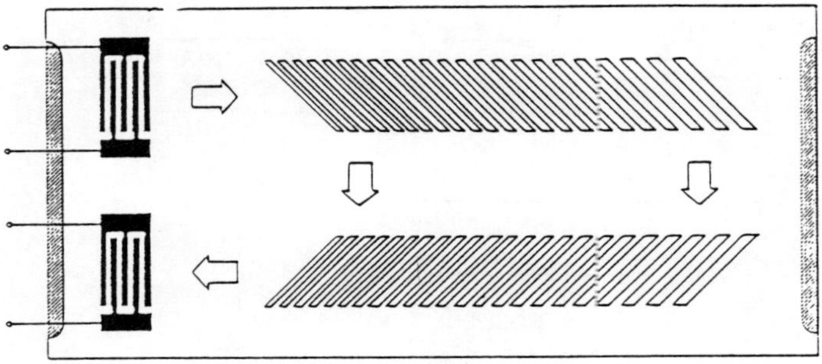

Figure 7 Reflective Array Matched Filter (After Morgan)

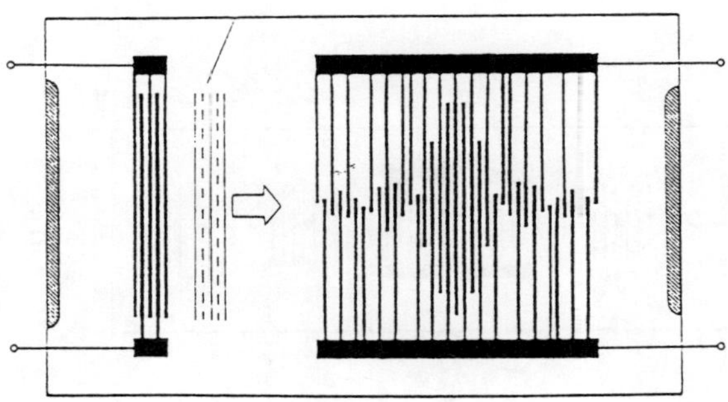

Figure 8 S A W Traversal Filter for Bandpass Filtering (After Morgan)

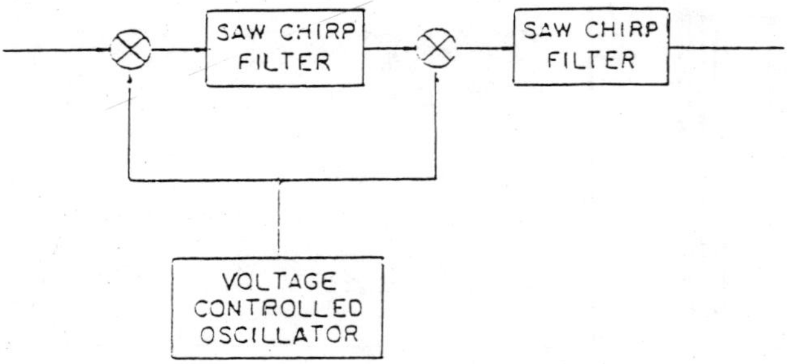

Figure 9 Variable Time Delay Filter (After Kino)

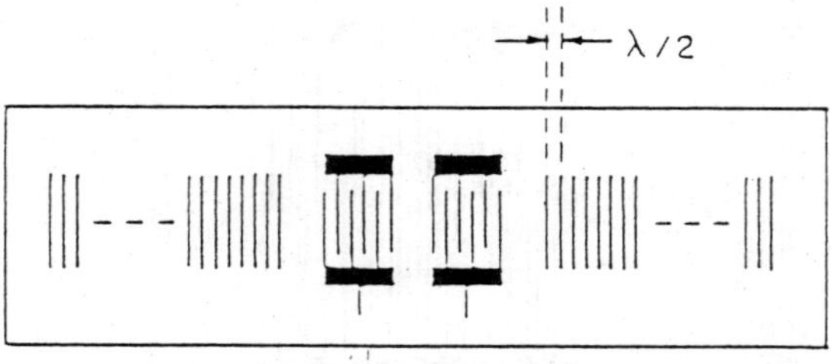

Figure 10 Resonator Filters (After Kino)

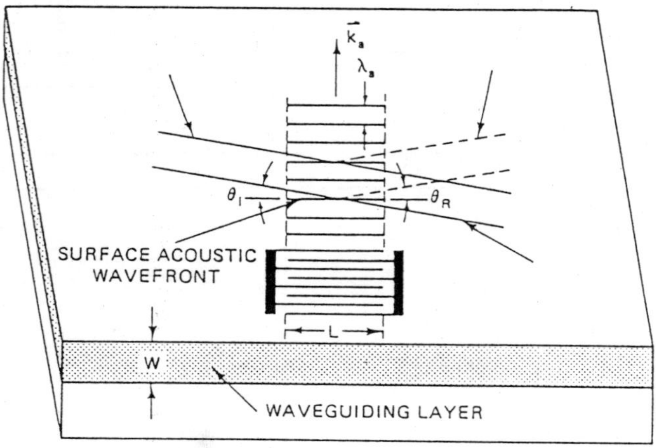

Figure 11 Acousto-optic Tapped Delay Line (After Kino)

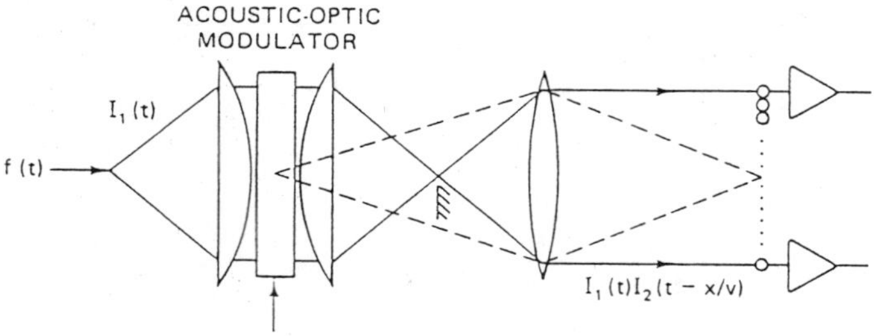

Figure 12 Time-integrating Correlator (After Kino)

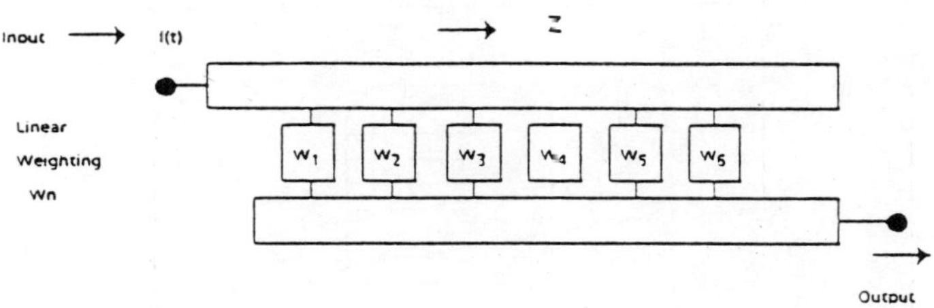

$$\text{Output} \sim \sum_n f\left(t - \frac{z_n}{V}\right) W_n$$

$$W_n = W(z_n)$$

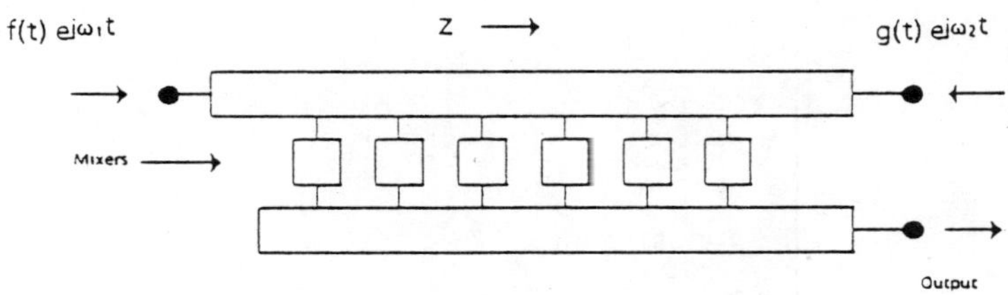

$$\text{Output} \sim \sum_n f\left(t - \frac{z_n}{V}\right) g\left(t - \frac{L - z_n}{V}\right) e^{j\omega_3 t} e^{j(\omega_1 - \omega_2)\frac{z_n}{V}}$$

$$\omega_3 = \omega_1 + \omega_2$$

(b)

Figure 13 Comparison of (a) Transversal Filter anc (b) Non Linear Convolver

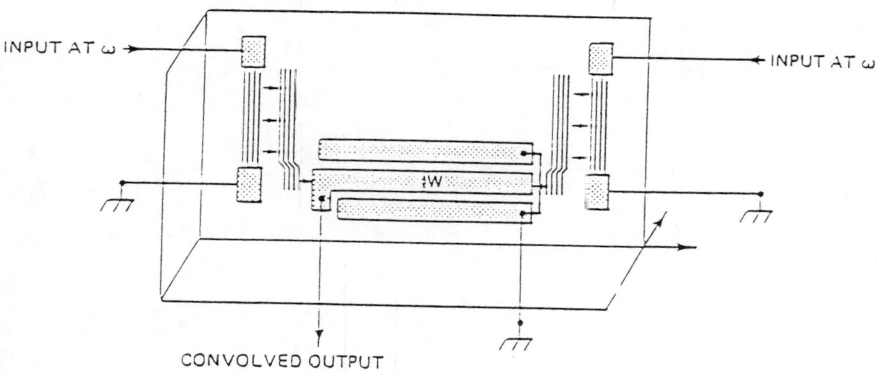

Figure 14 Waveguide Convolver (After Kino)

NONLINEARITIES IN ULTRASONIC SIGNAL PROCESSING

Jean-Jacques GAGNEPAIN

Laboratoire de Physique et Métrologie des Oscillateurs
associé à l'Université de Franche-Comté-Besançon
32, avenue de l'Observatoire - 25000 Besançon - France

ABSTRACT

After a brief recall of the microscopic models of chains of atoms where it is shown that nonlinearities are at the origin of some of the basic properties of solids, the general equation of elasticity are presented and the material constants are discussed for some of the most usual piezoelectric crystals. Two cases are considered in devices used for signal processing : nonlinearities are spurious effects (in resonators, filters, etc.) which must be minimized, or they are optimized for achieving convolution products or for developping sensors.

I - INTRODUCTION

All the physical systems experience, at some degree, nonlinearities. The observation of such a behaviour is essentially a question of the powers which are involved. The origin of the nonlinearities is in the interatomic forces, and their nonlinear potential. The microscopic theory of the vibration of a crystal lattice leads to the concept of phonons. In the harmonic approximation the different vibration modes of the lattice are independent and they cannot exchange energy. This enables to introduce the concept of specific heat. But due to the lattice anharmonicity phonon exchanges take place between the modes. These interactions are at the origin of fundamental properties like acoustic attenuation, thermal expansion, finite heat conductivity. In the macroscopic approach the induced nonlinearities (nonlinearities of deformation) appear in the equilibrium and constitutive equations. In the energy the intrinsic nonlinearities are characterized by the higher order fundamental constants, in relation with the elastic, dielectric, piezoelectric, ... properties of the medium.

The nonlinearities are the cause of harmonic generation, amplitude-velocity effect, intermodulation, ..., and also they are responsible of the sensitivity of acoustic waves to the environment, i.e. to temperature, forces, accelerations, electric fields, etc.

In acoustic devices used for signal processing, the nonlinear effects can be considered either as unwanted spurious effects or inversely as useful effects. For instance in resonators and filters nonlinearities must be reduced for a better frequency selection and a more precise control of the transfert function. The same comment could apply to transducers. At the opposite the various nonlinearities are advantageously used for analog convolvers or correlators, for parametric underwater wave generation, for acousto-optic interaction, for sensors.

II - MICROSCOPIC MODEL (a brief recall)

The simplest model is a one-dimension chain of N identical atoms with mass m. and mean distance a between two adjacent atoms.

Let u_n be the displacement of the nth atom with respect to its equilibrium position. Considering only interaction between the nearest atoms, with force F, in the linear approximation the dynamical equation reads

$$m\,\ddot{u}_n = F\,(u_{n+1} - 2u_n + u_{n-1}) \tag{1}$$

and the corresponding interaction energy is

$$H_2 = \frac{1}{2} \sum_n F\,(u_{n+1} - u_n)^2 \tag{2}$$

Considering a travelling wave with frequency ω and wave number k, the dispersion relation obtained from equation (1) is

$$\omega_k^2 = \frac{4F}{m}\, sin^2 \frac{ka}{2} \tag{3}$$

In term of normal coordinates Q_k the solution u_n can be written

$$u_n = \frac{1}{\sqrt{Nm}} \sum_k Q_k(t)\, e^{jkna} \tag{4}$$

and according to quantum mechanics, the total mean energy of the lattice is

$$<E> = \sum_{h} \hbar\, \omega_k\, [n_k + \frac{1}{2}] \tag{5}$$

where $\hbar\omega_k$ is the energy of one phonon.

The summation is performed over all normal modes of wave number k and frequency ω_k. n_k is the mean number of phonons stored in mode k.

The specific heat C is given by the mean energy temperature derivative. The mean number of phonons n_k follows the Bose-Einstein distribution

$$n_k = [e^{\hbar\omega_k/k_B T} - 1]^{-1} \tag{6}$$

where k_B is the Boltzmann constant and T the absolute temperature. Therefore

$$C = \frac{\hbar^2}{V k_B T^2} \sum_{h} \omega_k^2\, n_k (n_k + 1) \tag{7}$$

V is the total volume of the crystal.

To understand the other properties of the solid it is necessary to go beyond the harmonic approximation and to consider higher order terms in the atomic displacements. As a consequence of these anharmonicities are the interactions between phonons. Details can be found in the works of Klemens[1], Maradudin[2], and Maris[3].

In the Landau-Rumer theory[4] these interactions are represented by means of higher order Hamiltonian composed of products of creation and annihilation operators. The interactions must verify the energy conservation law. According to the involved process the quasi-momentum is conserved (N processes) or not (Umklapp processes). A consequence of the interactions is the finite life time ι of the phonons. The sound wave is considered as a beam of low energy phonons with frequency Ω, their number being largely in excess with respect to the equilibrium number. On account of the uncertainty principle, the finite life time of a phonon makes its energy uncertain by $\hbar/\iota$. Therefore the Landau-Rumer method should be valid when $\hbar/\iota \ll \hbar\Omega$, or $\Omega\iota \gg 1$. This corresponds to high frequency sound wave or to thermal phonons with large lifetimes, i.e. when the temperature is low.

A semi-classical approach due to Akhieser[5] is valid in the limit $\hbar\Omega \ll k_B T$, when the phonon mean free path is smaller than the sound wave length. This condition is verified at high temperature. In this theory the sound wave is treated macroscopically and the anharmonicities are introduced by stating that the frequency ω_k of a phonon depends upon the state of strains of the crystal.

The acoustic attenuation of a sound wave propagating in a solid is due to the exchange of its energy with the heat reservoir constituted by the cloud of thermal phonons. In average this exchange will correspond to a loss of the energy of the sound wave, given by the attenuation factor

$$\alpha = \frac{CT <\gamma^2_{eff}>}{2\rho s_o^3} \frac{\Omega^2 \tau}{1 + \Omega^2 \tau^2} \tag{8}$$

where $<\gamma^2{}_{eff}>$ is an effective Grüneisen constant[3], s_o is the sound wave mean phase velocity and ρ the crystal specific mass.

Heat diffusion also is due to a redistribution of phonons between the different modes of the lattice.The heat flow in a given direction is

$$q_i = \frac{1}{V} \sum_k N_k \hbar \omega_k \upsilon_i(k) \tag{9}$$

where $\upsilon_i(k)$ is the phonon group velocity. The difference of the number of phonons with respect to equilibrium $\Delta N_k = N_k - n_k$ is calculated with the Boltzmann transport equation and it is shown that the thermal conductivity coefficient takes the simpler form

$$\kappa_{ij} = C <\tau \upsilon_i \upsilon_j> \tag{10}$$

Thermal expansion has its origin in the same phenomena. The free energy A of the crystal is composed of the elastic deformation energy and the thermal phonon energy. The strain derivative of the free energy, at constant temperature, gives the thermodynamic tension and the thermal expansion coefficient $\alpha_{\gamma\delta}$ is obtained by derivating them with respect to temperature at constant tension.[2]

Thus

$$c_{\alpha\beta\gamma\delta}\, \alpha_{\gamma\delta} = - \frac{\hbar^2}{V k_B T^2} \sum_h \omega_{\alpha\beta}(k)\, \omega_k\, n_k (n_k + 1) \tag{11}$$

where

$$\omega_{\alpha\beta}(k) = \frac{\partial \omega_k}{\partial \eta_{\alpha\beta}} \tag{12}$$

$\eta_{\alpha\beta}$ represents the mechanical strain.

These results show that acoustic thermal conductivity, thermal expansion, attenuation and related velocity (or frequency) shifts all have their origin in the crystal lattice anharmonicities. The microscopic theory enables to relate these phenomena, which generally are considered as independent in the linearized macroscopic approach. Connection can be made between microscopic and macroscopic descriptions and for instance the Grüneisen constant was related to the third order elastic constants by Brügger[6].

III - MACROSCOPIC EQUATIONS

Basic and details of the general theory of elastic and electro-elastic solids can be found in the works of Thurston[7] and Tiersten[8]. The general equations and useful results will be summarized in this Section.

1) Initial and final states

Let a_j be the coordinates of a material point M of the solid at rest, corresponding to the initial state. After deformation the material point undergoes a displacement and takes a new position M' of coordinates x_j, denoting the final state (Fig. 1)

The components of the mechanical displacement are

$$u_j = x_j - a_j \tag{13}$$

and the instantaneous velocity of a particule is

$v_j = dx_j/dt = x_j$

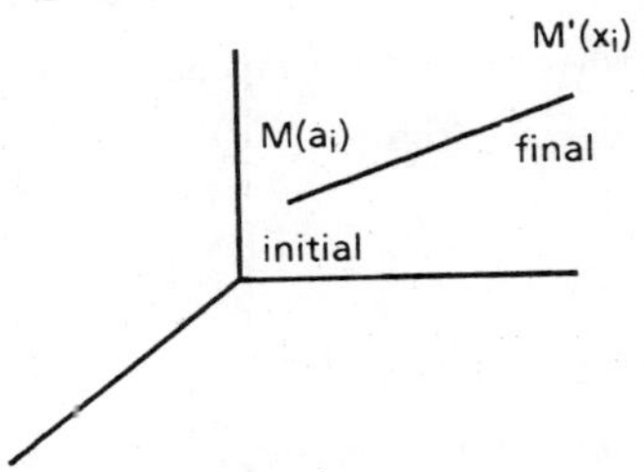

Fig. 1

2) Strains

A length element $d\ell_0$ of the solid at rest becomes $d\ell$ after deformation and

$$d\ell_o = da_i \, da_i \qquad d\ell = dx_j \, dx_j \tag{14}$$

The strain tensor η_{ij} is defined by

$$d\ell^2 - d\ell_o^2 = 2\,\eta_{ij}\, da_i\, da_j \tag{15}$$

therefore

$$\eta_{ij} = \frac{1}{2}\left(\frac{\partial x_k}{\partial a_i}\frac{\partial x_k}{\partial a_j} - \delta_{ij}\right) \tag{16}$$

δ_{ij} is the Kronecker symbol. By using (13) into (16) the strain tensor is given as a function of the displacements

$$\boxed{\eta_{ij} = \frac{1}{2}\left(\frac{\partial u_j}{\partial a_i} + \frac{\partial u_i}{\partial a_j} + \frac{\partial u_k}{\partial a_i}\frac{\partial u_k}{\partial a_j}\right)} \tag{17}$$

In this relation the strain tensor is given with respect to the initial state coordinates a_i. Similar relation could be given with respect to the final state. Relation (17) shows that the strain tensor is composed of a linear part, which corresponds to the usual definition in the approximation of infinitesimal deformations, and of a quadratic term which is nonlinear.

3) Stresses

Considering in the usual manner an elementary tetrahedron submitted to external surface forces Q_j, the induced stresses T_{ij} are given by

$$\boxed{Q_j = n_i\, T_{ij}} \tag{18}$$

where n_i is the component of the unit normal of the surface element, the surface forces Q_j are acting on.

4) Electrical forces

Following Tiersten[8], the electroelastic solid is composed of an electronic charge continuum coupled to the elastic continuum. The lattice continuum is assumed to have a positive charge density, and the electronic continuum a negative one. Relative motions of the two continua create electric dipoles. Each elementary volume element is submitted to an electric body force G_j and a body couple C following

$$G_j = P_i \frac{\partial E_j}{\partial x_i}$$

$$C = E_j \wedge P_j \tag{19}$$

E_j is the quasistatic Maxwell electric field and P_i the dipole density. The Maxwell electrostatic stress tensor T_{ij}^E is

$$T_{ij}^E = \varepsilon_o E_i E_j + P_i E_j - \frac{1}{2} \varepsilon_o E_k E_k \delta_{ij} \tag{20}$$

ε_0 being the permittivity of the vacuum. The electric displacement D_i is

$$D_i = \varepsilon_o E_i + P_i \tag{21}$$

5) Conservation Laws

5.1 - Conservation of the mass

Let ρ_0 and ρ be respectively the specific mass in the initial and final states. The mass dM of a volume element dV_o in the initial state, and dV in the final state is conserved, thus

$$\rho_o dV_o = \rho\, dV = dM \tag{22}$$

Introducing the Jacobian matrix J of the deformation

$$J = \lim_{dV_o \to 0} \frac{dV}{dV_o} = \left|\frac{\partial x_j}{\partial a_i}\right| \tag{23}$$

equation (22) is written

$$\rho J = J_o \tag{24}$$

5.2 - Conservation of the linear momentum

The resultant of the surface forces Q_j, the body forces F_j, and the electrical forces Gj equals the time derivative of the linear momentum

$$\int_s Q_j\, dS + \int_V (F_j + P_i \frac{\partial E_j}{\partial x_i})\, dV = \frac{d}{dt} \int_V \rho\, \upsilon_j\, dV \tag{25}$$

Using equation (18) in (25) and applying the Green Theorem, conservation of the linear momentum yields

$$\rho \upsilon_j = \frac{\partial}{\partial x_k} (T_{kj} + T^E_{kj}) + F_j \tag{26}$$

in the volume, and on the surface limiting the body in the case of a free surface

$$n_k (T_{kj} + T^E_{kj}) = 0 \tag{27}$$

5.3 - Conservation of the angular momentum

The sum of the forces couples acting on the solid equals the time derivative of its angular momentum

$$\int_s (x_j \wedge Q_j)\, dS + \int_V (x_j \wedge F_j + x_j \wedge P_i \frac{\partial E_j}{\partial x_i} + E_j \wedge P_j)\, dV = \frac{d}{dt} \int_V (x_j \wedge \rho \upsilon_j)\, dV \tag{28}$$

This conservation law, after some transformation, leads to

$$T_{ki} - T_{ik} = E_i P_k - E_k P_i \tag{29}$$

which indicates that the stress tensor is not symetrical in electroelastic solids. The symetry condition will be verified only in pure elastic solids (or in the linear approximation).

5.4 - Conservation of the energy

The rate of energy supplied to the solid equals the variation of its total energy.This last one is the sum of the internal energy per unit volume ρU and of the kinetic energy $(1/2)\rho\, v_k v_k$

$$\frac{d}{dt} \int_V \rho (\frac{1}{2} \upsilon_k \upsilon_h + U)\, dV = \int_S (Q_k \upsilon_k - n_k q_k)\, dS + \int_V F_k \upsilon_k\, dV + \int_V \sigma\, dV \tag{30}$$

$Q_k v_k$ and $F_k v_k$ correspond to the rate of working of the surface and body forces. $n_k q_k$ is the thermal energy flow and σ is related to the electric forces

$$\sigma = P_i \frac{\partial E_j}{\partial x_j} \upsilon_i + E_i \frac{dP_i}{dt} \tag{31}$$

The local form of equation (19) is obtained by applying the mass conservation law and Green theorem

$$\rho \frac{dU}{dt} = T_{ij} \frac{\partial v_j}{\partial x_i} + \rho \theta \frac{d\eta}{dt} + E_i \frac{dP_i}{dt} \tag{32}$$

where θ is the temperature and η the entropy per unit mass.

The energy conservation equation can be expressed also as a function of the electric field and the temperature, as variables rather than entropy and polarization, by introducing the enthalpy χ

$$\chi = U - \eta \theta - \frac{E_i P_i}{\rho} \tag{33}$$

This leads to

$$\boxed{\rho \frac{d\chi}{dt} = T_{ij} \frac{\partial a_m}{\partial x_i} \frac{d}{dt}\left(\frac{\partial x_j}{\partial a_m}\right) - P_i \frac{dE_i}{dt} - \rho \eta \frac{d\theta}{dt}} \tag{34}$$

Considering $\partial x_j / \partial a_m$, E_i, and θ as independent variables, i.e. $\chi = \chi\, (\partial x_i / \partial a_m, E_i, \theta)$, and identifying $d\chi/dt$ with equation (34) give the constitutive equations

$$T_{ij} = \rho \frac{\partial \chi}{\partial (\partial x_j / \partial a_m)} \frac{\partial x_i}{\partial a_m} \quad , \quad P_i = -\rho \frac{\partial \chi}{\partial E_i} \quad , \quad \eta = -\frac{\partial \chi}{\partial \theta} \tag{35}$$

It will be more practical to express the previous relations in the material (initial) coordinate system, as a function of the strains η_{ij}, the electric field W_k in the material state, and the temperature, following : $\chi = \chi\, (\eta_{ij}, W_k, \theta)$. If ϕ is the potential

$$W_k = -\frac{\partial \phi}{\partial a_k} = -\frac{\partial \phi}{\partial x_m} \frac{\partial x_m}{\partial a_k} = -E_m \frac{\partial x_m}{\partial a_k} \tag{36}$$

The constitutive equations become

$$T_{ij} = \rho \frac{\partial x_i}{\partial a_m} \frac{\partial x_j}{\partial a_k} \frac{\partial \chi}{\partial \eta_{mk}} + \rho \frac{\partial x_i}{\partial a_m} E_j \frac{\partial \chi}{\partial W_m} \tag{37}$$

$$P_i = -\rho \frac{\partial x_i}{\partial a_m} \frac{\partial \chi}{\partial W_m} \quad \text{and} \quad \eta = -\frac{\partial \chi}{\partial \theta} \tag{38}$$

IV - MATERIAL CONSTANTS

In elastic and piezoelectric solids, the internal energy is a function of mechanical strains, electric fields, and temperature. Other effects like piezomagnetic, pyroelectric, etc. effects will not be considered here. It is convenient to express the energy in a power series of the independent variables in the initial state

$$\begin{aligned}\rho_o X = & \frac{1}{2} c_{ijk\ell} \eta_{ij} \eta_{k\ell} + \frac{1}{6} c_{ijk\ell mn} \eta_{ij} \eta_{k\ell} \eta_{mn} \\ & + \frac{1}{24} c_{ijk\ell mnpq} \eta_{ij} \eta_{k\ell} \eta_{mn} \eta_{pq} + \dots \\ & - \frac{1}{2} \varepsilon_{mn} W_m W_n - \frac{1}{6} \varepsilon_{mnp} W_m W_n W_p \dots \\ & - e_{mij} W_m \eta_{ij} - \frac{1}{2} e_{m.ijk\ell} W_m \eta_{ij} \eta_{k\ell} \\ & - \frac{1}{2} e_{mn.ij} W_m W_n \eta_{ij} \dots \\ & + \text{higher order terms}\end{aligned} \tag{39}$$

$c_{ijk\ell}$, $c_{ijk\ell mn}$, $c_{ijk\ell mnpq}$ represent respectively the 2nd, 3rd and 4th order elastic constants (the order is the rank in the energy). ε_{mn}, ε_{mnp} are the 2nd and 3rd order dielectric constants. e_{mij} are the regular 2nd order piezoelectric constants. Finally $e_{m.ijk\ell}$ and $e_{mn.ij}$ correspondant to nonlinear 3rd order piezoconstants, which are respectively known as electroelastic and electrostrictive constants (the latter exist even in nonpiezoelectric media). These constants define the peculiar properties of each material. However, in the present paper the discussion will be mainly limited to the most usual piezoelectrics, which are quartz, lithium niobate, and lithium tantalate.

Elastic constants

The 2nd order elastic constants are generally available in the form of the compliance or of the stiffness constants. In a piezoelectric material distinction must be made between the elastic constants at zero electric field or at zero electric displacement.

The 3rd order elastic constants of quartz have been measured by Thurston et al[9]. There are 31 non zero 3rd order constants, and among them 14 are independent. The confidence one can have in the values of these constants is high according to the good agreement which is observed between theoretical evaluations and experimental data on the sensitivity of quartz resonators to mechanical forces. 3rd order constants have also been measured for many materials belonging to cubic, trigonal, tetragonal crystallographic systems. The values can be found in Tables of Ref. 10. But 4th order elastic constants are almost completely unknown. For quartz crystal only C_{6666}, C_{1111}, C_{3333} have been evaluated.[11]

Piezoelectric, electroelastic, and electrostrictive constants

The second order constants are also known for most of the piezoelectric materials. The electroelastic constants couple, in the energy, one component of the electric field with two components of the strain tensors, and the strain derivative of this quantity gives the stress tensor

$$T_{ij} \propto e_{m.ijk\ell} E_m \eta_{k\ell} \tag{40}$$

Therefore electroelastic effect appears as a nonlinear coupling between electric fields and mechanical strains. If, for instance, E_m is a static quantity (a DC field), and T_{ij} and $\eta_{k\ell}$ dynamic quantities $e_{m.ijk\ell} E_m$ in that case is equivalent to a correcting term of the 2nd order elastic constants. In this example electroelastic effect induces a direct change of the elastic constants under the application of a DC electric field.

The electrostrictive constants correspond in the energy to the coupling of two electric field components with one strain component in the term $(1/2)e_{mn.ij} E_m E_n \eta_{ij}$.

After strain derivation it follows

$$T_{ij} \propto e_{mn.ij} E_m E_n \tag{41}$$

which is a stress component proportional to a quadratic form of the electric field. This property, called electrostriction, can be observed in particular in materials which are not piezoelectric.

The electroelastic effect of quartz was studied by Hruska[12], who evaluated the values of linear combinations of electroelastic constants. Complete sets of the 8 independent constants (there are 23 nonzero constants) were determined by different authors. Kusters,[13] Brendel,[14] Hruska and Brendel[15] used frequency measurements of resonators under DC fields. Reider,[16] Kittinger,[17] determined the same constants with a pulse echo method. Large discrepancies appear in the values given by the different authors. This was commented by Hruska,[18] who showed that in the present state the values of the fundamental constants cannot be completely determined and only combinations of them are accessible.

The electrostrictive constants were measured for many materials and in particular for quartz[17], lithium niobate and lithium tantalate, in addition for these last materials of the electroelastic constants.[19-24]

Dielectric constants

Linear 2nd order and nonlinear higher order dielectric constants were measured in quartz and lithium niobate,[11,224] and many other materials.[10]

Electro-optic and piezo-optic constants

The electro-optic and piezo-optic interactions involve fundamental constants which already appeared in the previous sections of this paper. Electro-optic effect corresponds to the interaction of a light beam, i.e. an electromagnetic wave at very high frequency with a low frequency electric field. Even if this last one is in the microwave range it can be considered as a DC field in comparison to the light frequency. Since the mathematical representation of the interaction links the modified light wave to the initial light wave multiplied by the electric field, the electro-optic effect must involve in the expression of internal energy three components of the electric field (or electric displacement) and thus corresponds to the quantity (1/6) $\varepsilon_{mnp} E_m E_n E_p$. Therefore the electro-optic constants are similar to the 3rd order dielectric constants. If the 4th order constants are considered, they will lead to the Kerr effect.

When the crystal is piezoelectric, mechanical deformations occur which can be coupled with the light beam. The corresponding term in the energy involves two electrical components with one mechanical one.This is (1/2) $e_{mn.ij} E_m E_n \eta_{ij}$. At low frequency the piezo-optic constants will correspond to the electrostrictive constants. A rigourous presentation would have to make the difference between constants defined by taking stress or strain, and electric field or electric displacements, as independent variables in the definitions. These distinctions have not been made in this short presentation.

V - NONLINEAR PROPAGATION OF A FINITE AMPLITUDE WAVE

1) Travelling wave in an infinite medium

Considering the simpler case of a nonpiezoelectric elastic medium, combination of the mechanical equilibrium equation (26) with the constitutive equation (37), leads to the propagation equation, expressed with respect to the initial coordinate system by using the strain expression (17)

$$\rho_o \ddot{u}_j = c_{ijk\ell} u_{k,\ell i} + \gamma_{ijk\ell mn} (u_{k,\ell} u_{m,n})_{,i} + \delta_{ijk\ell mnpq} (u_{k,\ell} u_{m,n} u_{p,q})_{,i} \tag{42}$$

where the notation $\partial/\partial a_i = ,i$ has been used.

$\gamma_{ijk\ell mn}$ and $\delta_{ijk\ell mnpq}$ are nonlinear coefficients related to the 2nd, 3rd, and 4th order elastic constants following

$$\gamma_{ijk\ell mn} = c_{ink\ell} \delta_{jm} + \frac{1}{2} c_{ijn\ell} \delta_{km} + \frac{1}{2} c_{ijk\ell mn} \tag{43}$$

$$\delta_{ijk\ell mnpq} = \frac{1}{2} c_{iqn\ell} \delta_{km} \delta_{jp} + \frac{1}{4} c_{ijk\ell qn} \delta_{pm}$$

$$+ \frac{1}{4} c_{ijq\ell mn} \delta_{kp} + \frac{1}{2} c_{iqk\ell mn} \delta_{pj} + \frac{1}{6} c_{ijk\ell mnpq} \tag{44}$$

For the propagation of a plane wave the previous equation can be transformed into

$$\rho_o \ddot{U}_r = \lambda_r U_{r,ss} + \Gamma_r (U_{r,s})^2_{,s} + \Delta_r (U_{r,s})^3_{,s} \tag{45}$$

where λ_r, Γ_r, and Δ_r are effective elastic constants of the 2nd, 3rd, and 4th orders. For instance

$$\lambda_r = c_{ijk\ell} N_i N_\ell \ell_k^{(r)} \ell_j^{(r)} \tag{46}$$

where the N_i's are the director cosines of the propagation direction and the $\ell_k^{(r)}$'s the director cosines of the eigenvectors (expressed in the initial coordinate system). U_r is the total mechanical displacement and s the distance travelled by the wave. r indicates the wave mode. Equation (45) corresponds to modes which are not coupled. This is an approximation and in fact the three modes are coupled by the nonlinear terms.[25]

2) Examples

Short examples are given in this section. More details on harmonic generation, amplitude-frequency effect, intermodulation, convolution product can be found in references (26-37).

- *harmonic generation* is the direct consequence of the quadratic and cubic nonlinear terms of the propagation equation. Quadratic terms induce 2nd harmonic and a DC component. Cubic terms induce 3rd harmonic and components at the fundamental frequency. The coupling (quadratic) between the 2nd harmonic and the fundamental gives also 3rd harmonic and fundamental. And so on.

The components generated at the fundamental frequency modify the phase of the wave as a function of the travelled distance. This is equivalent to a dependence of the velocity on the amplitude, and is known as the velocity-amplitude effect, following the relation

$$V = V_o \left[1 + \frac{3}{4} \frac{\omega^2}{V_o^2} U_o^2 \left(\frac{\Gamma^2}{\lambda^2} - \frac{\Delta}{2\lambda} \right) \right] \tag{47}$$

where V_0 is the wave velocity at low amplitude. U_0 is the wave amplitude and ω its angular frequency. In a resonator the same effect appears as the dependence of the resonance frequency on the amplitude ; this is the amplitude-frequency effect

$$\frac{\omega_r - \omega_o}{\omega_o} = \frac{U_o^2 \pi^2}{128\, h^2} \left[- \frac{21\, \Gamma^2}{2\, \lambda^2} + \frac{9\Delta}{\lambda} \right] \tag{48}$$

2h is the thickness of the resonator.

- *intermodulation* : when the signal is composed of two frequencies ω_1 and ω_2, the quadratic nonlinearities generate components at the frequencies $\omega_1 + \omega_2$ and $\omega_1 - \omega_2$. The cubic nonlinearities induce $2\omega_1 + \omega_2$, $2\omega_2 + \omega_1$, $2\omega_1 - \omega_2$, and $2\omega_2 - \omega_1$.In the case of a resonator most of these components are again filtered out, except the intermodulation frequencies $2\omega_1 - \omega_2$ and $2\omega_2 - \omega_1$ which stand within the resonator bandwidth. This will induce spurious effects, which can be important in devices for example like front end filters.

- *convolution* : At the difference of the previous examples, nonlinearities can also be used for performing the products of signals with simultaneous integration as in the acoustic convolver.

The principle of the acoustic convolver is based on the nonlinear interaction of two surface waves generated on a piezoelectric substrate by two interdigital transducers and propagating in opposite direction as shown in Fig. 2.

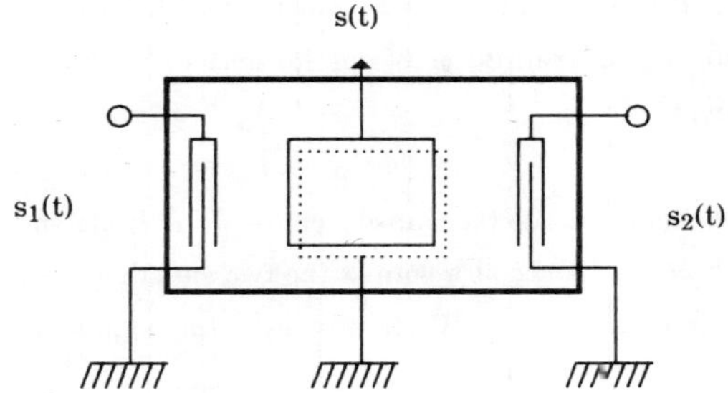

Fig. 2 : Degenerate acoustic convolver

The two waves can be represented by

$$s_1(t) = A_1 \cos(\omega_1 t - k_1 x)$$
$$s_2(t) = A_2 \cos(\omega_2 t - k_2 x) \quad (49)$$

The mixing of the two waves is due to the different nonlinear properties of the piezoelectric substrate involving the nonlinear elasticity, the electroelastic and electrostrictive properties, and the nonlinear dielectric permittivity. The product $s_1(t)\, s_2(t)$ is

$$s_1(t)\, s_2(t) = \frac{1}{2} A_1 A_2 \left\{ \cos[(\omega_1 + \omega_2)t - (k_1 - k_2)x] + \cos[(\omega_1 - \omega_2)t - (k_1 + k_2)x] \right\} \quad (50)$$

If the two waves are at the same frequency $\omega = \omega_1 = \omega_2$, and consequently $k_1 = \omega_1/V = k_2 = \omega_2/V_o = \omega/V_o = k$ equation (50) reduces to

$$s_1(t)\, s_2(t) = \frac{1}{2} A_1 A_2 \left\{ \cos 2\omega t + \cos 2kx \right\} \quad (51)$$

In addition are the two terms $s_1(t)^2$ and $s_2(t)^2$ and also the original waves $s_1(t)$ and $s_2(t)$.

The center electrode of Fig. 2 can detect only time dependent terms which are uniformly spatially distributed, i.e. independent of x, and this corresponds to $1/2\ A_1A_2 \cos 2\omega t$. In this case the convolver is called degenerate convolver, because the response is obtained when the two input frequencies are equal. If the frequencies are different the component with frequency $\omega_1 + \omega_2$ and wave number $k_1 - k_2$ can be detected by replacing the uniform electrode with an interdigital transducer of spatial periodicity $2\pi/(k_1 - k_2)$. This is the nondegenerate convolver.

The effect of the detecting electrode (or transducer) is to integrate the product of the two signals over its length $2L$, and since at a point x the two signals have relative time delays corresponding to $s_1\,(t - x/V)$ and $s_2\,(t + x/V)$, finally the output signal is proportional to

$$s(t) = \int_{-L}^{L} s_1\,(t - \frac{x}{v})\; s_2\,(t + \frac{x}{v})\, dx \qquad (52)$$

If the input signal have a limited duration so that the product occurs only on a spatial length which is smaller than the electrode length, and if introducing $\tau = t - x/V$, equation (52) becomes

$$s(t) \propto \int_{-\infty}^{+\infty} s_1(t)\; s_2\,(2t - \tau)\, dt \qquad (53)$$

which is just the convolution product of the two input signals (the factor 2 appears only as a time contraction).

The efficiency of the convolver depends on the powers of the input signals and on the values of the nonlinear constants of the material.This efficiency was improved in the waveguide convolver, where the electrode is used as a narrow waveguide which increases the energy density and thus the intensity of the product.[37] On the other hand materials with strong electromechanical coupling factors, such as lithium niobate, are selected. Other devices were also developped, which use the nonlinear effects of semiconductors[38] to improve the low efficiency of the acoustic convolver.

VI - WAVE PROPAGATION IN A BIASED MEDIUM

1) Propagation equation

(For simplicity the presentation is limited to the pure elastic case but this problem was extended to piezoelectric solids and details are given in references (39-41)).

If the solid is submitted to static deformations and stresses due to the application of forces, electric fields, ... superimposed on a high frequency vibration, three states are to be considered,

- the natural state, when the solid is at rest, with coordinates a_i
- the initial state, after application of the static deformation, with coordinates x_i
- the final state, when the vibration is superimposed on the static deformation, with coordinates X_i.

Let ρ_o, $\overline{\rho}$ and ρ be the specific mass respectively in the different states.

One distinguishes the static mechanical displacement $\overline{u}_i = x_i - c_i$, the dynamic one $\tilde{u}_i = X_i - x_i$ and the total displacement $u_i = X_i - a_i$. The same notation $\overline{\tau}_{ij}$, $\tilde{\eta}_{ij}$ and η_{ij} is used for the other quantities.

The mechanical equilibrium equation in the final state is

$$\rho \frac{dv_j}{dt} = \frac{\partial}{\partial X_i}(T^S_{ij} + T^{ES}_{ij}) = \frac{\partial \tau_{ij}}{\partial X_i} \tag{54}$$

In the initial state

$$\overline{\rho} \frac{dv_j}{dt} = \frac{\partial K_{sj}}{\partial x_s} \tag{55}$$

with

$$K_{sj} = J \frac{\partial x_s}{\partial X_i} \tau_{ij} \quad , \quad J = \overline{\rho} / \rho \tag{56}$$

and in the natural state

$$\rho_o \frac{dv_j}{dt} = \frac{\partial P_{j\ell}}{\partial a_\ell} \tag{57}$$

with

$$P_{j\ell} = \bar{J}\,\frac{\partial a_\ell}{\partial x_s}\,K_{sj} \quad , \quad \bar{J} = \rho_o / \bar{\rho} \tag{58}$$

In the expression of $P_{j\ell}$ the static and dynamic parts, respectively $\overline{P}_{j\ell}$ and $\widetilde{P}_{j\ell}$, are separated. By difference with the static equilibrium equation $\rho_o\,(d\bar{v}_j/dt) = \partial\overline{P}_{j\ell}/\partial a_\ell$ (which supposes that the static deformation is not affected by the vibration and this is true if the wave amplitude is small), the dynamic equation is obtained

$$\rho_o\,\frac{d\tilde{v}_j}{dt} = \frac{\partial\widetilde{P}_{j\ell}}{\partial a_\ell} \tag{59}$$

with

$$\widetilde{P}_{j\ell} = \overline{A_{j\ell kr}}\,\frac{\partial\tilde{u}_k}{\partial a_r} \tag{60}$$

and the boundary condition

$$N_j\,[P_{j\ell}] = 0 \tag{61}$$

The $\overline{A_{j\ell kr}}$ coefficients are considered as effective elastic constants.

$$\overline{A_{jk\ell r}} = c_{j\ell kr} + \hat{c}_{j\ell kr} \tag{62}$$

where $\hat{c}_{j\ell kr}$ appears as a perturbation of the elastic constant $c_{j\ell kr}$, which depends on the static bias. The expression of $c_{j\ell kr}$ limited to nonlinearities of the 3rd order is

$$\hat{c}_{j\ell kr} = (c_{jrmn}\,\delta_{\ell k} + c_{j\ell nr}\,\delta_{km} + c_{jnkr}\,\delta_{\ell m} + c_{j\ell krmn})\,\frac{\partial\overline{u_m}}{\partial a_n} \tag{63}$$

In a piezoelectric medium effective piezoelectric and dielectric constants will be defined, using the same method, as a function of the static mechanical displacement gradients $\partial\bar{u}_m/\partial a_n$ and of the static electric potential or field $\partial\bar{\phi}/da_n$.

2) Examples

The nonlinear coupling of the wave with a static bias is the cause of the sensitivity, of devices to external perturbations. Velocity (or frequency) is modified that way by temperature, accelerations, forces, pressures, electric fields, etc. For most of the applications these sentivities must be reduced in order to improve the stabilities.

Quartz resonators which are used in stable oscillators are temperature compensated by choosing appropriate cristallographic orientations known as AT cut, BT cut, etc. Studies were also made for compensating stresses, which can be induced by the influence of the supports, by thin film electrodes, or by temperature gradients. A simultaneous compensation of static temperature and stresses (or dynamic temperature) was made possible by using SC-cut.Today large effects are performed for reducing the sensitivity to accelerations for aero-space applications. Variations of the atmospheric pressure also induce frequency shifts. Sensitivities to irradiations, and more recently to magnetic fields are studies which are in progress. These different problems are important for surface acoustic wave devices as well as for bulk acoustic wave devices.

At the opposite the nonlinear coupling can be increased also by using particular cuts, and if the device is selectively sensitive of one given physical quantity it is used as a sensor. Such sensors transform the variation of the measured physical quantity into a frequency variation And since frequency can be measured with the highest accuracy such sensor can have the best performances. Studies are made on temperature probes, pressure sensors, accelerometers, but also on gas or liquid flow sensor, gas detectors, etc.

References

1) P.G. Klemens, "The thermal conductivity of dielectric solids at low temperatures", Proc. Roy. Soc. A 208, 108 (1951).

2) A.A. Maradudin, "Thermal expansion and phonon frequency shifts", Phys. Stat. Sol., 2, 1493 (1962).

3) A.J. Maris, "Interaction of sound waves with thermal phonons in dielectric crystals", Physical Acoustics, vol. VIII, 279, Academic Press, N.Y. (1971).

4) L. Landau, G. Rumer, Phys. Z. Sowjet Union, 11, 18 (1937).

5) A. Akhieser, "On the absorption of sound in solids", J. Phys. (USSR), 1 (1939).

6) R. Brugger, "Generalized Grüneisen parameters in the anisotropic Debye model", Phys. Rev. 137, N° 6A (1963).

7) R.N. Thurston, "Wave propagation in fluids and normal solids", Physical Acoustics, vol. I, part A, V.P. Mason, edit., Academic Press, New York (1964).

8) H.F. Tiersten, "On the nonlinear equations of thermo-electro-elasticity", Int. J. Engng. Sci., 9 (1971).

9) R.N. Thurston, J.J. McSkimin, P. Andreatch, "Third order elastic coefficients of quartz", J. Appl. Phys., 37 (1966).

10) Landolt-Börnstein, "Elastic, piezoelectric, and related constants of crystals", III, vol. 11, Springer-Verlag (1979).

11) J.J. Gagnepain, R. Besson, "Nonlinear effects in piezoelectric quartz crystals", Physical Acoustics, vol. XI, Academic Press (1975).

12) C.K. Hruska, "The electroelastic tensor and other second order phenomena in quasilinear interpretation of the polarizing effect with thickness vibrations of α-quartz plates", Proc. of the 31st AFCS, 159 (1977).

13) J. Kusters, "The effect of static electric fields on the elastic constants of a quartz", Proc. of the 24th AFCS (1970).

14) R. Brendel, "Material nonlinear piezoelectric coefficients for quartz", J. Appl. Phys., 54 (9), 5339-5346 (1983) ; 55 (2), 608 (1984).

15) C.K. Hruska, R. Brendel, "The material electro-elastic constants of quartz determined by the resonator method", 1er Forum Européen Temps-Fréquence, Besançon, LPMO, (1987).

16) G.A. Reider, E. Kittinger, J. Tichy, "Electroelastic effect in α-quartz", J. Appl. Phys., 53 (12), 8716-8721 (1982).

17) E. Kittinger, J. Tichy, W. Friedel, "Nonlinear piezoelectricity and electrostriction of α-quartz", J. Appl. Phys., 60 (4), 1465-1471 (1986).

18) C.K. Hruska, "On the disagreement in the order of magnitude of the electroelastic constants of α-quartz", Proc. of the 40th AFCS (1986).

19) R.B. Thomson, C.F. Quate, "Nonlinear interaction of microwave electric fields and sound in $LiNbO_3$", J. Appl. Phys., 42 (3) (1971).

20) A.I. Korobov, V.E. Lyamov, "Nonlinear coefficients of $LiNbO_3$", Sov. Phys. Sol. State, 17 (5) (1975).

21) B.A. Agishev, V.V. Lemanov, N.K. Yushi, "Electroacoustic coefficients of lithium tantalate and niobate crystals", Sov. Phys. Sol. State, 20 (9) (1979).

22) S.I. Chizhikov, N.G. Sorokni, U.S. Petrokov, "The elastoelectric effect in the noncentrosymmetric crystals", Ferroelectrics, 41 (1982).

23) A. Palma, L. Palmieri, G. Socino, E. Verona, "Nonlinear electroacoustic effect in Lamb wave propagation in $LiNbO_3$ plates", 1984 Ultrasonics Symp., IEEE cat 84 CH 2112-1.

24) A.K. Ganguly, K.L. Davis, "Nonlinear interaction in degenerate surface acoustic wave elastic convolvers", J. Appl. Phys., 51 (1980).

25) N. Daher, G.A. Maugin, "Intermodulation and generation of elastic and piezoelectric waves in anisotropic solids", to be published.

26) H.F. Tiersten, J.C. Baumhauer, "Second harmonic generation of surface waves in isotropic elastic solids", Ultrasonics Symposium, IEEE cat. 73 CHO 8078SU (1973).

27) P.J. Vella, G.I. Stegeman, U.M. Ristic, "Analysis of parametric mixing and harmonic generation of surface acoustic waves", J. Appl. Phys., 50, 81 (1979).

28) M. Planat, G. Théobald, J.J. Gagnepain, "Propagation non linéaire d'ondes élastiques dans un solide anisotrope" : I - Ondes de volume : l'Onde Electrique, 60, (8-9), 33 (1980), II - Ondes de surface : l'Onde Electrique, 60 (11), 61 (1980).

29) H.F. Tiersten, "Analysis of nonlinear resonance in rotated Y-cut quartz thickness-shear resonators", 29th A.F.C.S. (1975).

30) H.F. Tiersten, "An analysis of intermodulation in rotated Y-cut quartz thickness-shear resonators", 28thA.F.C.S. (1974).

31) M. Planat, G. Théobald, J.J. Gagnepain, P. Siffert, "Intermodulation in X-cut lithium tantalate resonators", Electronics Lett., 16 (5), 174 (1980).

32) M. Planat, "Nonlinear characteristics of SAW grooved resoators", Ultrasonics Symposium, IEEE 80 CH 1602.2 (1980).

33) A. Alippi, "Nonlinear acoustic propagation in piezoelectric crystals", Ferroelectrics, 42, 109-116 (1982).

34) G.S. Kino, S. Ludvik, A.J. Shaw, W.R. Shrere, J.M. White and D.K. Winslow, "Signal processing by parametric interactions in delay-line devices", IEEE Trans. MTT-21 (1973).

35) D.P. Morgan, "Surface wave devices for signal processing", Elsevier (1985).

36) P. Defranould, C. Maerfeld, "A SAW planar piezoelectric convolver", Proc. IEEE, 64 (1976).

37) M. Planat, G. Vanderbock, M. Gautier, C. Maerfeld, "Linear and nonlinear analysis of a piezoelectric waveguide convolver using a finite element method", IEEE Acoustics Symposium, CH (1982).

38) G.S. Kino, "Acoustoelectric interactions in acoustic surface wave devices", Proc. IEEE, 64 (1976).

39) J.C. Baumhauer, H.F. Tiersten, "Nonlinear electroelastic equations for small fields superposed on a bias", J. Acoust. Soc. Am., 54 (4) (1973).

40) G. Théobald, D. Hauden, J.J. Gagnepain, "Propagation des ondes élastiques en milieu perturbé", l'Onde Electrique, 65 (2) (1985).

41) J.J. Gagnepain, "Nonlinear constants and their significance", 41st Ann. Freq. Cont. Symposium, 87 CH 2427-3 (1987).

FIELD RECONSTRUCTION USING TOMOGRAPHIC AND FOURIER-OPTICAL TECHNIQUES

R. Reibold

Physikalisch-Technische Bundesanstalt
Braunschweig
Fed. Rep. of Germany

ABSTRACT

This object of lecture is to further the basic understanding of field reconstruction. What is attempted here is, by means of selected topics, to help develop a "feel" for some of the physical processes and numerical procedures involved.

An acousto-optic measurement method is introduced which, combined with a tomographic evaluation, allows the complex sound pressure to be determined in any cross-sectional plane of the ultrasonic field. Fourier-optical techniques are then applied to reconstruct the three-dimensional field by forward and backward projection of the respective pressure profile. Numerical examples and experimental results will be given.

Last year when Prof. Alippi asked me to give a talk about "Field Reconstruction" I immediately accepted his invitation. This happened when I was in a hurry to leave Ultrasonics International in order to catch my plane. Afterwards, during the flight when there was plenty of time to think about this, I probably would not have agreed to his request. Actually, what does "Field Reconstruction" mean? What is Prof. Alippi´s personal opinion on this subject? I am sure that I am not qualified to talk about "Field Reconstruction", but perhaps I can talk on a few selected topics concerning field reconstruction.

For this School, I have chosen a more specialized subject, the title of which is now: "Field Reconstruction Using Tomographic and Fourier-Optical Techniques". Perhaps some of you will be less familiar with some of the aspects I am discussing. For many of you, I will probably leave out the most important aspects of field reconstruction and cover less important topics. All of this illustrates the extremely broad scope of what is covered by the subject of field reconstruction.

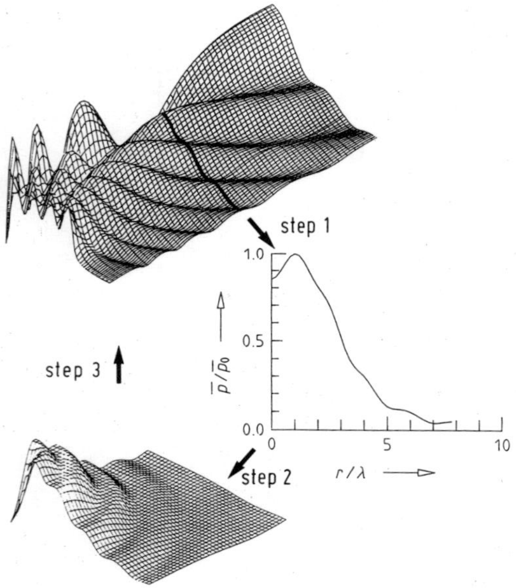

Fig. 1:
Schematic representation of the subject.
Step 1: Optical investigation.
Step 2: Tomographic reconstruction.
Step 3: Fourier-optical reconstruction.

The subject can be divided into three parts. It can most easily be made understood by means of Fig. 1. The upper part shows the sound field of a circular piston-like transducer which represents an unknown real sound field the object of interest. In the first step, any cross-sectional plane of this sound field, e.g. the plane marked by a thick line, is optically investigated. The measurement technique used gives the projection of the sound field at this particular plane which is shown on the right-hand side. In the second step, subsequent tomographic evaluation is applied to reconstruct the complex pressure profile of the respective plane given at the bottom. When the complex pressure profile at any cross-sectional plane is known, then the complete sound field can be reconstructed by means of numerical forward and backward-projection procedures. The technique used in this third step is known as the Fourier transform method of angular spectrum propagation.

1. LIGHT DIFFRACTION TECHNIQUE

The conventional light diffraction method is currently used for studying the sound-pressure amplitude of continuous ultrasonic waves by measuring the light intensity in the Fraunhofer diffraction orders. For the investigation of the amplitude and phase of the ultrasound, however, a method is required which enables the pressure waveform to be measured.

In 1982 Riley et al[1)] published a paper dealing with the study of ultrasonic waveform by means of a conventional schlieren system. If either the positive or negative diffraction orders are removed by the schlieren stop, a time variation of the light intensity in the image plane is observed. Moreover, this technique provides a good approximation of the pressure waveform for small amplitudes and a uniform frequency response.

In this paper, a measuring technique is described which is slightly modified in comparison with the arrangement proposed by Riley et al. This method enables the sound-pressure amplitude and phase in an essentially enlarged range of application to be determined. The technique is applicable to continuous waves[2)], standing waves[3)] and ultrasonic pulses[4)].

1.1 Continuous Waves Propagation[2)]

Let us consider a progressive ultrasonic wave in a liquid which is normally traversed by a parallel beam of monochromatic light (Fig.2).

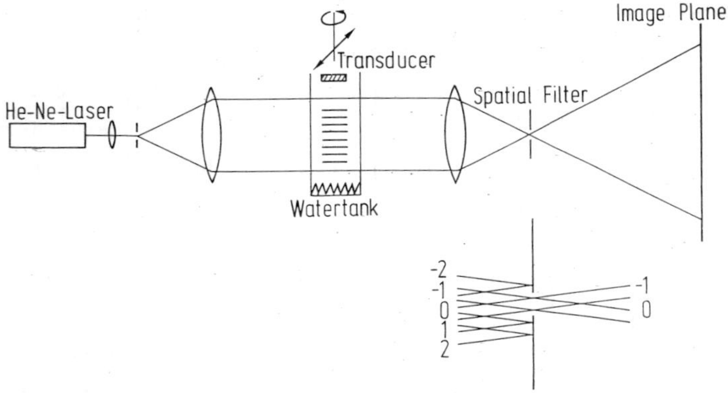

Fig. 2: Basic schlieren arrangement used for the measurement of the projection data.

The pressure at any arbitrary point in the sound field can be described by

$$p(x,y,z,t) = \hat{p}(x,y) \sin (\omega^* t - k^* z + \phi(x,y)) \tag{1}$$

Here, ω^* is the angular frequency and k^* is the wave number of the ultrasound. ϕ is the phase related to any reference plane which is in practice simulated by the exciting voltage of the transducer. Under Raman-Nath conditions, which means for small pressure amplitudes, the optical phase retardation, usually termed the Raman-Nath parameter, in the y direction relative to the undisturbed medium is

$$v(x,z,t) = k \frac{\partial n}{\partial p} \int \hat{p}(x,y) \sin (\omega^* t - k^* z + \phi(x,y))\, dy \tag{2}$$

$$= \hat{v}_c(x) \sin (\omega^* t - k^* z) + \hat{v}_s(x) \cos (\omega^* t - k^* z) \tag{3}$$

$$= \hat{v}(x) \sin (\omega^* t - k^* z + \bar{\phi}(x)) \tag{4}$$

with

$$\hat{v}_c(x) = k \frac{\partial n}{\partial p} \int \hat{p}(x,y) \cos \phi(x,y)\, dy \tag{5}$$

$$\hat{v}_s(x) = k \frac{\partial n}{\partial p} \int \hat{p}(x,y) \sin \phi(x,y)\, dy \tag{6}$$

$$\hat{v}^2(x) = \hat{v}_c^2(x) + \hat{v}_s^2(x), \quad \tan \bar{\phi}(x) = \hat{v}_s(x)/\hat{v}_c(x) \tag{7}$$

Here, k is the wave number of the light and $\partial n/\partial p$ the adiabatic piezo-optic coefficient. After some slight modifications, the final expression for the optical phase retardation is obtained:

$$\hat{v}(x)e^{i\bar{\phi}(x)} = k\,\frac{\partial n}{\partial p}\int \hat{p}(x,y)\,e^{i\phi(x,y)}dy \tag{8}$$

The real and imaginary parts of eq. (8) represent line integrals which are - as will be shown below - the basis of the tomographic back projection. On the left-hand side, the projection data $\hat{v}(x)$ and $\bar{\phi}(x)$ are given which are to be measured. On the right-hand side, the local data $p(x,y)$ and $\phi(x,y)$ are given which are to be calculated.

The amplitude of the light in the image plane (Fig. 2) for a given distance $z=z_o$ from the transducer may be written as

$$\begin{aligned} E(x,y,z_o,t) &= \exp\{i(\omega t - ky) - i\hat{v}(x)\sin(\omega^* t - k^* z_o + \bar{\phi}(x))\} \\ &= e^{i(\omega t - ky)}\sum_{n=-\infty}^{\infty} J_n(\hat{v}(x))e^{-in(\omega^* t - k^* z_o + \bar{\phi}(x))} \end{aligned} \tag{9}$$

where J_n is the n-th order Bessel function. In this expression the first term describes the incident plane light wave and the second term takes into account its phase retardation due to the ultrasonic field. If only the zero and the first positive diffraction order pass the spatial filter (Fig. 2), the series in eq. (9) is reduced to two terms only

$$E(x,y,z_o,t) = e^{i(\omega t-ky)}\{J_o(\hat{v}(x)) + J_1(\hat{v}(x))e^{-i(\omega^* t - k^* z_o + \bar{\phi}(x))} \tag{10}$$

For the light intensity it then follows that:

$$I(x,z_o,t) = J_o^2(\hat{v}) + J_1^2(\hat{v}) + 2J_o(\hat{v})J_1(\hat{v})\cos(\omega^* t - k^* z_o + \bar{\phi}(x)) \tag{11}$$

Fig. 3 shows the dc and the ac component of the normalized light intensity as a function of the Raman-Nath parameter $\hat{v}$. It can be seen that in the range up to the first zero of J_o, a unique determination of the phase $\bar{\phi}(x)$ is possible. The Raman-Nath parameter $\hat{v}$ is obtained by either measuring the dc component or the amplitude of the ac component.

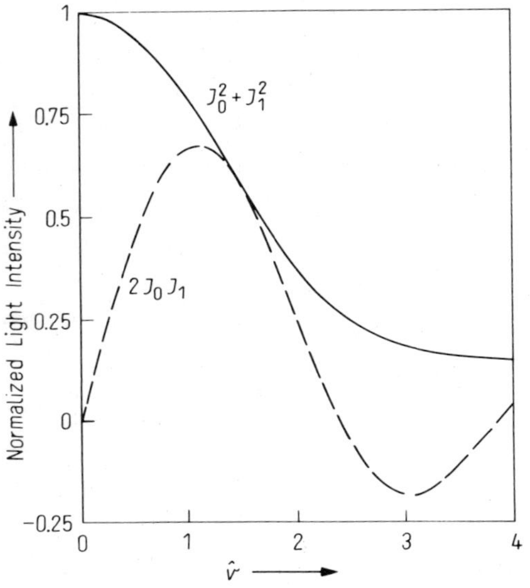

Fig. 3: Variation of the normalized light intensity with the Raman-Nath parameter in the image plane when the zero and positive first diffraction order pass the schlieren stop (harmonic waves).

In order to give an idea of the light intensity variation and phase variation over the cross section of an ultrasonic beam, Fig. 4 and 5 show some typical plots. In Fig. 4 the reproducibility of the measuring method for five successive cross-sectional scans under the same conditions and for a rather high pressure is shown.
For the measurement of the light intensity, the dc component (eq.(11)) was taken. Consequently, with increasing sound pressure towards the sound field axis, the light intensity decreases. Fig. 5 gives an example of a composite transducer with distinct intensity and phase variations under selected conditions (details will be given in section 2). In this case the ac term of eq. (11) was used.

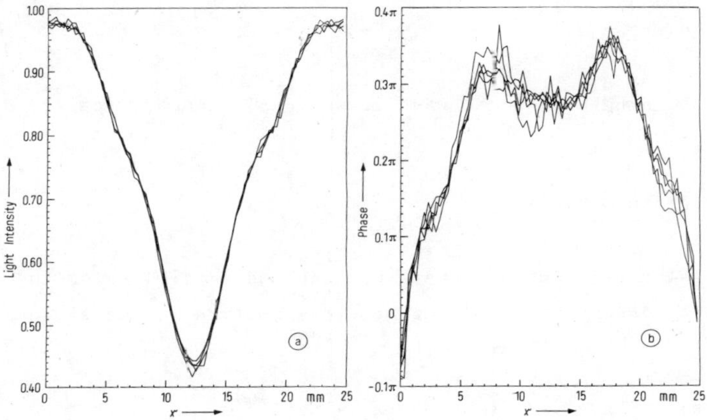

Fig. 4:
Successive scans of the normalized light intensity and phase taken under the same conditions (3.18 MHz, transducer diameter 20 mm, measuring distance 200 mm)

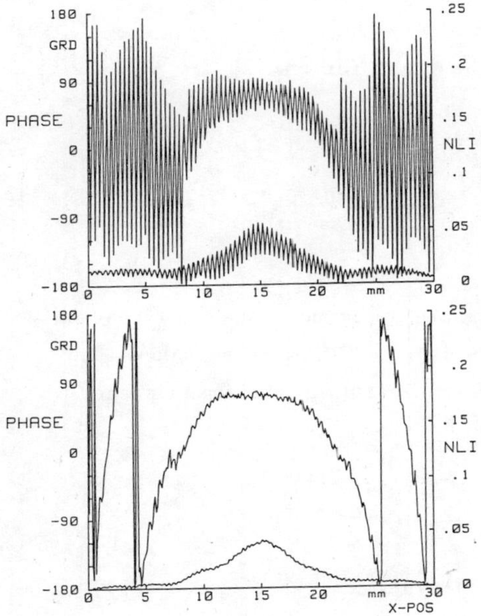

Fig. 5:
Normalized light intensity (lower curves) and phase (upper curves) of the sound field of a composite transducer at 0° (top, parallel to the cuts) and 45° (bottom).

1.2 Standing Waves[3)]

A plane standing ultrasonic wave in an ideal resonator can be described by

$$p(z,t) = \hat{p} \sin \omega^* t \cos k^* z \tag{12}$$

In the real case, however, the amplitude and possibly also the phase are locally dependent. A more general assumption is therefore required

$$p(x,y,z,t) = \hat{p}(x,y) \sin (\omega^* t + \phi(x,y)) \cos k^* z \tag{13}$$

Following the same procedure as used for progressive waves, the Raman-Nath parameter results in

$$v(x,z,t) = \frac{\hat{v}(x)}{2} \{\sin(\omega^* t - k^* z_o + \bar{\phi}(x)) + \sin(\omega^* t + k^* z_o + \bar{\phi}(x))\} \tag{14}$$

and with this the amplitude of the light for $z=z_o$

$$E(x,y,z_o,t) = e^{i(\omega t - ky)} \sum_{n=-\infty}^{\infty} J_n(\tfrac{\hat{v}}{2}) e^{-in(\omega^* t - k^* z_o + \bar{\phi}(x))} \sum_{l=-\infty}^{\infty} J_l(\tfrac{\hat{v}}{2}) e^{-il(\omega^* t + k^* z_o + \bar{\phi}(x))} \tag{15}$$

We now assume that the sound pressure amplitude is so low that only three partial waves of the series in eq. (10) must be taken into account. For the normalized light intensity in the image plane, it then follows that

$$\begin{aligned} I(x,z_o,t) = {} & J_o^4 + 4J_o^2 J_1^2 \cos (2\omega^* t + 2\bar{\phi}(x)) && \text{(0 DO)} \\ & + 2J_o^2 J_1^2 - 2J_o^2 J_1^2 \cos (2\omega^* t + 2\bar{\phi}(x)) && \text{(1 DO)} \\ & + 4J_o^3 J_1 \sin (\omega^* t + \bar{\phi}(x)) \sin k^* z_o && \text{(IT)} \end{aligned} \tag{16}$$

In this approximation which neglects terms containing J_1^3, we assumed that again, the zero and the first positive diffraction order pass the schlieren stop. In eq. (16) the first line describes the light intensity solely for the zero diffraction order, the second line that of the first order. The third line gives the interaction term of both diffraction orders. It should be stressed that in the components containing twice the ultrasonic frequency, the phase is also doubled and that the interaction term shows the typical form of a standing wave.

To obtain an estimate of the range of validity of the approach in eq. (16), the same procedure was carried out using five partical optical waves instead of three. The results are shown in Fig. 6. The solid lines give the interaction term (curve 1) and the interference term of the zero diffraction order (curve 2) according to eq. (16). The broken lines represent the corresponding shapes when five partial waves are considered. Finally, the dots indicate experimental results. These experiments were performed using the arrangement in Fig. 2 with the water tank replaced by an ultrasonic resonator.

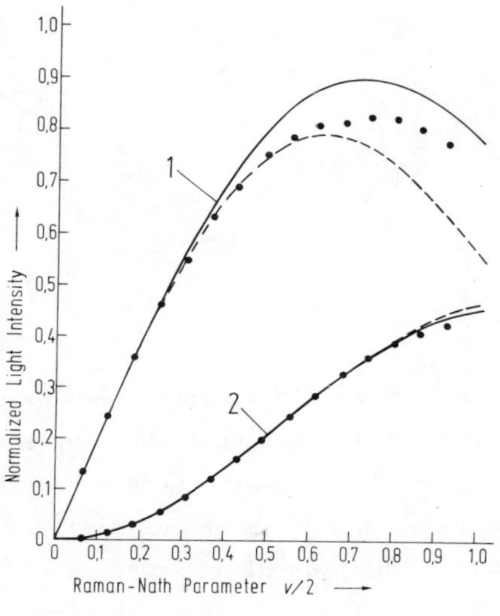

Fig. 6:
Normalized light intensity versus the Raman-Nath parameter for the interaction term (curve 1) and the interference term of the zero diffraction order (curve 2). The solid lines represent the theoretical shape according to eq. (16) and the broken lines when five partial waves are considered. The dots show experimental results.

An interesting insight into the performance of the light diffraction technique results from the transient behaviour of the standing waves in the resonator (Fig. 7). If only the zero diffraction order passes the schlieren stop, the laser beam "sees" a standing wave after 1.5 times the single travelling time in the resonator (Fig. 7a). It is recorded with twice the ultrasonic frequency. After each doubled travelling time the light intensity increases step by step up to the steady-state condition.

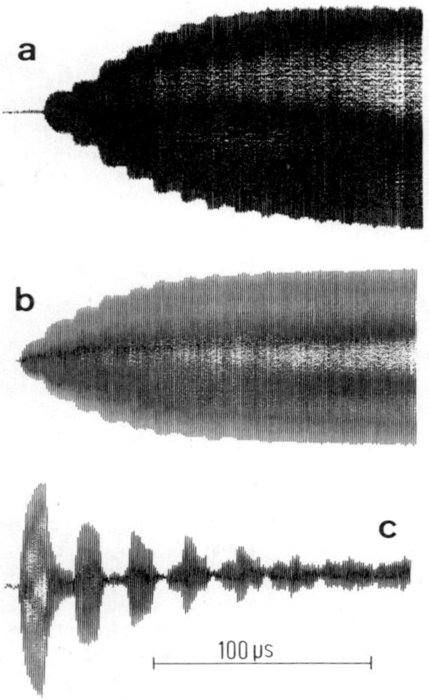

Fig. 7:
Transient behaviour of the standing wave field in the resonator.

a) $2\omega^*$ component of the interference term of the zero diffraction order,

b) interaction term when $\sin k^*z = 1$ in eq. (16),

c) progressive sound field components when $\sin k^*z = 0$ in (16).

For the interaction term, i.e. if the zero and first diffraction order pass the schlieren stop, a signal then occurs after half the single travelling time (Fig. 7b). Here, the laser beam "sees" the progressive wave front (see eq. (11)). Provided that $\sin k^*z = 1$ in eq. (16) (IT), the light intensity is doubled when the condition of a "standing wave" is reached, which means after 1.5 times the single travelling time from the beginning. Here, the light intensity alternates with the fundamental frequency.

Finally, in the last example (Fig. 7c) the photodiode is positioned so that $\sin k^* z = 0$ in eq. (16). In this position, only progressive sound field components can be recorded. Progressive components occur for odd multiples of the single travelling time but they vanish for even multiples of this.

2. FIELD RECONSTRUCTION FROM PROJECTIONS

2.1 Reconstruction Algorithm [5)]

The basic mathematical problem of image reconstruction from projections was solved by Radon in 1917. We assume a two-dimensional function f(x,y), which will be left unspecified for the time being. The integral of f(x,y) along a line through the xy plane is called a ray integral

$$P_\Theta(R) = \iint f(x,y)\, \delta(x \cos \Theta + y \sin \Theta - R)\, dxdy \qquad (17)$$

where δ(.) is the Dirac delta function. Θ is the angle formed by the direction of the ray and the y co-ordinate, and R is the perpendicular distance of a particular ray from the origin of the co-ordinates. A set of ray integrals along certain lines is called a projection. In practice, the data (line integrals) are sampled at equidistant sampling intervals by a linear shift of the transducer and its additional equiangular rotation after each projection.

The reconstruction algorithm is usually made with the following three relationships:

$$f(x,y) = \int_0^\pi Q_\Theta(x \cos \Theta + y \sin \Theta)\, d\Theta \qquad (18)$$

where $Q_\Theta(R)$ are called filtered projections for a given angle Θ. They are related to $P_\Theta(R)$ by

$$Q_\Theta(R) = \int P_\Theta(r)\, h(R-r)\, dr = P_\Theta(R) * h(R) \tag{19}$$

h(R) is referred to as the filter impulse response. If a band-limited projection with a limiting value W in the spatial frequency domain is assumed, h(R) can be described by the inverse Fourier transform of $|w|$

$$h(R) = \int_{-W}^{W} |w| \exp(i2\pi Rw)\, dw \tag{20}$$

The digital implementation of these equations allows an approximate calculation of f(x,y) to be made. The respective formulas for a suitable reconstruction algorithm are as follows.

In the first step the filtered projections $Q_\Theta(R)$ are calculated by means of discrete convolution

$$Q_\Theta(R) = \tau \sum_m P_\Theta(m\tau)\, h((n-m)\tau) \tag{21}$$

where τ is the sample interval and

$$h(n\ \tau) = \begin{cases} 1/(4\tau)^2 & \text{for } n = 0 \\ 0 & \text{for } n \text{ even} \\ -1/(n\pi\tau)^2 & \text{for } n \text{ odd} \end{cases}$$

In the second step the back projection is performed by summing up over all projection directions Θ_i :

$$f(x,y) \approx \frac{\pi}{M} \sum_{i=1}^{M} Q_{\Theta_i}(x \cos \Theta_i + y \sin \Theta_i) \tag{22}$$

where M is the number of angular positions.

2.2 Computer Simulation and Experimental Results

Computer simulation is used to demonstrate the efficiency of the reconstruction algorithm used. Fig. 8 shows the radial pressure and phase distributions (top, solid lines) of the sound field of a circular piston-like transducer (wavelength λ, radius a = 4 λ) at a distance of z=0 and z=4.5λ (last but one maximum on the axis) from the transducer surface. The corresponding projection profiles are given at the bottom. After tomographic back projection, the shapes represented by broken lines are obtained. The agreement between the original and the reconstruction is good in spite of the high spatial frequency components involved in this simulation. It should be stressed here that in accordance with eq. (8), the reconstruction procedure is to be performed twice, i.e. for the real and for the imaginary part separately.

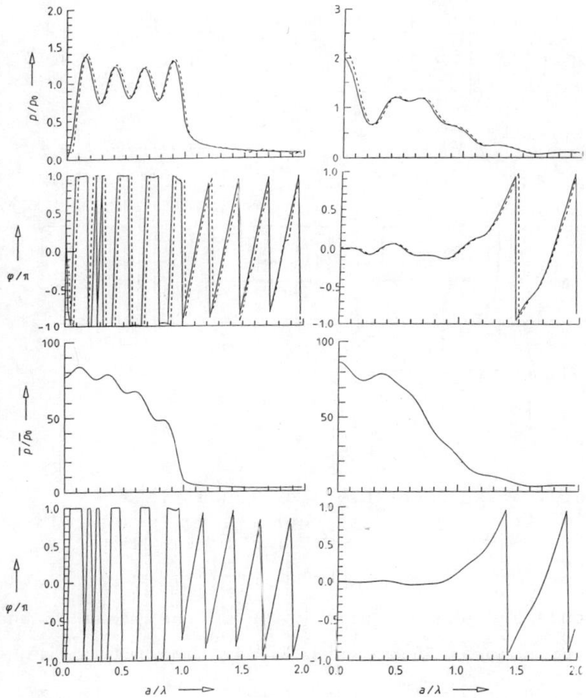

Fig. 8:
Radial pressure and phase distributions of the sound field of a piston radiator with radius a=4λ at z=0 (left) and z=4.5λ (right).
Upper part: Original distribution (solid lines) and after tomographic reconstruction (broken lines).
Lower part: Corresponding Projection profiles.

The experimental assembly used for data acquisition is shown in Fig. 9. As already mentioned, the optical part is that of the conventional schlieren system with a He-Ne laser as a light source and a photodiode positioned in the image plane. The necessary spatial resolution is obtained by magnified imaging of the sound field to the image plane and by using a small aperture photodiode. With the aid of a shift unit the transducer can be horizontally shifted in steps of 1/100 mm and rotated about its axis in steps of 0.18° (2000 steps per revolution). An additional adjusting device with two degrees of freedom allows the transducer to be finely adjusted with the centre of the transducer surface fixed in its position.

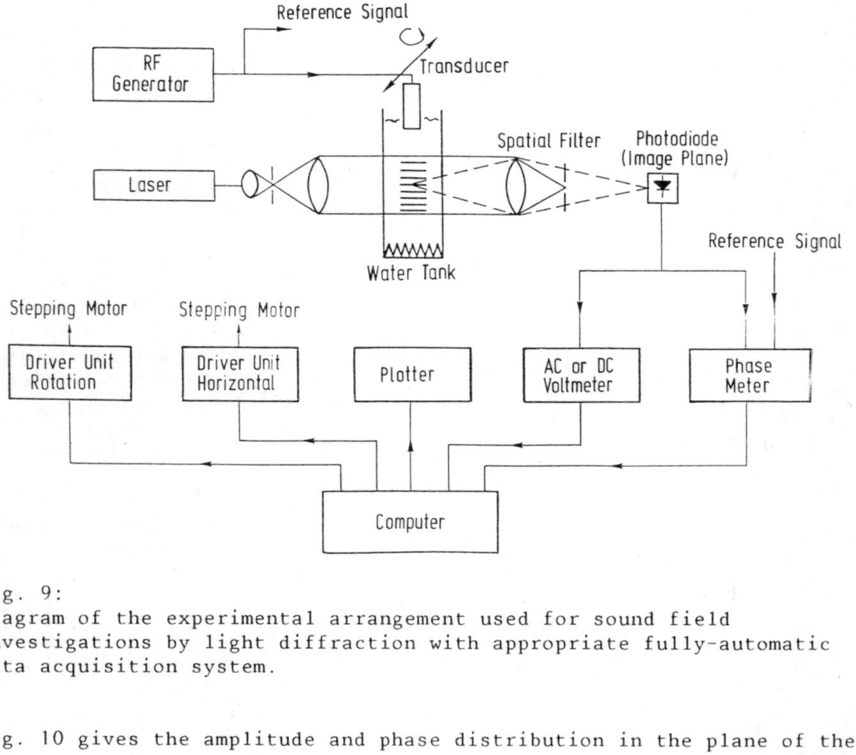

Fig. 9:
Diagram of the experimental arrangement used for sound field investigations by light diffraction with appropriate fully-automatic data acquisition system.

Fig. 10 gives the amplitude and phase distribution in the plane of the last maximum of a 3 MHz transducer 20 mm in diameter, the projection profile of which is shown in Fig. 4. The peak value of the sound pressure amplitude is 140 kPa. The phase varies between $\pm 0.2\pi$. The second example falls back upon the sound field of a composite transducer*, projections of which are shown in Fig. 5.

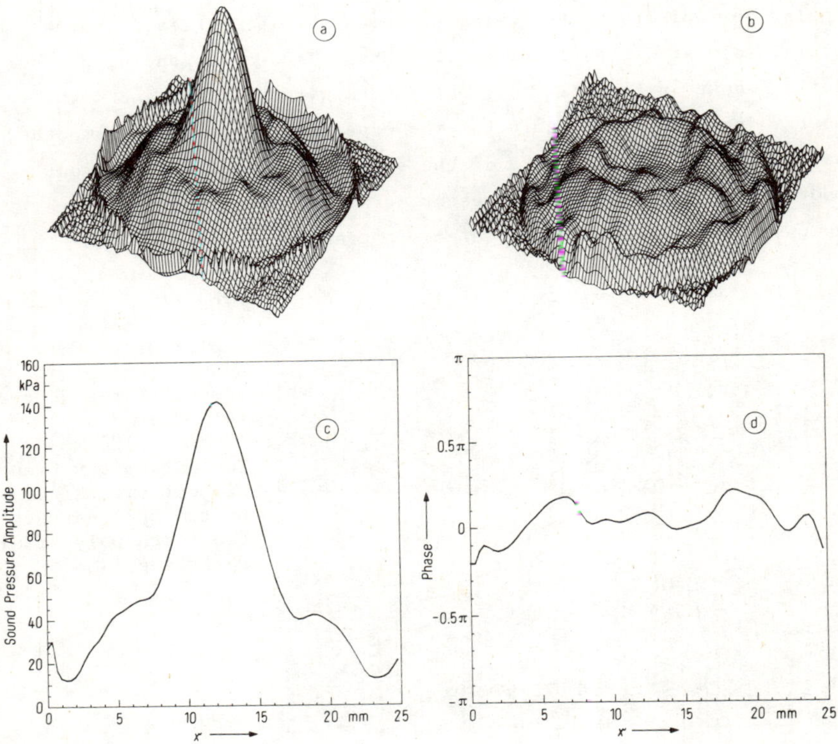

Fig. 10:
Filtered back projection of the sound-pressure amplitude (a) and the phase (b) by means of discrete convolution with corresponding sectional views (c,d). The measured values are sampled at a distance of z_o=200 mm from the transducer surface (nominal frequency: 3 MHz, excitation frequency: 3.18 MHz, diameter: 20 mm, applied voltage: U_{rms} = 5V). The amplitude is determined using the dc component in (11).

---

*The experiments on the composite material were performed at the PTB in colaboration with Dr. K. P. Richter from the Institute of Applied Biophysics of the Martin-Luther University in Halle, GDR.

The composite used consists of a two-dimensional periodic array of rectangular ceramic elements 0.2 mm in length and with a spacing of 0.2 mm. These composite materials show three principal types of resonance, one of which is the various lateral modes caused by the regular periodicity of the ceramic elements. Fig. 5 clearly reveals a lateral mode at 0° (parallel to the cuts). The corresponding pressure profile shown in Fig. 11 is reconstructed without (top) and with (bottom) polynomial smoothing of the experimental data. The smoothed shape shows the main feature of the sound field, whereas the non-smoothed shape might also represent details of the fine structure. In any case, both shapes show a defect in transducer manufacture.

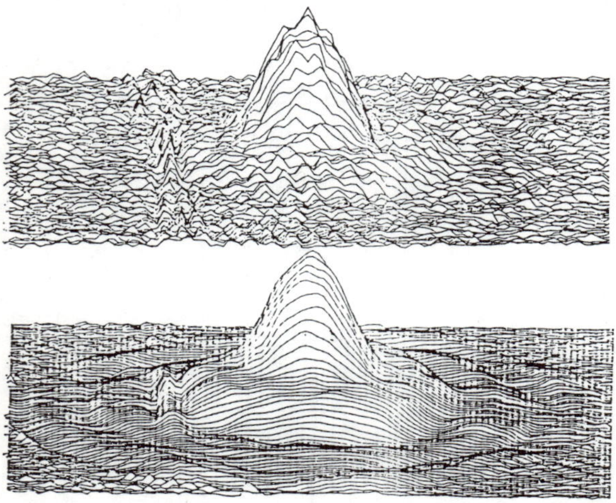

Fig. 11: Sound-pressure amplitude of the sound field of a composite transducer (4.2 MHz, diameter 10 mm, z=100 mm). Top: Reconstruction without smoothing. Bottom: Reconstruction with polynomial smoothing.

In the third example (Fig. 12) three pressure profiles of the sound field of a broad-band transducer are shown for various distances. We shall return to this example once again in connection with Fourier-optical sound-field calculations.

For the investigation of standing waves, an ultrasonic resonator with a 1 MHz transducer 20 mm in diameter is used. The reflector consists of a 3 mm thick brass plate. Measurements are performed using the zero and first diffraction order (interaction term in eq. (16)) and assuming axial symmetry of the sound field. Fig. 13 shows the measured amplitude and phase profiles for a distance of about 20 mm between the transducer and the reflector. It can be seen that the amplitude shows approximately a Gaussian distribution. The phase is almost constant in the inner region and decreases rapidly at the edge of the sound field.

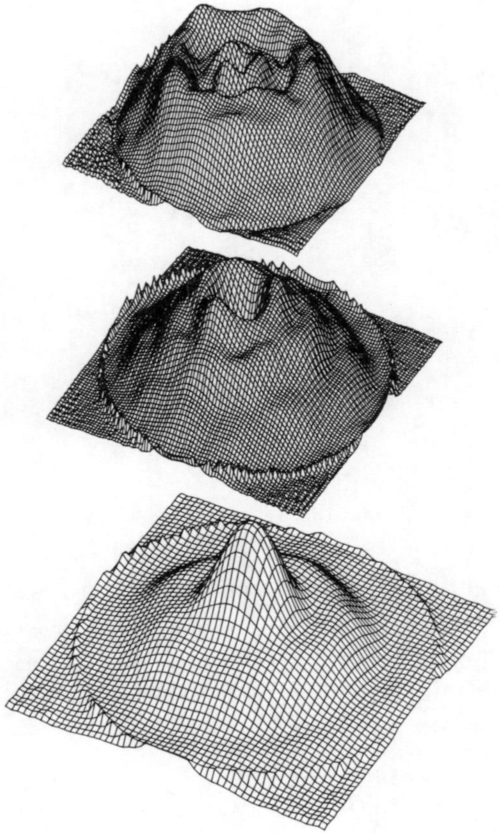

Fig. 12:
Pressure amplitude profiles of a broad-band transducer (3 MHz, 10 mm diameter) measured at z=5 mm (a), z=21 mm (b) and z=50 mm (c).

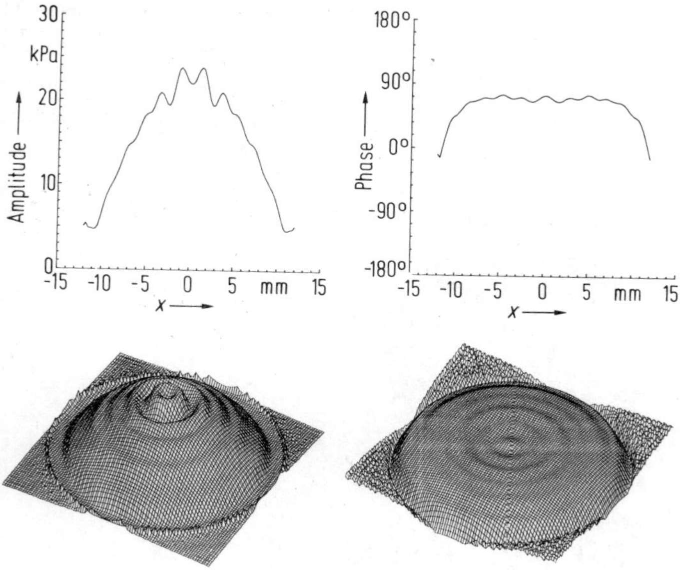

Fig. 13: Amplitude and phase profiles in the central cross-sectional plane of a resonator.

3. ANGULAR SPECTRUM FOURIER ANALYSIS

3.1 Basic Relationships[6)]

We take an ultrasonic beam propagating in the direction of the positive z-axis of a Cartesian coordinate system. Let the complex-valued pressure $p(x,y,z_1)$ in the cross-section at $z=z_1$ have a two-dimensional Fourier transform $P_1(f_x,f_y;z_1)$. The angular spectrum Fourier analysis then permits the corresponding profile in an arbitrary cross section (say $z=z_2$) to be calculated using the following relationship

$$p(x,y,z_2) = F^{-1}\{P_1(f_x,f_y;z_1)\,H_{pp}(f_x,f_y;z_2-z_1)\} \tag{23}$$

Here, H_{pp} is the transfer function and f_x and f_y are the spatial frequencies in the x and y direction, respectively. The transfer function

$$H_{pp}(f_x,f_y;z_1-z_1) = \frac{P_2}{P_1} = \exp\{ik^*(z_2-z_1)(1-\lambda^2 f_x^{\,2}-\lambda^2 f_y^{\,2})^{\frac{1}{2}}\} \tag{24}$$

is valid for all physical quantities obeying the Helmholtz equation. In this formula $f_x^2+f_y^2<1/\lambda^2$ is assumed, since evanescent waves can be neglected here.

Furthermore, by means of modified transfer functions the mapping of the z component of the particle velocity u_z can be calculated, provided the pressure profile is given and vice versa. The respective formulas read

$$H_{pu} = \frac{1}{\rho c} H_{pp}(f_x,f_y;z_2-z_1)(1-\lambda^2 f_x^2-\lambda^2 f_y^2)^{\frac{1}{2}} \tag{25}$$

$$H_{up} = \rho c\, H_{pp}(f_x,f_y;z_2-z_1)(1-\lambda^2 f_x^2-\lambda^2 f_y^2)^{-\frac{1}{2}} \tag{26}$$

c is the sound velocity and ρ is the density of the liquid at rest.

Once these field quantities are known, the axial component of the time-average intensity, i.e. the energy flux density, is given by

$$I_z = \frac{1}{2} \operatorname{Re}\{p^* u_z\} \tag{27}$$

3.2 Computer Simulation and Experimental Results [7)8)9)10)]

The model of a plane piston radiator - already applied in section 2.2 - is a useful means for theoretical considerations of sound field structures as well as for worst-case estimates of the spatial resolution of receivers. These calculations are usually performed in the spatial domain using the Huygens-Rayleigh diffraction integral. The Fourier-optical technique, in turn based on the two-dimensional complex FFT, makes possible a fast and simple evaluation of experimental data. Care must be taken, however, that no overlapping occurs in either the space or the frequency domain (sampling theorem), which also imposes some restrictions on the permissible distance from the transducer for a given number of NxN sampling points.

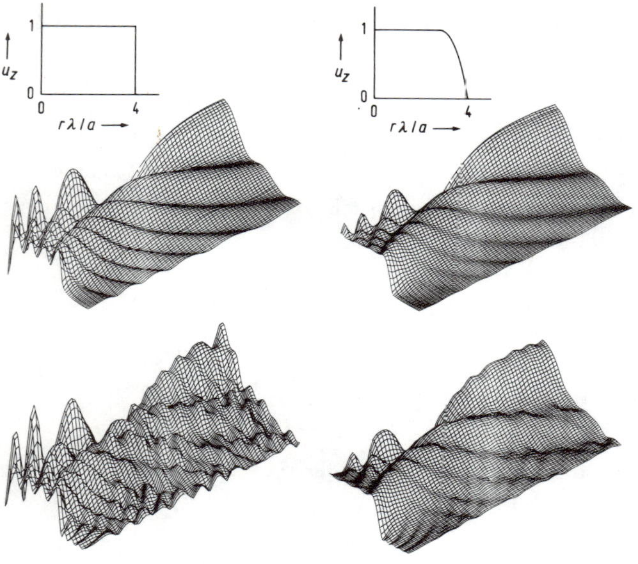

Fig. 14:
Calculation of the longitudinal pressure distribution using the Huygens-Rayleigh diffraction integral (top) and the Fourier-optical technique (bottom) for different particle velocity distributions. The diagram clearly reveals the limits of the Fourier-optical technique.

Fig. 14 shows an example of both the Huygens-Rayleigh principle based on spherical elementary waves (top) and the Fourier-optical principle based on plane partial waves (bottom). On the left-hand side a piston-like radiator is assumed and on the right-hand side a cosin-shaped particle velocity dependence at the edge is assumed. The rounding off of the particle velocity profile obviously results in a smoothing of the Fresnel zone and in a low pass filtering of the spatial frequency spectrum. Consequently, aliasing effects in the Fourier-optical sound field calculations are less pronounced in the band-limited case than for the piston radiator which is not band limited and cannot be truly described by a sampled function.

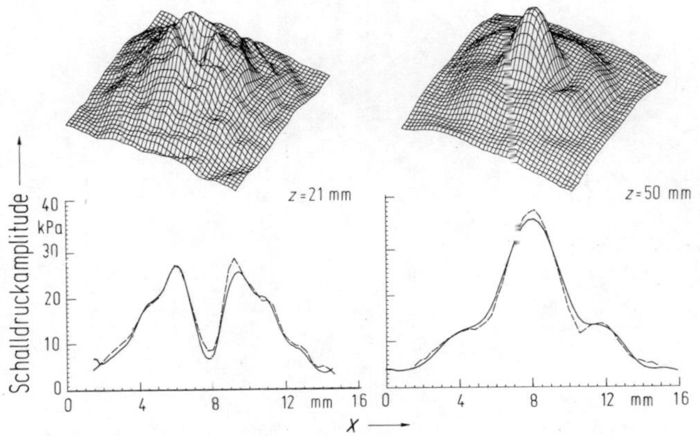

Fig. 15:
Pressure amplitude profiles at z=21 mm (a) and z=50 mm (b) calculated from the experimental data at z=5 mm (see Fig. 12). The corresponding sectional views (c) and (d) show the agreement between calculated (broken line) and experimental profiles (solid line).

The application of the Fourier-optical technique to experimental data is demonstrated in Fig. 15. Here, the data of the z=5 mm plane in Fig. 12 were taken to calculate the profile at z=21 mm and z=50 mm using a 128 x 128 point routine. The sectional views illustrate the good agreement between calculated and measured profiles. Fig. 16 gives the longitudinal plot in the xz plane for this example of application. Compared with the cicular piston in Fig. 14, the higher spatial frequency components which are the basis of the well-known highly structured interference pattern, are missing here due to the band limitation of the radiated pressure profile. It should be stressed that no aliasing effects were observed in any of the experimental applications.

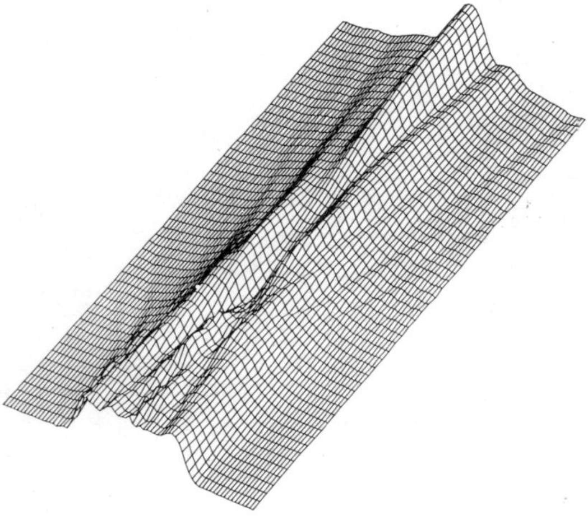

Fig. 16:
Longitudinal plot of the pressure profile in the xz plane of a broad-band transducer (see Fig. 12 and 15).

The use of eqs. (25) and (27) allows the component of the time-average intensity to be calculated (Fig. 17). The transducer used in the

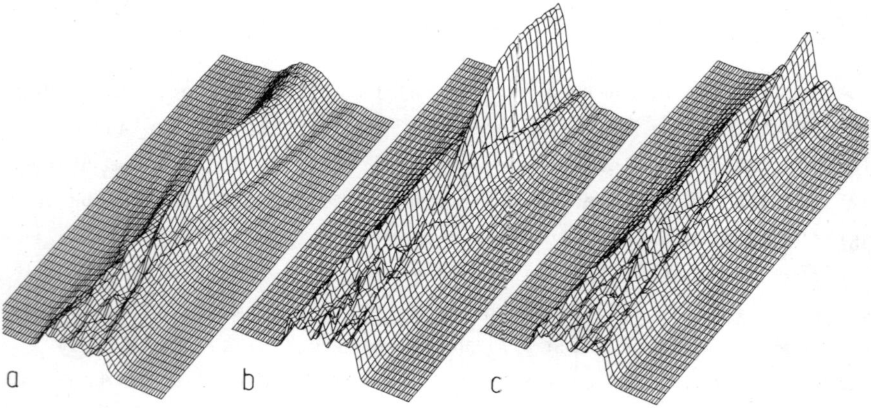

Fig. 17:
Longitudinal plots of the intensity fields of a broad-band transducer at 2.2 MHz (a), 3.0 MHz (b) and 3.8 MHz (c).

preceding case was again taken and it was excited at 2.2 MHz, 3.0 MHz and 3.8 MHz. For all frequencies studied, the intensity fields of this broad-band transducer show a rather uniform structure in the Fresnel zone, disregarding the length of the Fresnel zone which increases with increasing frequency.

4. CONCLUDING REMARKS

Although only a few theoretical and experimental examples of the subject field reconstruction could be given, it can be seen that the angular spectrum Fourier analysis offers a useful tool for obtaining a maximum of information about the sound field structure, and at the same time entails a minimum of experimental effort. This is convincingly demonstrated by Figs. 15, 16 and 17b showing cross-sectional and longitudinal pressure profiles together with the corresponding intensity profile, on the basis of one single experimental data set. Furthermore, the light diffraction method in combination with the second reconstruction technique, computer tomography, is distinguished by high spatial resolution and provides an excellent tool for those fields of application where conventional measuring techniques fail.

REFERENCES

|1| Riley, W. A., Love, L. A. and Griffith, D. W., "Observation of Raman-Nath optical diffraction in the phase grating plane", J. Acoust. Soc. Amer. 71, 1149 (1982).

|2| Reibold, R. and Molkenstruck, W., "Light diffraction tomography applied to the investigation of ultrasonic fields. Part I: Continuous waves", Acustica 56, 180-192 (1984).

|3| Reibold, R., "Light diffraction tomography applied to the investigation of ultrasonic fields. Part II: Standing waves", Acustica 63, 283-289 (1987).

|4| Reibold, R., "Investigation of ultrasonic pulses using light diffraction", Proceedings of the 5th. FASE-Congress, Lisboa, 19 - 22 (1987).

|5| Scudder, H. J., "Introduction of computer-aided tomography", Proc. IEEE 66, 628 (1978).

|6| Goodman, J. W., "Introduction to Fourier optics", Mc Graw-Hill, San Francisco (1968).

|7| Reibold, R., "Calibration of ultrasonic fields using light-diffraction tomography and Fourier-transform techniques", Ultrasonics International 85, Conference Proceedings, Butterworth & Co, Guildford, 686-691 (1985).

|8| Reibold, R. and Holzer, F., "Complete mapping of ultrasonic fields from optically measured data in a single cross section". Acustica 58, 11-16 (1985).

|9| Holzer, F. and Reibold, R., "Numerical analysis of ultrasonic transducer vibrations from optically measured beam profiles". Acustica 60, 236-245 (1986).

|10| Reibold, R., "Sound source reconstruction using Fourier optics". Acustica 63, 60-64 (1987).

SCANNING ACOUSTIC MICROSCOPY : GENERAL DESCRIPTION FIELD OF APPLICATIONS AND QUANTITATIVE MEASUREMENTS

J. ATTAL
Laboratoire de Microacoustique
Université des Sciences et Techniques du Languedoc
Place Eugène Bataillon, 34060 Montpellier cédex, France

1. INTRODUCTION

There have been a number of attempts to make an acoustic microscope since the idea of Sokolov to use high frequency acoustic waves in 1939. However the technology at that time did not allow the idea to be implemented, and no serious work was started before the late 1960s. At the Zenith Laboratoy in Chicago, L. Kessler and A. Korpel developed an acoustic system in which a scanned laser beam is used to read out the image from a plastic membrane deformed by an acoustic signal. Undeniably, the most extensive work in that field began at Stanford University with B. Auld who demonstrated and analyzed several systems of microscopic acoustic imaging and C.F. Quate who realized the first scanning acoustic instrument which is presented and discussed all along this paper (1). In opposition to other acoustic technics, acoustic waves are only used, and both the source and the detector are scanned relative to the specimen to be imaged, or conversely.

It is clear that in utilizing a form of radiation fundamentally different from light one will probe different properties of an object. Electron beams, X-rays, thermal waves and ultrasonic waves, all reveal different aspects of the matter, simply because the nature of the interations is basically different. Electron microscopy. X-rays and thermal wave microscopy, photo-acoustic, photo-termal and electron-acoustic microscopy, many forms of microscopy which are available nowadays and complete with optical and acoustic instruments.

2- SCANNING ACOUSTIC MICROSCOPE : FUNDAMENTALS

The first scanning acoustic microscope operated in transmission mode in 1974. An acoustic wave was brougth to a focus by a sapphire lens, and in absence of a specimen, was received by a similar lens arranged confocally. The object was then introduced into the waist of the common focal zone and mechanically scanned between them. The first pictures were taken from biological specimens and showed that acoustic waves are sufficiently scattered by the variations in elastic properties and good contrast is obtained without the need of any kind of staining.

Gradually interest began to shift from transmission microscopy to reflection microscopy with a single lens being used for both transmitting and receiving the acoustic signal. It was mainly because of the difficulty of aligning the transmitting and receiving lenses increasing as we push the frequency higher and higher. Another important reason is the interest of imaging subsurfaces of specimens much thicker than the path between the two lenses which does not justify to be crossed all the way through their thickness by ultrasonic waves.

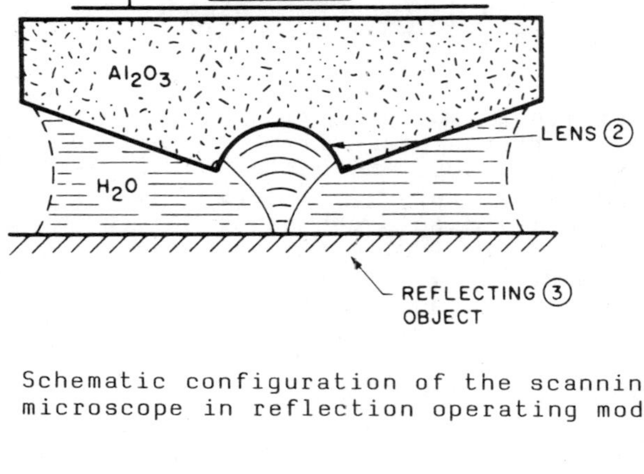

Fig. 1 : Schematic configuration of the scanning acoustic microscope in reflection operating mode.

In reflection mode, the acoustic microscope system (fig. 1) resembles a miniature radar or sonar system. The acoustic beam, after reflection from the sample immersed in a liquid (such as water), passes back through the lens and into the piezoelectric transducer that generated it. The transducer, now working as a receiver, detects the signal, whose amplitude and phase carry information about the sample. The acoustic lens moves in a raster pattern over the specimen (or vice versa) examining each point sequentially. The system requires several seconds to cover a single frame containing 512 x 512 pixels, each of which is typically 0.5 microns in size. The returning pulses modulate the intensity of the beam in a cathode-ray tube-monitor via a digital storage capability.

The lateral resolution of the microscope is set by the diameter of the spot size at focus, which is approximately given by $v/2f \sin\theta$, where v is the sound velocity in the liquid coupler, f is the operating frequency and θ is the half angular aperture. We see that the resolution increases with the frequency. However, whereas increasing the frequency, the total losses in the system increases too, which leads to a reduction in detection sensitivity. For classical liquids the absorption coefficient increases with f^2, so we have to deal with low velocity and low loss liquids for use near room temperature, to improve the resolution of the microscope. It turns out that water is one of the best liquid compatible with most objects with an absorption constant of about 200 dB per millimeter at 1 GHz.

2-1 ACOUSTIC LENS DESIGN

For regular applications, the acoustic lens consists of a spherical cavity ground and polished in a solid buffer rod (generally sapphire or fused quartz). The high velocity ratio beetween the solid and the liquid (around 8 with sapphire-water interface) causes the rays to converge to a focus very near to the centre of the spherical cavity : spherical aberration is therefore minimized leading to a nearly diffraction limited focus (fig.2). At the back of the rod a sputtered ZnO transducer is formed between two Cr-Au electrodes. This transducer structure is almost universal for high frequency lenses (above 300 MHz). At low frequencies a plate transducer (lithium niobate) is bonded to the buffer rod. The transducer converts the applied electromagnetic pulse to an acoustic pulse which travels along the rod in the form of a quasi-plane wave. The spherical lens produces the appropriate phase curvature on the wavefronts to bring the energy to focus at the focal plane. In the receiver mode, the lens recollimates the reflected acoustic energy and the transducer converts it back to an electromagnetic signal.

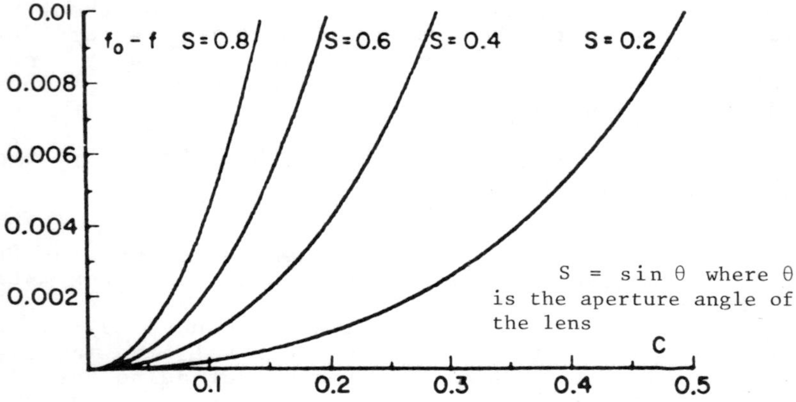

Fig. 2 : Spherical aberration in unit of radius of the lens as a function of the liquid/solid velocity ratio.

Although the high velocity ratio at the lens/couplant interface makes it possible to acheive a diffraction limited focus, it is unfortunately, in a majority of cases, associated with a large impedance discontinuity at the interface, which results in a large reflection loss (e.g. 9 dB for a sapphire/water interface for a plane wave at normal incidence). The loss can be greatly reduced by the inclusion of a quarter-wavelength matching layer. Sputtered glass can have the impedance needed to act as an impedance transformer.

Owing to high water losses at high frequencies, one tends to use lenses with smaller radii of curvature to minimize the path losses. For high frequencies we are concerned with lenses whose radii of curvature are much less than 1 millimeter. For a 1 GHz lens, a radius of curvature of about 100 μm is typical.

The length of the crystal rod is an important design parameter. It must satisfy two requirements. Firstly, it must be such as to provide a temporal separation between the echoes from the surface of the lens and the object echo. Normally the lens geometry is chosen so that the information pulse falls in a time-window between two successive echoes from the lens/water interface. Secondly to obtain a good focus, the lens should be illuminated as uniformly as possible. The top dot electrode deposited on the transducer determines the diameter of the collimated acoustic beam. The lateral distribution of amplitude at various planes parallel to the transducer is shown schematically in Fig.3. Between the transducer and the Fresnel focal length there are large fluctuations in the amplitude distribution. For best performance, it is usually desirable to illuminate the lens aperture with a phase and amplitude distribution as uniformly as possible.

2-2 ELECTRONIC SYSTEM

The operation of a reflection microscope is somewhat similar to that of a pulsed radar system (fig. 4). The output of a CW signal generator is pulsed by means of a series of cascaded PIN switches. The width of the pulses is generally of the order of 25 ns for frequencies above 500 MHz. The minimum on/off ratio is dictated by the maximi-

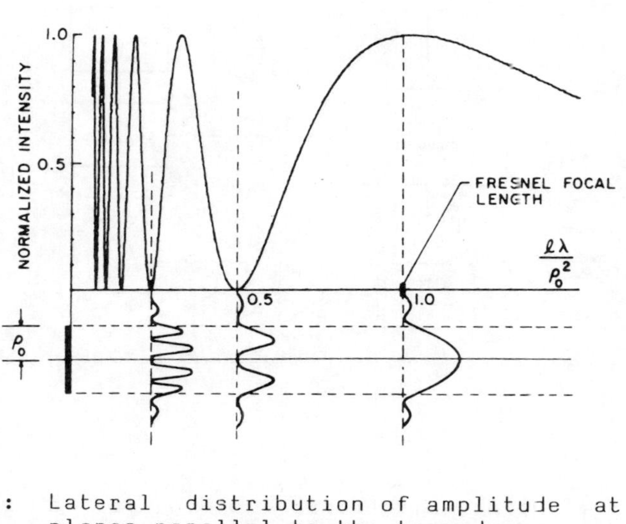

Fig. 3 : Lateral distribution of amplitude at various planes parallel to the transducer.

mum loss allowable through the lens system. Typically 80 dB is currently achieved which leaves 40 dB of dynamic range for imaging with an input electrical power of a few tenths of a watt. No special signal processing is used which can drastically slow down the acquisition time of one frame. Successful attempts have been done with real time processing using a pulse expansion and compression technique in the low frequency range.

A time gating system suppresses the echoes from the surface of the lens and other unwanted signals. Finally after suitable detection and amplification, the outcoming pulses are fed to a sample-and-hold or a box car integrator circuit which provides the video signal used in the display system. A digital frame storage system is used for this purpose associated to a scan converter which allows the information to be written-in at the mechanical scanning rates and read out at the TV rate. The informations are therefore digitized and stored in an electronic memory, during a frame scan.

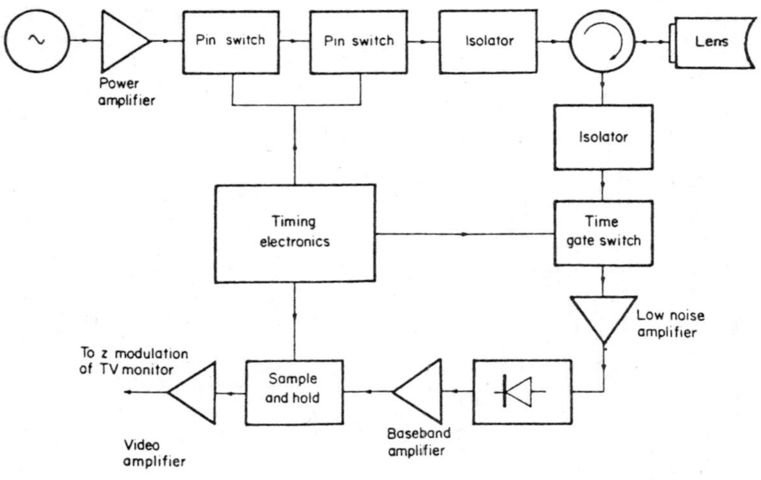

Fig. 4 : Typical electronics set up for a reflection microscope.

3- SURFACE APPLICATIONS

The scanning acoustic microscope covers a wide range of applications from surface imaging such as in micro-electronic components, metal, composites, ceramics, to in-depth applications with inspection of bonding between two materials, adhesion, flaws, cracks which do not open to the surface.

3-1 SURFACE IMAGING

The main feature of the acoustic microscope is generally the higher contrast of the images compared to Optics. This can be surprising if we consider the expression of the reflection coefficient at the interface given in normal incidence by :

$$R = \frac{Z_2 - Z_1}{Z_2 + Z_1}$$

For a typical solid-water interface Z_1 is about 30 Mrayls and Z_2 = 1.5 Mrayls, so R = 0.9. Now, if we consider a change of say 30 per cent in the impedance of the solid, this would induce a difference of 5 per cent in the amplitude of the reflected wave. The coefficient of reflection is so nearly unity that its departure from unity is only weakly dependent on the impedance of the solid. Fortunately, this simple consideration of the reflection coefficient at normal incidence does not consider the strong change of this function around the Rayleigh angle. The theory for this effect will be developed further.

Let us first list some typical applications of surface imaging selected in different fields.

- MICROELECTRONICS

In that field most applications can be devided in two parts: device inspection and material characterization (2) : delamination of thin passivating films, adhesion of silicon layer to the sapphire substrate in silicon on sapphire (SOS) devices (3), non uniformity in the adhesion of the aluminium tracks in RAM chip, masking misalignement, topographic variations on the surface, hillocks shorting two layer metallizations, defects in epi-silicon (spikes, dislocations in unetched epi-silicon), minute thickness variations.

- METALLURGY

The applications range from the study of strength of materials to fracture mechanics. Study of diffusion bonds, characterization of grain structure and orientation in polycristalline metals, locating and imaging surface and near surface defects, are all areas where reflection microscopy has provided fruitful results (4,5). As in microelectronics, the acoustic microscope facilitates the study of the adhesion of metal thin films, a disbonded region would lead to 100% reflection. For instance the adhesion of protective layers of TiN on cutting tool (6) has been studied and it has been found stress-relief cracks induced in the hard material during the manufacturing of the coating.

- COMPOSITES AND CERAMICS

Pores in ceramics can be readily imaged in the acoustic microscope (7,8). A fluid filled porous ceramic has large impedance variations in its structure and can give high contrast. When a polycrystalline material is imaged, different orientation of the crystallites give rise to contrast between the grain. In glass fibre (9) or carbon-carbon composite optical images give little contrast in opposition with acoustic images where the components have quite different elastic properties. Microcracks are easily visible such as the distribution of the fiber in the matrix, and their elastic properties.

3-2 SURFACE AND MATERIAL CHARACTERIZATION

3-2-1 General consideration

As said before, the high contrast observed in using a reflective scanning microscope cannot be interpreted in terms of reflection coefficient near normal incidence. The second feature observed is that this contrast is strongly dependent on the defocus. We observe a succession of contrast reversals as we move the specimen toward the lens (in z direction) so we cannot simply say that a brighter area corresponds to a higher (or lower) density or to a greater (or smaller elastic modulus. Interpretation (10,11) must be based on the variation of the signal intensity (V, the voltage generated at the transducer) with defocus (z). Its expression for an object placed at the focus (z = 0) may be written :

$$V(0) = \int_0^\infty P(\theta)\, R(\theta)\, 2\pi\, r\, dr \qquad (1)$$

where $P(\theta)$ is the composite pupil function for the lens and transducer together, $R(\theta)$ the reflectance function of this object with respect to incidence angle θ, r is the radial coordinate relative to the transducer.

If f is the focal length sustituting $f\sin\theta = r$, $f.\cos\theta\, d\theta = dr$, equation (1) yields.

$$V(0) = 2\pi\, f \int_0^{\pi/2} P(\theta)\, R(\theta)\, \sin\theta \cos\theta\, d\theta \qquad (2)$$

If we moved the surface of the object towards the lens by a displacement $-z$ from the focus, the phase of the wave incident at a given point on the surface will advance by $k.z$ (where k is the wavevector in the fluid) and waves returning to the lens will advance their phase by twice this, and the signal at the transducer is now :

$$V(z) = 2\pi\, f \int_0^{\pi/2} P(\theta)\, R(\theta)\, e^{-2ikz\cos\theta} \sin\theta \cos\theta\, d\theta \qquad (3)$$

If we substitute $t = 2k\cos\theta$
$dt = -2k\sin\theta\, d\theta$

$$\text{then } V(z) = 2\pi\, f \int_0^{2k} P'(t/k)\, R'(t/k)\, e^{-itz}\, \frac{t}{4k}\, dt \qquad (4)$$

where P' and R' are suitably defined for the new argument and the integral may be taken over infinite limits provided $R' = 0$ unless $0 \lesssim t/k \lesssim 2$:

$$V(z) = \frac{\pi f}{2k} \int_{-\infty}^{\infty} P'(t/k)\, R'(t/k)\, e^{-itz}\, t\, dt \qquad (5)$$

This expression can be recognized as a Fourier transform with t and z as the transform pair.

The reflectance function for waves in a fluid incident on the surface of an isotropic solid derives from the boundary conditions using the impedances of the two medias in oblique incidence. Snell's law relates the incidence angle to longitudinal and shear angles according to the following equations :

$$\frac{\sin\theta}{c} = \frac{\sin\theta_1}{c_1} = \frac{\sin\gamma_1}{b_1} \tag{6}$$

$$Z = \frac{\rho c}{\sin\theta} \qquad Z_\ell = \frac{\rho_1 c_1}{\cos\theta_1} \qquad Z_t = \frac{\rho_1 b_1}{\cos\gamma_1} \tag{7}$$

where Z, ρ , θ , c are respectively the impedance, density, incidence angle and velocity of longitudinal waves in liquid and solid (with the subscript 1) and b being the incidence angle and velocity for shear waves in solid according to Brekhovskikh's notation (12).

In order to keep valid the formulation of the reflectance function in normal incidence, we define a total impedance for the solids by the equation :

$$Z_{tot} = Z_\ell \cos^2 2\gamma_1 + Z_t \sin^2 2\gamma_1 \tag{8}$$

Then the reflectance function is :

$$R(\theta) = \frac{Z_{tot} - Z}{Z_{tot} + Z} \tag{9}$$

As an example we show the reflectance function for acoustic waves incident from water onto glass (fig.5). $R(\theta)$ is a complex function, so the modulus is refered to the left ordinate, and the phase to the right ordinate. starting from zero angle of incidence,the modulus of the reflectance function shows up its first major fluctuations at around 15 which corresponds to the critical angle for longitudinal wave excitation in the glass. Above the shear wave critical angle no energy can be propagated into the solid, so the modulus of the reflectance function must be unity. The phase of the reflectance function exhibits small fluctuations around the longitudinal critical angle, but its most dramatic behaviour occurs a few degrees beyond the shear wave critical angle. This is centered around the angle θ_R at which waves in the liquid can couple into Rayleigh surface acoustic wave on the solid, and as can be seen the phase changes by almost 2π over a fairly small change in the angle of incidence. This phase change dominates the behaviour of V(z).

In order to understand the curve V(z) of fig. 6, let us

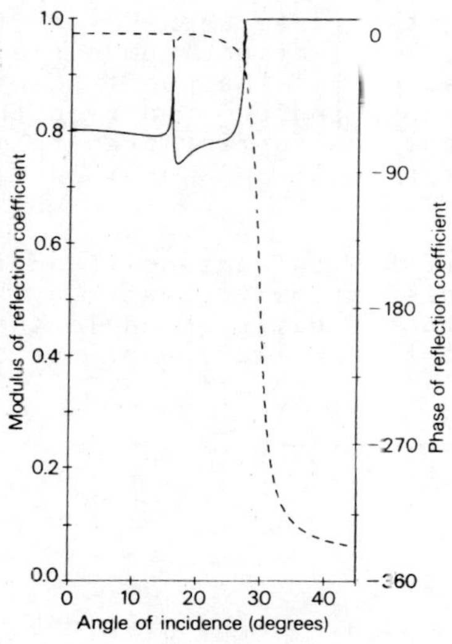

Fig. 5 : Acoustic reflectance functior of water glass interface as function of the ircidence angle.

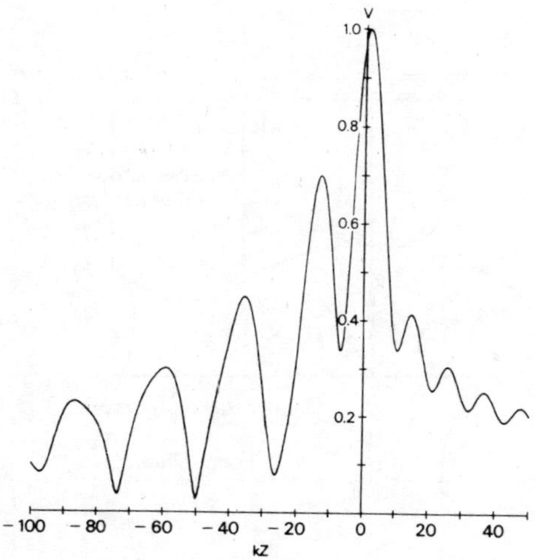

Fig. 6 : V(z) curve for glass with water coцplant.

consider first the case when R = 1 for all angles of incidence (i.e. reflection from a rigid surface or from a boundary with vacuum) if the pupil function were unity (or constant) up to some angle and zero thereafter, then V(z) is not far from the Fourier transform of a rectangular window which for small θ is a sinc^2 function ($\text{sinc}\,\theta$ is equal to $\sin\theta/\theta$).

If now we insert the reflectance function of a real material the phase discontinuities at $t = 2k\cos\theta_R$ and $t = 2k$, corresponding to the Rayleigh angle and normal incidence $\theta = 0$, give rise to oscillation of periodicity :

$$\Delta z = \frac{2\pi}{2k - 2k\cos\theta_R} \tag{10}$$

in the Fourier domain.

This Fourier analysis emphasizes the role of two waves : incident at the Rayleigh angle and normally. So it is possible to predict the periodicity of the oscillations in V(z) by using a simple ray model illustrated in fig. 7.

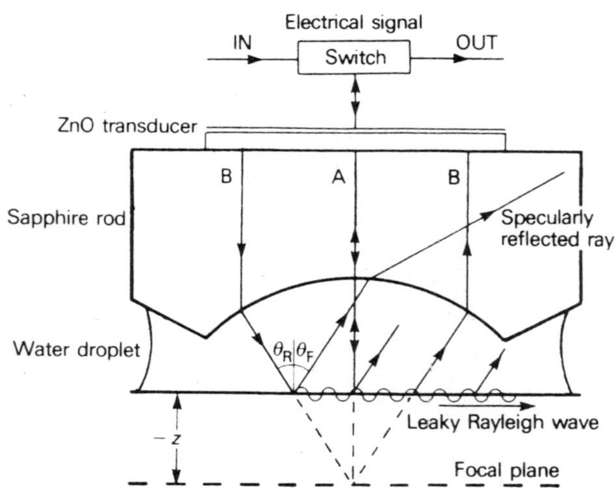

Fig. 7 : Simple ray approach to explain the V(z) mechanism.

Compared with focus the normal ray A undergoes a change in path length of 2z, whereas the ray incident at the Rayleigh angle B has its path in water shortened by $2z/\cos\theta_R$ while it excites a Rayleigh wave on the specimen surface. This propagates along the surface at a velocity $c/\sin\theta_R$ which gives an equivalent acoustic path in the fluid of $2z \,\mathrm{tg}\, \theta_R \sin\theta_R$.

The phase difference between the two rays is therefore :

$$\Delta\phi = (2z - \frac{2z}{\cos\theta_R} + 2z \,\mathrm{tg}\, \theta_R \sin\theta_R) \cdot k + \pi \tag{11}$$

The extra phase change of π is associated with the excitation of the Rayleigh wave as seen on the reflectance function.

$$\Delta\phi = 2\,kz\,(1 - \cos\theta_R) + \pi \tag{12}$$

This will give oscillations of periodicity

$$\Delta z = \frac{2\pi}{2k(1 - \cos\theta_R)} \tag{13}$$

In conclusion, the contrast in the scanning acoustic microscope is generally described by the V(z) formalism for surface and subsurface observation (13). The dominant feature is the behaviour of the reflectance function around the Rayleigh angle.

The efficiency of the conversion depends on density ratio, the velocity ratio between liquid and solid. According to Dransfeld and Salzmann's formula (14) the re-emission factor of Rayleigh wave into longitudinal wave in liquid is given by:

$$\alpha_\ell \lambda_w = (\frac{\rho w}{\rho s}) (\sin\theta_R)^2 \tag{14}$$

where α_ℓ is the attenuation of the Rayleigh wave resulting from re-radiation into water, λ_w the wavelength in water, and ρw, ρs the densities of the water and the solid.

Using (14) we can have a simple estimation of the amplitude of the leaky wave voltage relative to the voltage produced by the geometrically reflected ray field found by integration over the transducer :

$$= \frac{2\pi^{1/4}}{F^{3/4}} \frac{\rho w}{\rho s} (\sin\theta_R)^{5/2} (f/R_T)^{3/2} \quad (15)$$

where R_T is the radius of the transducer and F the ratio of the separation between the transducer and back focal plane to the Fresnel distance of the transducer. In typical lens design F near unity.

It is essential to use a lens of numerical aperture large enough to include the Rayleigh angle. For low Rayleigh velocity such as in high density materials or plastics, the Rayleigh angle is greater than the opening angle of the acoustic lens acheivable or sometimes does not exist in real form since $\sin\theta_R > 1$ and no typical V(z) curve should be recorded. In fact in some case, one can observe a typical V(z) curve which may be attributed to longitudinal skimming waves since the surface wave velocity measured corresponds rather to that of longitudinal mode in the solid (fig.8). This has been seen in materials such as platics, copper, arsenic trisulfide, where sometimes Rayleigh waves can coexist with skimming longitudinal waves causing a faint modulation.

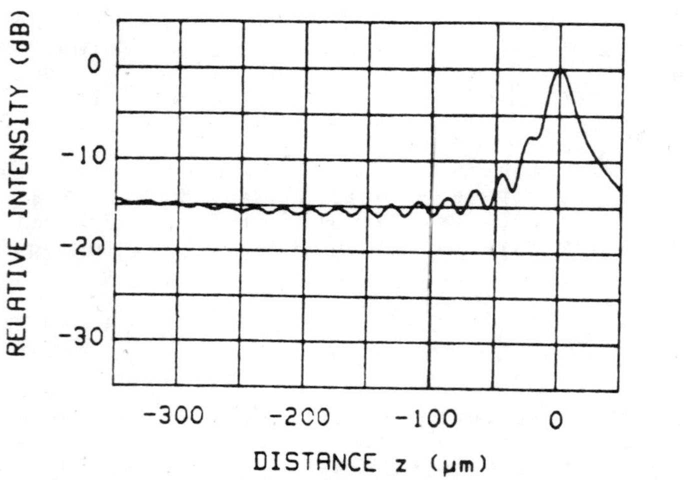

V(z) curves on water/PMMA boundary measured at 225 MHz.

Fig. 8 : V(z) curve for compressionnal skimming modes.

3-2-2 Surface velocity measurements

The simplest application of this technique called Acoustic Material Signature (AMS) tchnique is the direct calculation of the surface velocity by measuring the periodicity of the curves (15). It is an easy method for microprobing the elastic properties of the surface of materials without special preparation.

3-2-3 Anisotropy of materials

Anisotropy gives rise to different surface wave velocities, which depend on the lateral directions. A spherical lens launches surface waves radially in all directions : the surface wave velocity which gives rise to the V(z) dips is therefore velocity averaged over all directions. However, one can devise a cylindrical lens which generates the surface waves in one direction only (fig.9) (16). With this scheme V(z) curves at different angles give different periodicities and therefore the complete velocity surface can be determined. This complete determination of surface velocities is unfortunately related to a reduction in the lateral resolution due to the type of lens used.

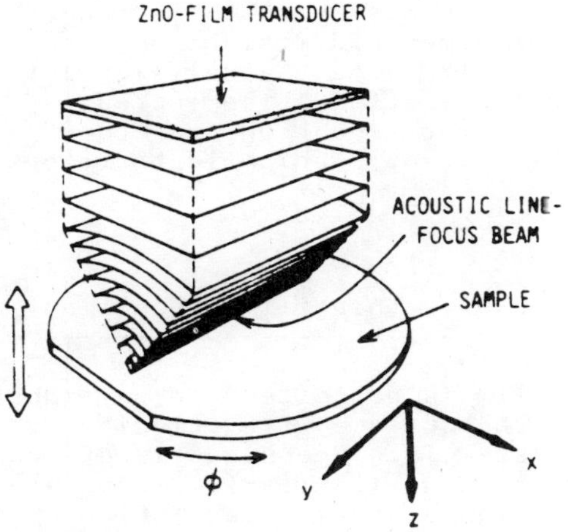

Fig. 9 : Schematic configuration of a cylindrical lens.

A consequence of this anisotropy is that it is now necessary to specify not only the zenithal angle θ in the reflectance function, but also the azimuthal angle ϕ and the Miller indices (h, k, l) of the surface, so that the reflectance function must be written R_{hkl} (θ, ϕ).

Larger deviations from isotropic behaviour are observed when ϕ is changed. A first phase change of 2π is observed for a given ϕ corresponding to the excitation of a Rayleigh wave. A second phase change of 2π is observed for a larger angle corresponding to a pseudo surface wave, a phenomenon which occurs only in anisotropic materials. This double phase change might be expected to have a marked effect on V(z). The pseudo surface wave sometimes dominates the reflectance function, so that the Rayleigh wave introduces only small fluctuations in the V(z) curve.

In order to observe the V(z) that would be produced by each of these reflectance functions, the cylindrical lens must be used.

3-2-4 Surface attenuation

The presence of porosity and grain structure in a material causes scattering and attenuation of an ultrasonic wave in that material. By measuring the attenuation coefficient, we can deduce some informations on, for example, the density or grain size (17). As the Rayleigh path length is increased, the surface attenuation increases correspondingly with an attendant reduction in the peak-to-peak amplitude in the V(z) curves, which is therefore a measure of the attenuation.

3-2-5 Influence of a thin film deposition

When a thin film is deposited on a substrate, we notice a change of the surface wave velocity as a function of the film thickness. This technique is of considerable value for two basic reasons : Firstly, the measurement is carried out with high lateral resolution, the film thickness can be recorded point by point. Secondly, generating the surface waves by acoustic lens does not require the specimen to be piezoelectric. Examination of thin films used in microelectronics as well as for thicker, plated layers used for protection, have been carried out using this technique.

The theory for calculating the reflectance function for a multilayered structure is given by Brekhovskikh (12). The case of a single layer that is thin compared with a Rayleigh wavelength may be approximately treated by considering the layer as a perturbation on the elastic properties of the substrate. Under this condition the relative change in the Rayleigh velocity of the substrate can be expressed in the linear form defined (18) by :

$$\frac{\Delta V}{V_R} = S_o \left(\frac{2\pi h}{\lambda_R} \right) \tag{16}$$

where S_o is the initial slope of the dispersion curve, and h/λ_R the normalized thickness of the layer in termes of wavelength of the substrate. S_o is a function of electric parameter of the substrate and the film. When h/λ_R is greater than 1 or when the substrate is highly anisotropic, large deviation form this perturbation theory can be observed.

4- IN DEPTH IMAGING

Reflection acoustic microscopy has been successfully applied to imaging microstructures below the surface of optically opaque solids (19,20). The maximum operating frequency is limited by the attenuation characteristics of the materials. When the wavelength becomes comparable to the grain size, the grain boundary scattering loss is the dominant factor in the attenuation coefficient.

If one attempts to focus an ultrasonic beam below the surface of a solid, two problems are immediately encounters. Firstly, the high velocity discontinuity between the couplant and the object causes a large spherical aberration which degrades the formation of the subsurface focus divided itself into longitudinal and shear focii (fig.10). Secondly, high reflection loss at this interface tends to mask the relatively smaller reflection from the plane of interest, which arrives only shortly after the surface echo. This makes the separation and detection of the image echo difficult. Different workers have resorted to various techniques to overcome these problems and in this section, we shall look at these techniques and their relevant applications.

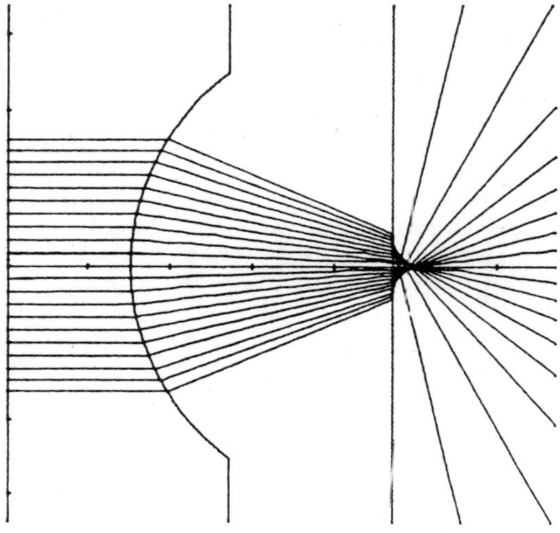

Fig. 10 : Spherical aberrations relative to in-depth imaging.

4-1 LIQUID METAL COUPLANTS

The use of water as a liquid couplant in the acoustic microscope causes high impedance discontinuity at the water/solid interface and allows only a small fraction of the energy to penetrate below the surface. The two-way reflection loss at this interface can be as high as 25 dB depending on the impedance of the object. By employing a liquid metal couplant, we cumulate two advantages : the higher density results in a larger impedance and the interface impedance discontinuity is then reduced and the object becomes more transparent ; the ultrasonic absorption in liquid metals is less than that in water which makes the couplant path loss smaller and higher operating frequency (i.e. higher resolution) can be achieved.

As far as the spherical aberration is concerned, gallium performs better than mercury, for which the velocity is very close to that in water (20). The reason is that the shear velocity in most solids is close to the longitudinal velocity in gallium. The shear waves generated at the interface can be brought to a nearly diffraction limited focus within the solid.

The use of liquid metals is not, however, without their problems. They can readily form amalgam with metals such as gold, tin lead, indium which can feature in some objects of interest. Furthermore, liquid metals do not easily wet most solids. Mercury will be prefered to gallium for all these wetting problems and the possibilities to prevent from amalgaming with specific care (19).

4-2 REDUCED APERTURE AND ASPHERICAL LENSES

The spherical aberration at an interface can be thought of as a distortion of the spherical wavefronts in the liquid. One can approach diffraction limited focussing with a spherical lens with a relatively small angular aperture. A typical reduced aperture lens produces a focus half angle of about 10° in water corresponding to 30-40° in the solid. One of the remarkable advantages of a reduced aperture lens is that a long range of subsurface depth can be probed, with a nearly diffraction limited focus. We see that, over a substantial range, the focusing performance is good (21), however the fact of using longitudinal waves in solids with small opening angle lead to a consequent decrease in the resolution.

Another way of imaging is using aspherical lenses which allow a better focusing inside the specimen. Longitudinal (22) or transverse (23) modes can be highly focused at a given depth by means of these type of lenses which can be suitable for inspection and repetitive examinations. Specific lens profiles can be made for either longitudinal or transverse focusing. Longitudinal modes are easier to focus because they correspond to paraxial incoming rays. Unfortunately the velocity ratio between liquid and object is high and the actual focus does not go very far under the surface. Transverse modes are more difficult to focus accurately since they are not generated in the object in normal incidence. Nevertheless, the lower velocity ratio lead to a better penetration in solids with an increase in resolution.

5- CRACKS AND DISCONTINUITIES

This section gives specific results on this subject which, in fact, could have been included in paragraphs 3 and 4. These problems deal with interference fringes seen in acoustic microscopy when a crack or discontinuity is present in the sample.

Three typical situations can occur :

1- The crack runs normal to the surface. The situation is described in figure 11, in which the lens is over a crack. If the crack is thin, the incident waves close to the normal (ray A) will scarcely be affected by it, and will be reflected almost as if there were no crack. However, waves close to the Rayleigh angle (ray B) will excite Rayleigh waves which may then strike the crack from the side. Such a wave will undergo strong reflection and scattering even when the crack is very thin. When a Rayleigh wave is normally incident on a surface breaking crack, there is a reflection back along the surface of original propagation, transmission onto the second surface of the wedge or the crack surface, and diffraction into bulk waves. In the case of a crack of finite depth, some energy is transmitted all around the surface to continue propagating in the original direction. In the presence of fluid, there may be diffraction into the fluid and also transmission directly across the crack. It is assumed here that reflection along the original surface is the dominant effect (this would correspond to an infinitely deep crack filled with vacuum).

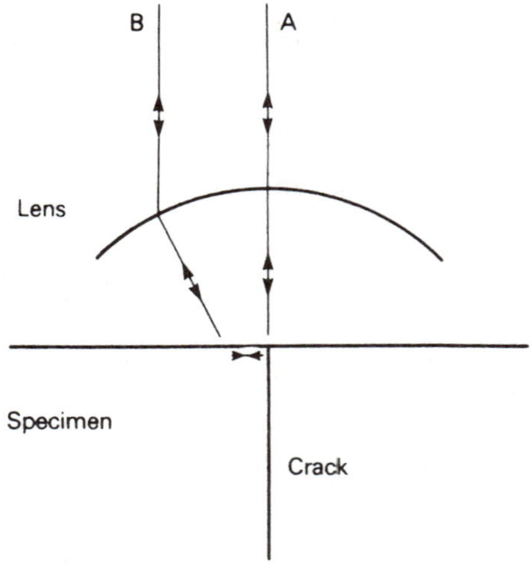

Fig. 11 : Geometrical ray configuration when the lens is directly over a normal crack.

The reflection involves a change of amplitude and phase which may be approximately described by a Rayleigh reflection coefficient (24) :

$$R = 0.4\, e^{i\,0.6} \tag{17}$$

This result applies to Rayleigh waves incident normally on the crack. This situation can be usually acheived by taking a cylindrical lens with the axis of the cylinder parallel to the line of intersection of the crack with the surface. However, one can observe a similar feature with a spherical lens focusing on a disc. Because of the symmetry of the situation, surface waves which are reflected by the edge of a disc interfere with the other rays responsible for the V(z) curve. A ringing pattern is therefore obtained as shown in figure 12 which represents a composite fiber. The fringe periodicity is $\lambda_R/2$ and can be eliminated by restricting the aperture of the lens so that it does not include the Rayleigh angle.

In order to make a quantitative approach, it is necessary to isolate the component of the reflectance funtion which corresponds to excitation of Rayleigh waves. The reflection coefficient may be expressed as :

$$R(\theta) = R_o(\theta)\,\frac{k_x^2 - k_o^2}{k_x^2 - k_p^2} \tag{18}$$

where k_x is the tangential component of the incident wave vecteur, k_o and k_p are determined by finding the imaginary incident wave vectors for which the numerator and denominator respectively of the analytical expression for $R(\theta)$ are zero. The reflectance function may be separted into two terms :

$$R(\theta) = R_o(\theta) + R_o(\theta)\,\frac{k_p^2 - k_o^2}{k_x^2 - k_p^2} \tag{19}$$

The second term on the right is associated with the singularity in the vicinity of $k_x = + k_p$ for waves incident around the Rayleigh angle. In the lossless case considered here $k_o = k_p^*$, the real part of k is approximately equal to the Rayleigh wave number for a free surface, and the imaginary part is due to re-radiation into the fluid. After reflection right over a fine crack or a surface boundary, a local effective reflectance function may be written :

$$R_{eff}(\theta) = R_o(\theta)\left\{ 1 + R_R\,\frac{k_p^2 - k_o^2}{k_x^2 - k_p^2} \right\} \tag{20}$$

In the case of cylindrical lens this expression may be inserted into the V(z) formula given by (25) :

$$V(z) = \int_0^{\pi/2} P^*(\theta)\, R_{eff}(\theta)\, e^{-2ikz\cos\theta} \cos\theta \, d\theta \qquad (21)$$

Comparaison with a crack-free material shows a strong difference between the two signatures.

2- Subsurface oblique cracks can also generate interference fringes in acoustic microscopy. An explanation of these fringes is illustrated in figure 13. It is assumed that two rays are important : ray A which strikes the surface normally and is reflected back to the lens and ray C, which is refracted at the specimen surface so as to strike the crack normally, and is also reflected back along its own path. As the lens is scanned parallel to the surface, the path followed by ray A remains constant, but that of ray C varies. Since the transducer sums the amplitude of the two contributions, periodic interference results, leading to the observed fringes. The periodicity is :

$$\Delta x = \frac{\lambda}{2 \sin \theta_c} \qquad (22)$$

where θ_c is the running direction of the crack with respect to the surface.

The two series of interferences which have been described can be easily identified by the dissymmetry of the fringes apart from the crack in the present case whereas it stays symmetrical when the phenomenon is governed by surface acoustic waves.

3- An other source of interference outcomes when one involves the phase of V(z) curve (fig.14). This is particularly interesting when imaging surface or subsurface topography and is very sensitive to the quality of the scanning system. To achieve this, one adds a third wave to the V(z) phenomenon which can be a fixed echo in the lens delay line or a leakage through the switch generating the RF pulse. This third signal must not vary during the scanning so it can be considered as a reference signal which is mixed to the two others.

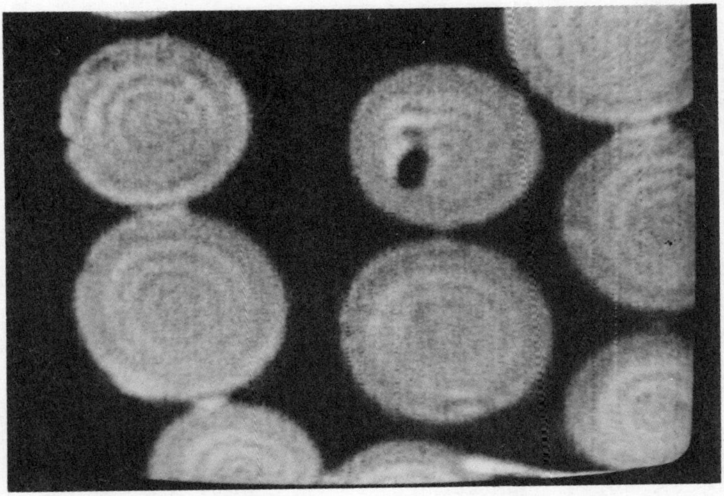

Fig. 12 : Ringing patterns in a composite fiber.

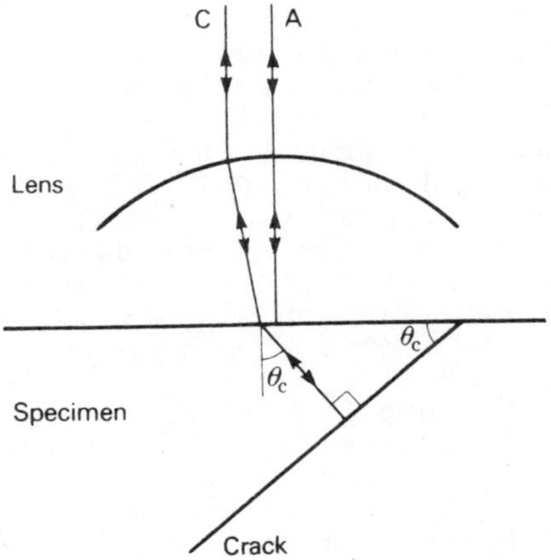

Fig. 13 : Interference fringes generated by an oblique subsurface crack.

Fig. 14 : Interference in topography involving the phase of V(z). (pyramidal indentation).

6- CONCLUSION

More than Optics, Acoustics in focused beam give a great deal of informations about the sample to be analyzed. Rayleigh waves, for instance, play a vital role in the contrast in subsurface examination. This can be correlated to the reflectance function well defined for a uniform surface, although the most significant informations come from discontinuities and grain boundaries. Surface wave velocity measurements tell about the mechanical properties of the surface, especially adhesion or hardness.

Focusing in depth longitudinal or transverse waves is another dominant feature of the acoustic microscope. Imaging interface of bonding between two opaque materials and volumetric quality control can be performed and give exciting results in the field of material science.

Several commercial scanning acoutic microscopy are now available and with the much wider access, it will be possible to use this technique for on-line process control and for research which will become much more widespread.

REFERENCES

1 Lemons R.A. and Quate C.F., Acoustic microscope, Scanning version. Appl. Phys. Lett. 16, 737-8 (1974).

2 Bennnett S.D., I C applications for acoustic microscopy. Proceedings of the International Conference on the quality of Electronic Devices, Bordeaux (1982).

3 Miller A.J. "Acoustical Imaging" (E.A. Ash and C.R. Hills Eds) Vol. 12, pp 67-68, Plenum, New-York (1982).

4 Briggs G.A.D., Ilett C. and Somekh M.G., "Acoustical Imaging" (E.A. Ash and C.R. Hills Eds) Vol. 12, pp 88-89, Plenum, New York (1982).

5 Nikoonahad H. and Ash E.A., "Acoustical Imaging" (E.A. Ash and C.R. Hills Eds), Vol. 12 pp 47-60, Plenum, New York (1982).

6 Briggs G.A.D., Somekh and Ilett, "Acoustic Microscopy in material sciences", Proceedings of the SIRA Conference, London (1982).

7 Yue G.Q., Nikoonahad M. and Ash E.A., Electr. Lett. 18, pp 767-769 (1981).

8 Nongaillard B., Grosmhaire J. and Rouvaen J.M., Zooming lens for ultrasonic microscopy and analysis of NDT images. Ultrasonics Symposium, pp 772-776 (1985).

9 Hollis R.L., Hammer R. and Al-Jaroudi M.Y., Subsurface imaging of glass fibres in a polycarbonate composite by acoustic microscopy. J. Mater. Sci. 19, pp 1897-903 (1984).

10 Atalar A., A physical model for acoustic signatures, J. Appl. Phys. 50, pp 8237-9 (1979).

11 Wickramasinghe H.K., Scanning Acoustic microscopy : a review. J. Microsc. 129, 63-73 (1983.

12 Brekhovskikh L.M., Waves in layered media (2nd ed.). Academic Press, New-York (1980).

13 Bertoni H., Ray Optical Evaluation of V(z) in the Reflective Acoustic Microscope, IEEE Trans. on Sonic and Ultras. SU 31, pp 105-116 (1984).

14 Dransfeld K. and Salzmann E., Physical Acoustics Vol.8 (W.P. Masson and R.N. Thurston Ed.) New-York, pp 260-263 (1970).

15 Weglein R.D. "Scanned Image Microscopy" (E.A. Ash Ed.) pp 127-136, Acad. Press, London and New-York (1980).

16 Kushibiki J., Ohkubo A. and Chubachi N., Linearly focused acoustic beams for acoustic microscopy, Electron. Lett. 17, 520-2 (1981).

17 Morsh A. and Arnold W., Signal Processing in Acoustic Microscopy with Applications to NDT, International School of Physical Acoustics, Nice (1988).

18 Weglein R.D., Acoustic Micro-metrology, IEEE Trans or Son. and Ultras. SU 32, pp 225-234 (1985).

19 Attal J. "Scanned Image Microscopy" (E.A. Ash ed.), pp 97-118, Academic Press, London and New-York (1980).

20 Jipson V.B., ppl. Phys. Lett. 35, pp 385-87 (1979).

21 Nikoonahad M., Yue G.Q. and Ash E.A., "Review of Progress in quantitative NDE" (D.O. Thompson and D.E. Chimenti Ed.) Vol. 2B pp 1611-1623, Plenum Press, New-York (1982).

22 Pino F., Sinclair D.A. and Ash E.A., Scanning Acoustic Microscopy of solid objects using aspheric lenses. Proceedings of 11th International Conference on Acoustical Imaging, Monterey (1981).

23 Attal J., Saied A., Saurel J.M. and Ly C.C., "Acoustic Microscopy : Deep focusing inside materials at gigahertz frequencies". Acoustical Imaging, 17th Internatiqonal Symposium, Sendai 1988 (to be published).

24 Achenbach J.D., Gautesen A.K. and Mendelsohn D.A., "Ray analysis of surface wave interaction with an edge crack", IEEE Trans. SU-27, 124-9 (1980).

25 Kushibiki J., Ohkubo A. and Chubachi N., "Theoretical analysis for V(z) obtained by acoustic microscope with line focus beam", Electron. Lett. 18, 663-4 (1982).

Physical Acoustoelectronics: Interaction of Acoustic Waves with Free Electrons in Semiconductors

Yu.V. Gulyaev

Institute of Radioengineering and Electronics
Academy of Sciences of the USSR, Moscow, USSR

Abstract

The general linear theory of electron absorption and amplification of acoustic waves (AW) in semiconductors is presented. The emphasis is put on the criteria of applicability of different existing theories for various regions of AW frequencies, semiconductor properties and external conditions. The drag of free electrons by AW is considered, which leads to acoustoelectric effect, acoustomagnetoelectric effect, acoustothermal effect etc. The nonlinear effects of acoustoelectronic interaction are considered, such as the dependence of AW absorption coefficient and its velocity on the amplitude of AW, the nonohmic behaviour of current-voltage curve of the specimen, the generation of harmonics etc. The specific features of acoustoelectronic interaction for surface acoustic waves (SAW) are pointed out. The comparison of the theory of acoustoelectronic interaction with experimental data is briefly given and some devices based on this interaction are described.

1. General Linear Theory of Electron Absorption and Amplification of Acoustic Waves in Semiconductors

There are two main mechanisms of interaction of AW with free electrons (acoustoelectronic (AE) interaction) in semiconductors: via deformation potential and (in piezoelectric semiconductors) via piezoeffect. The interaction via deformation potential is present in all solids and is connected with the change of energy of electrons in the conduction band due to the deformation of the crystal lattice by the acoustic wave:

$$\delta\epsilon = \frac{\partial}{\partial x_i}(\Lambda_{ij} u_j) \ . \tag{1}$$

Here Λ_{ij} are the components of the deformation potential tensor, u_j is the component of the mechanical displacement in the acoustic wave. The corresponding coupling constant is

$$K_d^2 = \frac{\epsilon q^2}{4\pi\rho v_s^2}\left(\frac{\Lambda}{e}\right)^2 \propto f_{AW}^2 \tag{2}$$

where ϵ and ρ are the dielectric constant and the density of the cristal, q and f_{AW} are the wave vector and the frequency of the AW, v_s its velocity and e is the charge of electron.

The interaction via piezoeffect is connected with the electric field (potential) produced by the AW in piezoelectric semiconductors. It is described by the electric field contribution to the stress tensor T_{ij} and the deformation contribution to the electric displacement D_i:

$$T_{ij} = c_{ijkl} u_{kl} - \beta_{kij}\frac{\partial\phi}{\partial x_k} \tag{3}$$

$$D_i = -\epsilon_{ij}\frac{\partial\phi}{\partial x_j} - 4\pi\beta_{ijk}\frac{\partial u_j}{\partial x_k} \tag{4}$$

Here c_{ijkl} is the elastic moduli tensor, β_{ijk} is the piezoelectric tensor, u_{kl} is the deformation tensor, ϕ is the electric potential of AW. The corresponding coupling constant is

$$K_p^2 = \frac{4\pi\beta^2}{\epsilon c} \ . \tag{5}$$

Comparison of coupling strength for AE interaction in typical semiconductors shows that for AW frequencies

$$f_{AW} \leq 10\ GHz \qquad K_d^2 < K_p^2 \tag{6}$$

Thus, the deformation potential interaction usually becomes important only for high frequencies, i.e. for $f_{AW} > 10\ GHz$. Due to high viscous absorption of AW at these frequencies, practical acoustoelectronic devices work at frequencies below $2\ GHz$, where the deformation potential interaction is negligably small. So in future consideration we shall mainly concentrate on piezoelectric semiconductors, such as CdS, GaAs, InSb, the generalization to nonpiezoelectric semiconductors with deformation potential interaction being made, if necessary, very easily.

We shall consider the case when AW wavelength is much larger than the lattice constant, $\lambda_s \gg \sigma$, i.e. we regard the crystal as a continous mediium from the acoustic point of view.

From the physical point of view there are two different cases:

1. **Classical case**

$$\lambda_s \gg \lambda_e\,, \quad i.e. \quad k \gg q \tag{7}$$

where $\lambda_e = 2\pi k^{-1}$ is the de Broglie wavelength of electron and $\lambda_s = 2\pi q^{-1}$ the acoustic wavelenght respectively. In this case the electron may be considered as a point particle moving in the classical field of the AW. This case is realized usually in semiconductors at high temperatures, high values of electron effective mass m^* and low acoustic frequencies. For example for InSb at $77^o K$, $m^* = 0.015\ m_0$ and for AW frequency $f = 26\ Hz$ one has $\lambda_e \simeq 10^{-5} cm$ and $\lambda_s \simeq 10^{-4} cm$, so $\lambda_e \ll \lambda_s$, a classical case.

2. **Quantum case**

$$\lambda_s \leq \lambda_e\,, \quad i.e. \quad k \leq q \tag{8}$$

In this case the interaction of the electron with the acoustic wave must be considered by quantum mechanical methods as diffraction of electron de Broglie wave on the AW. In other words, in the process of interaction with electrons the AW must be considered in this case as a flux of coherent phonons. For the same InSb crystal at $T = 4.2^o K$ and for $f = 10\ GHz$ we have $\lambda_e \simeq 5 * 10^{-5} cm$ and

$\lambda_s \simeq 2 * 10^{-5} cm$, i.e. $\lambda_e \approx \lambda_s$, a typical quantum case.

In this lecture I restrict myself to the classical case only, since the physical sense of phenomena which will be considered is the same for both cases. The classical case, in turn, may be divided into 3 subcases.

When

$$\lambda_s \gg L_e \qquad \omega_s \ll \tau_e^{-1} \tag{9}$$

where L_e and τ_e are the characteristic length and time required to establish local quasi-equilibrium Fermi distribution in the electron gas, we have the so called *hydrodynamic* case, in which the electron gas is characterized by local equilibrium kinetic coefficients.

When

$$\lambda_s \ll l \quad \omega_s \gg \tau^{-1} \quad (ql \gg 1, \omega\tau \gg 1) \tag{10}$$

where l and τ are the mean free path of electrons and their momentum relaxation time, we have the so called *collisionless* case, in which electron collisions may be neglected.

When

$$L_e \gg \lambda_s \neq l \quad \tau_e^{-1} \ll \omega_s \ll \tau^{-1} \tag{11}$$

we have the *intermediate frequency* case, in which electron collisions must be explicitly taken into account.

Methods of calculation for these three cases are quite different, the last case being, of course, the most difficult one.

It is useful to note here that for room temperatures and AW frequencies $f_s < 0.5\ GHz$ the hydrodynamic case is usually valid for all semiconductors.

1. Hydrodynamic case

Under the conditions $\lambda_s \gg L_e$, $\omega_s \ll \tau_e^{-1}$ at each point in space and at each moment the local quasi-equilibrium Fermi distribuition for electrons is valid,

$$f_{el} = \frac{1}{e^{\frac{\epsilon - \varsigma(x,t)}{kT}} + 1} \tag{12}$$

where the chemical potential is

$$\varsigma(x,t) = \varsigma_0 + \varsigma_1(x,t) \tag{13}$$

$$\varsigma_1(x,t) \propto e^{i(\omega t - qx)} \tag{14}$$

and ω and q are the frequency and wave vector of AW propagating in OX-direction.

What are the quantities L_e and τ_e? Let's first determine them from qualitative considerations. Suppose that the electron gas is put off the thermodynamic equilibrium. In the simplest case its relaxation to equilibrium is characterized by the two times τ (momentum relaxation time) and τ_ϵ (energy relaxation time). For elastic collisions $\tau_e \gg \tau$, so in fact the relaxation of the electron gas to equilibrium is described by the longest time τ_e introduced above.

There are several characteristic lengths which describe the electron gas relaxation in space. The mean free path l characterizes electron momentum relaxation. The energy relaxation diffusion length $L_\epsilon \doteq \sqrt{D\tau_\epsilon}$ (D is the diffusion coefficient) is the distance along which electrons *diffuse* in Brownian-like motion during the energy relaxation time. Usually, $L_\epsilon \gg l$. If there is directed drift of electrons, one more length $L_d = v_d\tau_\epsilon$ appears (v_d is the mean drift velocity of electrons), which characterizes the distance over which the electron travels due to drift during the energy relaxation time τ_ϵ. So the length L_e above introduced is the longest of the three lengths L_ϵ, L_d and l:

$$L_e = Max\left\{l, \sqrt{D\tau_\epsilon}, v_d\tau_\epsilon\right\} \tag{15}$$

Thus, the conditions which must be satisfied for the validity of hydrodynamic approach are:

$$T \gg \tau_\epsilon \ , \ \lambda_s \gg Max\left\{l, \sqrt{D\tau_\epsilon}, v_d\tau_\epsilon\right\} \tag{16}$$

(T is the period of AW), or

$$\omega_s\tau_\epsilon \ll 1 \ , \ Max\left\{ql, q\sqrt{D\tau_\epsilon}, qv_d\tau_\epsilon\right\} \ll 1 \tag{17}$$

The conditions (16) and (17), of course, strictly follow from the kinetic theory of AW interaction with electrons (see below). Here it is necessary to note that only the condition $ql \ll 1$ is often considered as necessary for the hydrodynamic case. In the majority of cases this is incorrect. Let us take for instance a typical example:

$$n - InSb, \ T = 77^o K, \ \mu = 10^6 \ cm^2/vs, \ D = 7 * 10^3 \ cm^2/s,$$

$$\tau = 10^{-12} \ s, \ \tau_\epsilon = 10^{-10} \ s, \ v_T = 10^7 \ cm/s, \ v_d = 10^6 \ cm/s,$$

From these data it follows that

$$\sqrt{D\tau_\epsilon} \simeq 10^{-3} \ cm, \ v_d\tau_\epsilon \simeq 10^{-4} \ cm, \ l \simeq 10^{-5} \ cm,$$

so $\sqrt{D\tau_\epsilon} \gg l$ and $v_d\tau_\epsilon \gg l$, and the condition (17) can by no means reduce to $ql \ll 1$.

We shall consider further the simplest case of piezoelectric semiconductor with one type of carriers (electrons), not taking into account trapping or heating of electrons. Then the basic system of equations is (for detailed references see the review papers by Gurevich (1968), Pustovoit (1969) and by Gulyaev ad Kozorezov (1977)):

$$\rho\frac{\partial^2 u_i}{\partial t^2} = \frac{\partial T_{ij}}{\partial x_j} \tag{18}$$

$$\nabla \vec{D} = 4\pi e(N - N_d) \tag{19}$$

$$\nabla \vec{j} = e\frac{\partial N}{\partial t} \tag{20}$$

$$T_{ij} = c_{ijkl}\frac{\partial u_k}{\partial x_l} - \beta_{kij}\frac{\partial \phi}{\partial x_k} + \Lambda_{ij}(N - N_d) \tag{21}$$

$$D_i = -\epsilon_{ij}\frac{\partial \phi}{\partial x_j} - 4\pi\beta_{ijk}\frac{\partial u_j}{\partial x_k} \tag{22}$$

$$j_i = \sigma_{ik}\left[E_{0k} - \frac{\partial \phi}{\partial x_k} - \frac{\Lambda_{kj}}{e}\frac{\partial^2 u_l}{\partial x_l \partial x_j}\right] - eD_{ik}\frac{\partial n}{\partial x_k} \tag{23}$$

Here N and N_d are the concentrations of free electrons and compensating positive charges.

In the one-dimensional case, for $\vec{q} \parallel \vec{E_0} \parallel OX$, the system of equations (18)-(23) reduces to the following one:

$$\rho\frac{\partial^2 u}{\partial t^2} = c\frac{\partial^2 u}{\partial x^2} - \beta\frac{\partial^2 \phi}{\partial x^2} + \Lambda\frac{\partial N}{\partial x} \tag{24}$$

$$\epsilon\frac{\partial^2 \phi}{\partial x^2} + 4\pi\beta\frac{\partial^2 u}{\partial x^2} = -4\pi e(N - N_d) \tag{25}$$

$$e\frac{\partial N}{\partial t} + \frac{\partial j}{\partial x} = 0 \tag{26}$$

$$j = \sigma(N)\left[E_0 - \frac{\partial \phi}{\partial x} - \frac{\Lambda}{e}\frac{\partial^2 u}{\partial x^2}\right] - eD(N)\frac{\partial N}{\partial x} \tag{27}$$

The kinetic coefficients σ and D depend, generally speaking, on the electron concentration N and on the scattering mechanism.

Let us now perform the linearization procedure by putting

$$N = N_0 + n \ (|n| \ll N_0 = N_d), \ j = j_0 + j_\sim \ (|j_\sim| \ll j_0), \left|\frac{\partial \phi}{\partial x}\right| \ll E_0 \tag{28}$$

$$u, \phi, n, j_\propto \sim exp[i(\omega t - qx)] \tag{29}$$

Then we have

$$\sigma(N) = \sigma(N_0) + \frac{\partial \sigma_0}{\partial N_0}n \ , \ D(N) = D_0(N_0) + \frac{\partial D_0}{\partial N_0}n \tag{30}$$

Since $\sigma(N) = e \cdot N \cdot \mu(N)$, μ is the mobility of electrons,

$$\frac{\partial \sigma_0}{\partial N_0} = e\mu_0(N_0)\left[1 + \frac{\partial \ln \mu_0}{\partial \ln N_0}\right] \tag{31}$$

For example, for degenerate semiconductors we have:

$$\mu = \frac{e\tau}{m^*}$$

where $\tau = \tau(\varsigma)$ is the momentum relaxation time, m^* is the effective mass, ς is the Fermi energy, $\mu = \mu(\varsigma)$, $\varsigma = (3\pi^2)^{\frac{2}{3}}\frac{\hbar^2 N^{\frac{2}{3}}}{2m^*}$, so we have $\mu = \mu(N)$. After linearization we have:

$$-eN_0\mu_0\frac{\partial\phi}{\partial x} - eD_0\frac{\partial n}{\partial x} - N_0\mu_0\Lambda\frac{\partial^2 u}{\partial x^2} \tag{32}$$

The second term in the right hand side of (32) describes the motion of space charge bunches n under the drift velocity

$$v_d^* = \frac{1}{e}\frac{\partial\sigma_0}{\partial N_0}E_0 \equiv v_d r \tag{33}$$

which is different from the average drift velocity of electrons $v_d = \mu_0 E_0$ by a factor

$$r = \frac{d\ln\sigma_0}{d\ln N_0} \equiv 1 + \frac{d\ln\mu_0}{d\ln N_0} \tag{34}$$

which is, generally speaking, not equal to 1.

Neglecting at this point the deformation potential coupling, we have from (26) and (32):

$$n = -\frac{e\phi}{kT}\frac{N_0}{1 - \frac{i\omega}{q^2 D}\left(1 - \frac{v_d^*}{v_d}\right)} \tag{35}$$

Thus one sees that the phase shift between n and ϕ depends on the sign of the factor $(1 - \frac{v_d^*}{v_s})$: it changes from positive to negative at the point where the drift velocity of electron bunches v_d^* (not the average drift velocity v_d!) becomes larger than the sound velocity v_s.

After linearization the dispersion equation for the system of equations (24)-(27) is found, if one puts

$$q = q_0 + \Delta q + i\alpha_e \ , \ |\Delta q|, |\alpha| \ll q_0 = \frac{\omega}{v_{s0}} \ , \ v_{s0} = \sqrt{\frac{c}{\rho}} \ , \ \frac{\Delta q}{q_0} = -\frac{\Delta v_s}{v_{s0}} \tag{36}$$

Here α_e is the electron absorption (amplification) coefficient of AW and $\frac{\Delta v_s}{v_{so}}$ is the relative electron correction to sound velocity.

One obtains:

$$\alpha_e = \frac{K_p^2 + K_d^2}{2} q_0 \frac{\omega\tau_M \left(1 - \frac{v_d^*}{v_s}\right)}{\omega^2\tau_M^2 \left(1 - \frac{v_d^*}{v_s}\right)^2 + (1 + q^2 r_D^2)^2} \tag{37}$$

$$\frac{\Delta v_s}{v_{s0}} = \frac{1}{2} \frac{K_p^2 \left[\omega^2\tau_M^2 \left(1 - \frac{v_d^*}{v_s}\right)^2 + q^2 r_D^2 (1 + q^2 r_D^2)\right] - K_d^2 (1 + q^2 r_D^2)}{\omega^2\tau_M^2 \left(1 - \frac{v_d^*}{v_s}\right)^2 + (1 + q^2 r_D^2)^2} \tag{38}$$

Here r_D is Debye length:

$$r_D = \begin{cases} \sqrt{\frac{\epsilon k T}{4\pi e^2 N_0}} & \text{- nondegenerate case} \\ \sqrt{\frac{\epsilon \frac{2}{3}\varsigma_0}{4\pi e^2 N_0}} & \text{- degenerate case} \end{cases} \tag{39}$$

$$\tau_M = \frac{\epsilon}{4\pi\sigma_0} \text{ - Maxwellian relaxation time} \tag{40}$$

For the non degenerate case $\frac{d\mu_o}{dN_o} = 0$, $r = 1$ and if one puts $K_d^2 = 0$ than the (37) is the well known expression for α_e in White's theory. For the degenerate case $r = 1 + \frac{2}{3}\frac{\partial \ln \tau(\varsigma)}{\partial \ln \varsigma}$, where $\tau(\varsigma)$ is momentum relaxation time. If $\tau(\varsigma) \propto \varsigma^\nu$, then

$$r = 1 + \frac{2}{3}\nu \text{ , (gain condition is } v_d > \frac{v_s}{r}) \tag{41}$$

For different electron scattering mechanisms one has:

$$\begin{cases} \nu = -\frac{1}{2} \text{ (def. pot. scatt.)} & r = \frac{2}{3} \quad \text{gain cond. } v_d > 1.5 v_s \\ \nu = +\frac{1}{2} \text{ (piezo. scatt.)} & r = \frac{4}{3} \quad \text{gain cond. } v_d > 0.75 v_s \\ \nu = +\frac{3}{2} \text{ (impurity scatt.)} & r = 2 \quad \text{gain cond. } v_d > 0.5 v_s \end{cases} \tag{42}$$

So, for example, for impurity scattering the acoustic wave amplification starts already when the electron drift velocity v_d equals half of the acoustic wave velocity. At exactly that point, as it was already said before, the phase shift between the electric potential ϕ in the acoustic wave and the electron concentration wave n changes its sign (electron bunches begin to *overtake* sound). The dependence of the electron absorption (amplification) coefficient α_e and of its velocity v_s on the drift electric field E_o is shown in Fig.1.

The physical explanation of why the electron absorption of sound turns into amplification at the point $v_d = \frac{v_s}{r}$, where the phase shift between ϕ and n changes its sign, is illustrated by Fig. 2.

2. Collisionless case (classical): $ql \gg 1$

In this case from one collision to another electrons fly through many AW wavelength (see Fig. 3), so the local connection between electron density and electric potential of AW is no longer valid.

Thus to calculate the acoustoelectronic interaction one must use kinetic equations. The basic system of equations now is:

$$\begin{cases} \rho\dfrac{\partial^2 u}{\partial t^2} = c\dfrac{\partial^2 u}{\partial x^2} - \beta\dfrac{\partial^2 \phi}{\partial x^2} \\ \\ \epsilon\dfrac{\partial^2 \phi}{\partial x^2} + 4\pi\beta\dfrac{\partial^2 u}{\partial x^2} = -4\pi e n \\ \\ \dfrac{\partial f_{\vec{p}}}{\partial t} + \vec{v}\,\nabla_{\vec{r}} f_{\vec{p}} + (e\vec{E} + \frac{e}{c}[\vec{v}\times\vec{H}])\nabla_{\vec{p}} f_{\vec{p}} = \hat{S}\{f_{\vec{p}}\} \end{cases} \tag{43}$$

Here $f_{\vec{p}}$ is the electron distribution function, $\vec{H}$ is the magnetic field (if any) and $\hat{S}\{f_{\vec{p}}\}$ is the collision integral.

The connection between amplitudes of n and ϕ now is to be found from the solution of kinetic equations (43). In the linear case considered here, it has the general form:

$$n(\vec{q},\omega) = K(\vec{q},\omega)\phi(\vec{q},\omega) \tag{44}$$

$$K(\vec{q},\omega) = \frac{iq\sigma(\vec{q},\omega)}{e^2 v_s} \tag{45}$$

Here the nonlocality of interaction of electrons with AW reflects in the dependence of complex conductivity $\sigma(\vec{q},\omega)$ on the wavevector $\vec{q}$ and frequency ω.

By putting $K = K' + iK''$ and solving the linearized equations (43) with the use of (44) in a usual manner, we obtain for the electron absorption coefficient of AW the expression:

$$\alpha_e = \frac{K_p^2 q}{2} \frac{\frac{4\pi e^2}{\epsilon_0 q^2} K''(\vec{q},\omega)}{\left(1 - \frac{4\pi e^2 K'(\vec{q},\omega)}{\epsilon_0 q^2}\right)^2 + \left(\frac{4\pi e^2 K''(\vec{q},\omega)}{\epsilon_0 q^2}\right)^2} \tag{46}$$

Now the problem is reduced to finding the complex conductivity of the specimen, $\sigma(\vec{q},\omega)$, which has spatial and temporal dispersion.

In the collisionless case $\hat{S}\{f_{\vec{p}}\} = 0$. So for the one-dimensional case $\vec{q} \parallel \vec{E}_o \parallel OX$; putting $f = f_o + f_1$ (f_o is electron distribution function in the d.c. electric field), we have from (43):

$$f_1 = \frac{e}{m^*} \frac{1}{\omega - q v_x} \frac{\partial f_0}{\partial v_x} q\phi \tag{47}$$

$$n = \frac{2}{(2\pi\hbar)^3} \int f_1(\vec{p}) d^3\vec{p} \equiv K(\vec{q},\omega)\phi \tag{48}$$

where

$$K(\vec{q},\omega) = \frac{2e}{(2\pi\hbar)^3} \int d^3\vec{v} \frac{q\frac{\partial f_0}{\partial v_x}}{\omega - q v_x} \tag{49}$$

Performing the integration around the pole at $v_x = \frac{\omega}{q}$ in the same manner as in the theory of Landau damping of plasma waves, we have:

$$K(q,\omega) = -\frac{eN_0}{kT}\left[1 + i\sqrt{\pi}\frac{m^*\omega\left(1 - \frac{\mu E_0}{v_s}\right)}{q\sqrt{2m^*kT}}\right] \tag{50}$$

For the electron absorption coefficient α_e from (46) one immediately obtains:

$$\alpha_e = \frac{K_p^2 \omega\sqrt{\pi}\left(1 - \frac{\mu E_0}{v_s}\right)}{q^2 r_D^2 \left(1 + \frac{1}{q^2 r_D^2}\right)^2 \sqrt{\frac{2kT}{m^*}}} \tag{51}$$

This is the well known formula of Spector theory.

The main contribution to the integral in (49) comes from the pole region at $v_x = \frac{\omega}{q}$, i.e. from electrons which have the projection of their thermal velocity $\vec{v_T}$, on the sound propagation direction equal to the sound velocity, v_s. Since v_T is usually much greater then v_s, this means that only electrons moving almost perpendicular to the AW direction contribute to collisionless AW absorption or to AW amplification.

The dependence of α_e on drift electric field E_o is presented in Fig.4.

3. **Intermediate frequency case.**

Here $L_e \gg \lambda_s \neq l$, $\tau_e \ll \omega_s < \tau^{-1}$, which includes the point $\lambda_s \simeq l$, *i.e.* $ql \simeq 1$. Usually at that point $\omega\tau \simeq ql\frac{v_s}{v_t} \ll 1$.

In this case electron collisions must be taken into account explicitly and for the calculation one must use the kinetic equation approach (43) with collision integral $\hat{S}\{f_{\vec{p}}\} \neq 0$. For simplicity let us put $\vec{H} = 0$ (no d.c. magnetic field). Then kinetic equation (43) is:

$$\frac{\partial f_{\vec{p}}}{\partial t} + \vec{v}\nabla_{\vec{r}} f_{\vec{p}} + e\vec{E}\nabla_{\vec{p}} f_{\vec{p}} = \hat{S}\{f_{\vec{p}}\} \tag{52}$$

Let us put

$$f = F + f^h + \Psi \tag{53}$$

where

$$F = f_0 - e\vec{E}_0\vec{v}\tau\frac{\partial f_0}{\partial \epsilon} \tag{54}$$

is the electron distribution function in d.c. drift electric field E_0, f_0 being the quasiequilibrium Fermi distribution function in the presence of the acoustic wave

$$f_0 = \frac{1}{e^{\frac{\epsilon-\varsigma}{kT}} + 1}, \qquad \varsigma = \varsigma_0 + \varsigma_1(x,t) \tag{55}$$

By f^h we denote that part of the electron distribution function which leads to the result of hydrodynamic White's theory. So if we pose $f = F + f^h$ then from the kinetic equation we obtain $K^h(\vec{q},\omega)$, which, substituted into (46), gives the well known White's expression for the electron absorption coefficient α_e. Ψ is that part of electron distribution function which gives deviations from White's theory.

By solving the kinetic equation (52) in the hydrodynamic case one easily obtains f^h in the form

$$f^h = f_+^h + f_-^h$$

where the symmetric and the antisymmetric parts are:

$$f_+^h = -\varsigma^{(1)} \frac{\partial f_0}{\partial \epsilon} \tag{56}$$

$$f_-^h = -i\vec{q}\vec{v}\tau \left\{ (\varsigma^{(1)} + e\phi) \frac{\partial f_0}{\partial \epsilon} + e\vec{E}_0 \vec{v}\tau \varsigma^{(1)} \frac{\partial^2 f_0}{\partial \epsilon^2} \right\} \tag{57}$$

Here

$$\varsigma^{(1)} = -\frac{q^2 r_D^2}{q^2 r_D^2 - i\omega' \tau_M} e\phi, \qquad \omega' = \omega - q v_d \tag{58}$$

As it was already said, by using $f = F + f^h$ from (54)-(56) (57) one obtains the result of White's theory. Now a question arises: what are the conditions in which Ψ is enough small for applying the White's theory?

Let's divide also Ψ into symmetric and antisymmetric parts

$$\Psi = \Psi_+ + \Psi_- \tag{59}$$

Then for Ψ_+ and Ψ_- from kinetic equation we obtain the following equations:

$$\begin{cases} -i\omega\Psi_+ + i\vec{q}\vec{v}\Psi_- + e\vec{E}_0 \frac{\partial}{\partial \vec{p}} \Psi_- + \hat{S}\{\Psi_+\} = \Pi_+ \\ -i\omega\Psi_- + i\vec{q}\vec{v}\Psi_+ + e\vec{E}_0 \frac{\partial}{\partial \vec{p}} \Psi_+ + \hat{S}\{\Psi_-\} = \Pi_- \end{cases} \tag{60}$$

where

$$\begin{cases} \Pi_+ = i\omega f_+^h - i\vec{q}\vec{v} f_-^h + \left(\frac{\varsigma^{(1)}}{kT} + 2\frac{E_\sim}{E_0} + \frac{E_\sim}{E_0} \frac{\varsigma^{(1)}}{e\phi} \right) \hat{S}\{F\} \\ \Pi_- = i\omega f_-^h \end{cases} \tag{61}$$

where $E_\sim = -\nabla\phi$ By putting for estimates $\frac{\partial}{\partial p} \simeq p^{-1}$, $\vec{S}\{\Psi_-\} \simeq \frac{\Psi_-}{\tau}$, $\vec{S}\{\Psi_+\} \simeq \frac{\Psi_+}{\tau_\epsilon}$ we have from (60)-(61)

$$\Psi_+ \simeq \frac{\Pi_+ \tau_\epsilon - \left(iql + \frac{v_d}{v_T} \right) \Pi_- \tau}{1 + q^2 D\tau_\epsilon - i\omega' \tau_\epsilon} \tag{62}$$

and similar expression for Ψ_{-}. By putting further

$$K(\vec{q},\omega) = K^{h}(\vec{q},\omega) + K^{\psi}(\vec{q},\omega) \tag{63}$$

we have

$$K^{\psi}(\vec{q},\omega) \simeq \frac{\left[i\omega' \tau_{\epsilon} q^2 r_D^2 - q^2 r_D^2 \left(\frac{v_d}{v_T}\right)^2 \frac{\tau_{\epsilon}}{\tau} - q v_d \tau_{\epsilon} \omega' \tau_M\right] \frac{n_0}{kT}}{(1 + q^2 D\tau_{\epsilon} - i\omega\tau_{\epsilon})(q^2 r_D^2 - i\omega' \tau_M)} \tag{64}$$

Now, the condition of applicability of White's theory obviously must be

$$\left|\Im K^{\psi}\right| \ll \left|\Im K^{h}\right| \quad and \quad \left|\Re K^{\psi}\right| \ll \left|\Re K^{h}\right| \tag{65}$$

or, after using calculated expression,

$$q\sqrt{D\tau_{\epsilon}} \ll 1 \ \ , \ \ q v_d \tau_{\epsilon} \ll 1 \ \ , \ \ \omega\tau_{\epsilon} \ll 1 \tag{66}$$

which coincide with conditions (17), obtained from intuitive considerations.

How to solve kinetic equation (52) when Φ is not small and conditions 66 are not satisfied? For $ql \geq 1$, $f_{\vec{p}}$ has a pole structure

$$f_{\vec{p}} \simeq \frac{1}{\left(\cos\theta - \frac{m v_S}{\hbar q}\right) - \frac{i}{ql}} \quad , \quad \theta = \widehat{\vec{p},\vec{q}} \tag{67}$$

and the approximation of collision integral

$$\hat{S}\{f_{\vec{p}}\} \simeq -\frac{1}{\tau}\left[f_{\vec{p}} - f_{\vec{p}0}\right] \tag{68}$$

is too rough. One must take into account all higher harmonics of $f_{\vec{p}}$:

$$f_{\vec{p}} = f_S(\epsilon_{\vec{p}}) + \sum_{n=1}^{\infty} f_n(\epsilon_{\vec{p}}) P_n(\cos\theta) \quad , \quad \cos\theta = \frac{P_x}{P} \quad , \quad \vec{q} \parallel OX \tag{69}$$

So the collision integral takes the form:

$$\hat{S}\{f_{\vec{p}}\} = \hat{S}_S\{f_S(\epsilon_{\vec{p}})\} - \sum_{n=1}^{\infty} \frac{f_n(\epsilon_{\vec{p}}) P_n(\cos\theta)}{\tau_n\left(\epsilon_{\vec{p}}\right)} \tag{70}$$

where each harmonic $f_n(\epsilon_{\vec{p}})$ has its own relaxation time $\tau_n(\epsilon_{\vec{p}})$ which is expressed in terms of scattering transition probability $W_n(\epsilon_{\vec{p}})$:

$$W(|\vec{p}-\vec{p'}|)_{p=p'} = \sum_n W_n(\epsilon_{\vec{p}}) P_n(\cos \widehat{\vec{p'},\vec{p}}) \tag{71}$$

Having these considerations in mind, we shall put

$$f_{\vec{p}} \;=\; F + f_p^{(1)} \tag{72}$$

and linearize kinetic equation (52) in $f_p^{(1)}$. It is possible to present it in observable form, if the conditions, quite opposite to those for applicability of White's theory, (66) are satisfied:

$$q\sqrt{D\tau_\epsilon} \gg 1 \quad , \quad qv_d\tau_\epsilon \gg 1 \quad , \quad q^2 D\tau_\epsilon \gg \omega\tau_\epsilon \tag{73}$$

Under these conditions we have

$$\frac{\partial f_{\vec{p}}^{(1)}}{\partial t} + v_x \frac{\partial f_p^{(1)}}{\partial x} - \hat{S}\{f_{\vec{p}}^{(1)}\} = e\nabla\phi \frac{\partial F}{\partial \vec{p}} \tag{74}$$

where (see (54))

$$\hat{S}\{f_{\vec{p}}^{(1)}\} = -\frac{f_{\vec{p}}^{(1)}}{\tau_0} + \sum_{\vec{p'}} W(\vec{p},\vec{p'}) f_{\vec{p'}}^{(1)} \tag{75}$$

$$\tau_0^{-1} = \sum_{\vec{p'}} W(\vec{p},\vec{p'}) \tag{76}$$

$$F = f_0 - e\vec{E}_0\vec{v}\tau \frac{\partial f_0}{\partial \epsilon} \; . \tag{77}$$

For elastic and isotropic scattering one obtains

$$\tau_0 \sum_{\vec{p'}} W(\vec{p},\vec{p'}) f_{\vec{p'}}^{(1)} = \frac{\int dO' S(\theta'\phi',\theta\phi) f_{\vec{p'}}^{(1)}(\theta')}{\int dO' S(\theta'\phi',\theta\phi)} \tag{78}$$

where $S(\theta'\phi',\theta\phi)$ is the corresponding scattering cross-section, the angles θ,ϕ are related to $\vec{p}$, the angles θ',ϕ' are related to $\vec{p'}$, and we have posed

$f_{\vec{p}} \propto \exp[iqx - i\omega t]$. After that the kinetic equation (74) is reduced to the linear Friedholm integral equation of the second kind:

$$f_{\vec{p}}^{(1)}(z) = \lambda \int_{-1}^{1} d\varsigma K(z,\varsigma) f_{\vec{p}}^{(1)}(\varsigma) + \Phi(\varsigma) \tag{79}$$

Here $z = \cos\theta,\ \varsigma = \cos\theta'$,

$$K(z,\varsigma) = -\frac{1}{w - z}\tilde{S}(z,\varsigma) \tag{80}$$

$$\tilde{S}(z,\varsigma) = \tilde{S}(\cos\psi) = \frac{1}{2\pi}\int_0^{2\pi} d\phi \int_0^{2\pi} d\phi' S(\theta'\phi', \theta\phi) \tag{81}$$

$$\Psi = \left(\widehat{\vec{p}, \vec{p}'}\right) \tag{82}$$

$$\Phi(z) = -\frac{e\phi}{w_0 - z}\left[\left(z - \frac{v_d^\epsilon}{v}\right)\frac{\partial f_0}{\partial \epsilon} - mv_d^\epsilon v z^2 \frac{\partial^2 f_0}{\partial \epsilon^2} - \nu \frac{mv_d^\epsilon v}{\epsilon} z^2 \frac{\partial f_0}{\partial \epsilon}\right] \tag{83}$$

$$\omega = \frac{i\omega\tau - 1}{iql} \quad , \quad \omega_0 = \frac{i\omega\tau_0 - 1}{iql_0} \quad , \quad l = v\tau \quad , \quad l_0 = v\tau_0 \tag{84}$$

$$v_d^\epsilon \equiv v_d(\epsilon) = \frac{eE_0\tau(\epsilon)}{m^*} \quad , \quad \tau(\epsilon) \propto \epsilon^\nu \tag{85}$$

$$\lambda = \frac{2\pi}{iql_0 \int dO' S(\theta'\phi', \theta\phi)} \tag{86}$$

We already suggested that for typical scattering mechanisms $\tau(\epsilon) \propto \epsilon^\nu$. In this case the equation (79) can be solved explicity (analytically). Here

$$\tilde{S}(\psi) = \frac{(1 - \cos\psi)^{\frac{3}{2} - \nu}}{(1 + x(\epsilon) - \cos\psi)^2} \tag{87}$$

where $x(\epsilon) = \dfrac{\hbar^2}{4m^*\epsilon r_D^2}$. Substituting into (80), (81) one obtains:

$$n = K(\vec{q}, \omega) e\phi \tag{88}$$

$$\begin{aligned} K(\vec{q},\omega) &= \frac{3N_0}{2}\frac{(2m)^{\frac{3}{2}}}{3\pi^2\hbar^3 n_0} \cdot \\ &\cdot \int_0^\infty \epsilon^{\frac{1}{2}} \frac{\partial f_0}{\partial \epsilon}\left[1 - \frac{Q_0(w_0) i\omega' \tau_0}{Q_0(w_0) + iql_0 + h(\epsilon)} + \frac{v_d^\epsilon}{v}\frac{1}{iql}\left(\nu - \frac{1}{2}\right)\right] d\epsilon \end{aligned} \tag{89}$$

Here

$$h(\epsilon) = iql_0 \left[-1 + Q_0(w_0)(w_0 + B)\right] \tag{90}$$

$Q_0(w_0)$ is the Legendre function of the second kind, B is the infinite continued fraction:

$$\beta = \cfrac{1}{\beta_{11} - \cfrac{2^2}{\beta_{22} - \cfrac{3^2}{\beta_{33} - \cdots}}} \tag{91}$$

$$\beta_{kk} = -(2k+1)w_k \quad , \quad w_k = \frac{i\omega\tau_k - 1}{iql_k} \tag{92}$$

$$\tau_k^{-1} = \tau_0^{-1}\left[1 - \frac{\int_{-1}^{1} d\varsigma S(\varsigma) P_k(\varsigma)}{\int_{-1}^{1} d\varsigma S(\varsigma)}\right] \equiv \tau_0^{-1}(1-\lambda_k) \tag{93}$$

$$l_k = v\tau_k \quad , \quad \tau_1 = \tau_\epsilon \tag{94}$$

Finally, with the help of (88)-(94) one obtains:

$$\alpha_e = K^2 q \cdot \frac{q^2 r_D^2 \left\langle \frac{1}{\epsilon}\right\rangle^{-1} \left\langle \frac{1}{\epsilon}\beta'' \left[1 - \frac{v_d^\epsilon}{v_s}\left(1 - \frac{v_s}{v}\frac{\nu - \frac{1}{2}}{ql\beta''}\right)\right]\right\rangle}{\left(\left\langle \frac{1}{\epsilon}\right\rangle^{-1} \left\langle \frac{1}{\epsilon}(1-\beta')\right\rangle + q^2 r_D^2\right)^2 + \left(\left\langle \frac{1}{\epsilon}\right\rangle^{-1} \left\langle \frac{1}{\epsilon}\beta'' \left[1 - \frac{v_d^\epsilon}{v_s}\left(1 - \frac{v_d^\epsilon}{v_s}\frac{\nu - \frac{1}{2}}{ql\beta''}\right)\right]\right\rangle\right)^2} \tag{95}$$

where

$$\beta = \beta' + i\beta'' = \frac{1}{1 + \frac{v}{v_S}B} \tag{96}$$

$$\langle \cdots \rangle = \frac{(2m)^{\frac{3}{2}}}{3\pi^2\hbar^3 N_0}\int_0^\infty \{\cdots\}\frac{\partial f_0}{\partial \epsilon}\epsilon^{\frac{3}{2}} d\epsilon \tag{97}$$

The criterium of amplification of acoustic wave by electron drift has the form

$$v_d > \frac{v_S}{r} \tag{98}$$

$$r = \frac{\left\langle \frac{1}{\epsilon}\tau(\epsilon)\beta''(\epsilon)\left(1 - \frac{v_S}{v}\frac{\nu - \frac{1}{2}}{ql(\epsilon)\beta''(\epsilon)}\right)\right\rangle}{\langle \tau(\epsilon)\rangle \left\langle \frac{1}{\epsilon}\beta''(\epsilon)\right\rangle} \tag{99}$$

Physically, here $r \neq 1$ means that at different frequencies and for different scattering mechanisms the main contribution to electron absorption (amplification) of the acoustic wave comes from groups of electrons with different energetic levels. Their *partial* drift velocities

$$v_d^\epsilon \equiv v_d(\epsilon) = \frac{e\tau(\epsilon)}{m^*}E_0 \tag{100}$$

generally speaking, are different from one another and from the mean drift velocity of electrons v_d.

Formulae (95)-(100) solve the problem in a general way. For each particular scattering mechanism of electrons the infinite fraction B can be summed. For example, for optical phonons scattering below Debye temperature ($\nu = 0$) one obtains

$$\beta(\epsilon) = \cfrac{1}{x + \cfrac{2^2}{2x + \cfrac{3^2}{3x + \cdots}}} = \frac{i}{2\sqrt{1+x^2}F(1, -p, 1-p, \frac{-1}{n})} \tag{101}$$

where

$$x = \frac{1}{ql_0} - iO \tag{102}$$

$$P = P(x) = -\frac{1}{2} + \frac{1}{2}\frac{x}{\sqrt{1+x^2}} \tag{103}$$

$$n = \left(x + \sqrt{1+x^2}\right)^2 \tag{104}$$

F is the Gauss hypergeometrical function,

$$1 \geq F \geq \frac{\pi}{4} \tag{105}$$

Thus, for α_e from (95) one obtains:

$$\begin{aligned} \alpha_e \quad &= \quad K^2 q \frac{2q^2 r_D^2}{(1+q^2 r_D^2)^2} \left\langle \frac{1}{\epsilon} \right\rangle^{-1} \left\langle \frac{1}{\epsilon} \frac{v_s}{v} \sqrt{1 + \frac{1}{q^2 l_0^2}} \left(1 - \frac{v_d^\epsilon}{v_s}\right) \times \right. \\ &\left. \times F \left(1; \frac{\sqrt{1+q^2 l_0^2} - 1}{2\sqrt{1+q^2 l_0^2}}; \frac{3\sqrt{1+q^2 l_0^2} - 1}{2\sqrt{1+q^2 l_0^2}}; \frac{q^2 l_0^2}{\left(1 + \sqrt{1+q^2 l_0^2}\right)^2} \right) \right\rangle \end{aligned} \tag{106}$$

Other scattering mechanisms ($\nu = -\frac{1}{2}$ is the acoustic fhonon scattering via deformation potential; $\nu = \frac{1}{2}$ is the optical phonon scattering above Debye temperature, or acoustic phonon scattering via piezopotential; $\nu = \frac{3}{2}$ is the ionized impurity scattering, etc.) can be considered in a similar manner.

This theory satisfactorily describes all known experimental data concerning acoustic waves electronic absorption and amplification for intermediate (and high!) frequencies $ql \neq 1$ including region $ql \simeq 1$. Example of this matching is given on Fig. 5.

4. Surface acoustic waves case.

In the uniform piezoelectric semiconductor the interaction of surface wave (SAW) with electrons can be considered in a way similar to the one we followed in previous chapters. So we shall not do it. Much more interesting is the case of layered structures piezoelectric-semiconductor (see Fig.6) which are of pratical importance. Here piezoelectric and semiconductor media are separated and so can be optimized in a better way. To find the SAW electric absorption (amplification) coefficient, α_e, one must write the equations of (18)-(23) type for each medium and use proper boundary conditions.

For the hydrodynamic case and thin dielectric layers

$$h \ll \lambda_{SAW} \tag{107}$$

one obtains:

$$\alpha_e = -\frac{K^2 q}{2}\frac{\epsilon_p}{1+\epsilon_p}\frac{\gamma\frac{\omega_c}{\omega}\frac{\epsilon_S}{1+\epsilon_p}th(q_R d)}{\gamma^2+\left[F+\frac{\omega_c}{\omega}\frac{\epsilon_S}{1+\epsilon_p}th(q_R d)\right]^2} \tag{108}$$

Here $\gamma = 1 - \frac{v_d}{v_R}$ is the drift parameter, v_R and q_R are the velocity and the wave vector of Raylaigh SAW, $F = F(q_R, d, \sigma)$ is the diffusion factor, $\omega_e = \frac{\sigma}{\epsilon_0 \epsilon_S}$, σ is the conductivity of a semiconductor, v_d is the drift velocity of electrons, $K^2 = 2\frac{v_0 - v_\infty}{v_0} F_1(q_R, h, d, \epsilon_p, \epsilon_d, \epsilon_S,)$ is the effective electromechanical coupling constant, v_0 and v_∞ are the velocities in the piezoelectric medium under consideration along the free and the metallized surface respectively, F_1 is of the order of unity. The excellent matching of experimental data for $LiNbO_3 - SiOx - InSb$ SAW amplifier with theoretical calculations in accordance with formula (108) shows the relevance of the theory (see Fig.7-9).

To conclude this section, it is possible to say that the theory of electronic absorption and amplification of acoustic waves in solids is developed quite well for different structures and types of waves, different material parameters and frequencies and it is in reasonably good agreement with all known experimental data.

2. The Effects Connected with the Drag of Electrons by AW in Semiconductors

The acoustic wave propagating in a semiconductor and losing its energy due to electronic absorption loses also some of its momentum which partly can be transferred to electrons, thus producing d.c. *drag* force applied to electrons. This drag force may, in principle, produce d.c. kinetic phenomena similar to those produced by temperature gradient ∇T or electron concentration gradient Δn: d.c. longitudinal current (or voltage in open circuit case), which is called acoustoelectric effect, d.c. transverse current (or voltage) in transverse d.c. magnetic field, which could be called acoustomagnetoelectric effect etc. This section is devoted to consideration of some of these effect.

2.1 Acoustoelectric Effect.

The value of the acoustoelectric (AE) current or voltage in the simplest one-dimentional case of isotropic semiconductors (see Fig. 10) can be obtained from the momentum conservation law. Let W be the acoustic wave intensity (energy flux), measured in W/cm^2. The energy and momentum conservation laws for the interaction of AW with electrons give

$$\frac{1}{v_s}\frac{\partial W}{\partial x} = -\alpha_e \frac{W}{v_s} \tag{109}$$

$$-\frac{1}{v_s}\alpha_e W = -m^* v_d \nu N_0 \tag{110}$$

Here ν is the frequency of electron collisions, $\nu = \tau^{-1}$. Introducing the mobility of electrons $\mu = \frac{e\tau}{m^*} = \frac{e}{m^*\nu}$ from (110) we have

$$-\frac{1}{v_s}\alpha_e W = -\frac{ev_d N_0}{\mu} \equiv \frac{j_{ae}}{\mu} \tag{111}$$

From the above equation we immediately obtain for the AE current j_{ae}

$$j_{ae} = \frac{\mu}{v_s}\alpha_e W \tag{112}$$

The previous equation constitutes the well known Weinreich relation for the short-circuited case.

For the open-circuited case, by putting $j_{ae} = \sigma E_{ae} \equiv e\mu N_0 E_{ae}$, we obtain

$$E_{ae} = \frac{\alpha_e W}{eN_0 v_s} \tag{113}$$

or, for acoustoelectric voltage in the case of weak AW absorption,

$$V_{ae} = \frac{\alpha_e W}{eN_0 v_s} L \tag{114}$$

which are the very well known Weinreich relations for AE field and voltage. Usually Weinreich relations in such a simple form are valid only in some particular cases. Indeed, for example in hydrodynamic case the AW interacts not with individual electrons, but with electron bunches and, hence, drags these bunches. But their drift velocity, which is used in the derivation of Weinreich relations (111) and (112), is different, as it was already shown above, from the average drift velocity of electrons by the factor r (see (33) and (34)). So, strictly speaking, in Weinreich relations (112)-(114) instead of v_d one must substitute $v_d^* = v_d r$, where r is the same as in amplification condition $v_d r > v_s$ (41). So the correct Weinreich relations are

$$j_{ae} = \frac{\mu}{v_s}\alpha_e W r \tag{115}$$

$$E_{ae} = \frac{\alpha_e W}{eN_0 v_s} r \tag{116}$$

$$V_{ae} = \frac{\alpha_e W}{eN_0 v_s} L r \tag{117}$$

where $r = \frac{d\ln\sigma_0}{d\ln N_0} = 1 + \frac{d\ln\mu_0}{d\ln N_0}$; for degenerate semiconductor $r = 1 + \frac{2}{3}\frac{d\ln\tau(\varsigma)}{d\ln\varsigma}$
The relations (115)-(117), of course, follow from the strict theory (see below).

Let us consider first the phenomenological theory for hydrodynamic case in one-dimensional isotropic configuration.

The current j, in the presence of the AW, is given by the expression

$$j_{\sim} = \sigma\left(-\frac{\partial\phi}{\partial x}\right) - eD\frac{\partial n}{\partial x} \tag{118}$$

(see (27) when $\Lambda = 0$, $E_0 = 0$)

$$\phi\ ,\ n\ \propto u_0 e^{i(\omega t - qx)} \tag{119}$$

u_0 is the AW amplitude, $\sigma = \sigma_0 + \frac{\partial\sigma_0}{\partial N_0}n$, $\sigma_0 = e\mu_0(N_0)N_0$

$$\frac{\partial\sigma_0}{\partial N_0} = \frac{\sigma_0}{N_0}\frac{\partial\ln\sigma_0}{\partial\ln N_0} \equiv \frac{\sigma_0}{N_0}r \tag{120}$$

So, from (118) one has:

$$j_{\sim} = -\sigma_0\frac{\partial\phi}{\partial x} - \frac{\sigma_0}{N_0}rn\frac{\partial\phi}{\partial x} - eD_0\frac{\partial n}{\partial x} \tag{121}$$

We are interested in the d.c. part of this current which may only come from the nonlinear term $-\frac{\sigma_0}{N_0}\cdot rn\cdot\frac{\partial\phi}{\partial x}$. So averaging over the AW period we have

$$\overline{j_{\sim}} \equiv j_{ae} = -r\frac{\sigma_0}{N_0}\overline{n\frac{\partial\phi}{\partial x}} = -r\frac{\sigma_0}{N_0}\frac{1}{2}\Re\left[n\left(\frac{\partial\phi}{\partial x}\right)^*\right] \tag{122}$$

From the basic equations (24)-(27) one has:

$$\left\{\begin{array}{lcl} \frac{\partial\phi}{\partial x} & = & \frac{4\pi\beta}{\epsilon}(iqu)\frac{q^2r_D^2 - i\omega\tau_M\left(1-\frac{v_dr}{v_s}\right)}{(1+q^2r_D^2) - i\omega\tau_M\left(1-\frac{v_dr}{v_s}\right)} \\ n & = & \frac{\beta q^2}{e}u\frac{1}{(1+q^2r_D^2) - i\omega\tau_M\left(1-\frac{v_dr}{v_s}\right)} \end{array}\right. \tag{123}$$

Introducing the acoustic wave intensity

$$W = \frac{c\omega}{q}\overline{\left(\frac{\partial u}{\partial x}\right)^2} \tag{124}$$

from (122) and (123) we obtain

$$j_{ae} = \frac{\mu_0}{v_s}\alpha_e W r \tag{125}$$

$$r = \frac{\partial \ln \sigma_0}{\partial \ln N_0} = \frac{1}{e\mu_0}\frac{\partial \sigma_0}{\partial N_0} \tag{126}$$

Formula (125) is the *corrected* Weinreich relation (115) which we have derived above from qualitative momentum concervation considerations. The other Weinreich relations, for example, $E_{ae} = \frac{\alpha_e W}{eN_0 v_s} r$ follow immediately.

Let us give some numerical estimates. From (116) one has (order of magnitude:

$$E_{ae} \simeq \frac{K^2 q W}{eN_0 v_s} \tag{127}$$

For InSb with $N_0 = 10^{12} cm^{-3}$, $v_s = 3 \cdot 10^5 cm/s$, $K^2 = 10^{-2}$, $c = 10^{12} dyn/cm^2$ and AW with $W = 1mW$, $f = 200MHz$ ($q = 4.2 \cdot 10^3 cm^{-1}$, $\omega = 1.26 \cdot 10^9 s^{-1}$) we have $qu_0 = \left\{\frac{W}{cv_s}\right\}^{\frac{1}{2}} = 1.8 \cdot 10^{-7}$ and hence from (127) $E_{ae} \simeq 1V/cm$.

2.2 Acoustomagnetoelectric, acoustothermal and other drag effect.

Let us first look at Fig.11. At it was already shown, in this geometry either an AE current will flow (short-circuited case), or an AE voltage will appear (open-circuited case) between contacts 1 and 2. Moreover the AW, acting as a dragging force on electrons, may produce, as it was also already said, other **drag** effects. Let us consider some of them.

(a) - **Acoustoelectric Hall effect** - This effect appears when contacts 1 and 2 are short-circuited and the AE current j_{ae} flows in the X-direction. Then in a transverse magnetic field $\vec{H}$ a Hall voltage between contacts 3 and 4 will obviously appear (or a Hall current will flow, if contacts 3 and 4 are short-circuited). The expression for *AE Hall* effect in the simplest case may be easily obtained from usual Hall effect formulae, where the ordinary electric current must be substituted by the acoustoelectric current, j_{ae}.

(b) - Acoustomagnetoelectric effect - This effect by definition is the appearance of a transverse voltage between contacts 3 and 4 in the presence of a transverse magnetic field $\vec{H}$, when contacts 1 and 2 are open-circuited, so that there is no longitudinal current. In order to obtain the value of the acoustomagnetoelectric (AME) effect, let us consider phenomenological expressions for averages (over period) of the AW electric current in the presence of both the longitudinal electric and transverse magnetic fields and AW (compare (122)):

$$\begin{aligned} j_{ae}^{(x)} &= \overline{\sigma_{xx}E_x} + \overline{\sigma_{xy}E_y} - eD_{xx}\overline{\frac{\partial n}{\partial x}} \\ &= \sigma_{xx}^{(0)}E_{ae}^{(x)} + \sigma_{xy}^{(0)}E_{ae}^{(y)} + \frac{\partial\sigma_{xx}^{(0)}}{\partial N_0}\overline{n\left(-\frac{\partial\phi}{\partial x}\right)} \end{aligned} \tag{128}$$

$$\begin{aligned} j_{ae}^{(y)} &= -\overline{\sigma_{xy}E_x} + \overline{\sigma_{xx}E_y} + eD_{xy}\overline{\frac{\partial n}{\partial x}} \\ &= -\sigma_{xy}^{(0)}E_{ae}^{(x)} + \sigma_{xx}^{(0)}E_{ae}^{(y)} - \frac{\partial\sigma_{xy}^{(0)}}{\partial N_0}\overline{n\left(-\frac{\partial\phi}{\partial x}\right)} \end{aligned} \tag{129}$$

Remembering (see (122)-(125)) that

$$-\overline{n\frac{\partial\phi}{\partial x}} = -\frac{1}{2}\Re\left[n\left(\frac{\partial\phi}{\partial x}\right)^*\right] = -\frac{\alpha_e W}{ev_s} \tag{130}$$

we have

$$\left\{\begin{aligned} j_{ae}^{(x)} &= \sigma_{xx}^{(0)}E_{ae}^{(x)} + \sigma_{xy}^{(0)}E_{ae}^{(y)} - \frac{\partial\sigma_{xx}^{(0)}}{\partial N_0}\frac{\alpha_e W}{ev_s} \\ j_{ae}^{(y)} &= -\sigma_{xy}^{(0)}E_{ae}^{(x)} + \sigma_{xx}^{(0)}E_{ae}^{(y)} + \frac{\partial\sigma_{xy}^{(0)}}{\partial N_0}\frac{\alpha_e W}{ev_s} \end{aligned}\right. \tag{131}$$

The components of the conductivity tensor σ_{ik} are well known:

$$\sigma_{xx}^{(0)} = \frac{e^2N_0}{m^*}\left\langle\frac{\tau}{1+\omega_M^2\tau^2}\right\rangle \quad , \quad \sigma_{xy}^{(0)} = \frac{e^2N_0}{m^*}\left\langle\frac{\omega_M\tau^2}{1+\omega_M^2\tau^2}\right\rangle \tag{132}$$

where the symbol $\langle\cdots\rangle$ was given by formula (97).

$$j_{ae}^{(x)} = 0 \quad , \quad j_{ae}^{(y)} = 0 \tag{133}$$

By using here (131) we immediately have:

$$\left\{\begin{aligned} E_{ae}^{(x)} &\equiv E_{ae} &= \frac{\alpha_e W}{eN_0v_s}r_\parallel \\ E_{ae}^{(y)} &\equiv E_{AME} &= \frac{\alpha_e W}{eN_0v_s}r_\perp \end{aligned}\right. \tag{134}$$

where the factors $r_\parallel, r_\perp$ are:

$$\left\{ \begin{array}{lcl} r_\parallel & = & \dfrac{1}{2}\dfrac{\partial \ln\left(\sigma_{xx}^{(0)2}+\sigma_{xy}^{(0)2}\right)}{\partial \ln N_0} \\ r_\perp & = & -\dfrac{\sigma_{xx}^{(0)}\sigma_{xy}^{(0)}}{\sigma_{xx}^{(0)2}+\sigma_{xy}^{(0)2}}\dfrac{\partial \ln\left(\dfrac{\sigma_{xy}^{(0)}}{\sigma_{xx}^{(0)}}\right)}{\partial \ln N_0} \end{array} \right. \tag{135}$$

One can easily see that in the absence of magnetic field $\sigma_{xy}^{(0)} = 0, \quad \sigma_{xx}^{(0)} = \sigma_0$ $r_\parallel = \dfrac{\partial \ln \sigma_0}{\partial \ln N_0} \equiv r$ introduced above and $r_\perp = 0$ so E_{AME} and E_{ae} is given by Weinreich relation (116)

For degenerate electron gas, putting $\tau(\epsilon) \sim \epsilon^\nu$ we have for AME field

$$E_{AME} = \frac{\alpha_e W}{eN_0 v_s}\frac{2}{3}\nu\frac{\frac{\mu_0 H}{c}}{1+\left(\frac{\mu_0 H}{c}\right)^2} \tag{136}$$

where c is the light velocity.

One may see that for different electron scattering mechanisms the AME effect may have different sign ($\nu = \frac{1}{2}$ for acoustic phonon scattering and $\nu = \frac{3}{2}$ for impurity scattering, for example). So the AME effect may by very good tool to determine electron scattering mechanisms in semiconductors, which is very important for semiconductor electronics.

The physical reason for appearance of the AME effect is the following. *Drag forces* applied to electrons of different energies from the acoustic wave and from compensating AE field, $E_{ae}^{(x)}$, due to bunching of electrons and the dependence of the *partial* mobility of electrons $\mu(\epsilon) = \dfrac{e\tau(\epsilon)}{m^*}$ on their energy, are generally speaking, different. They do not compensate one another for each individual electron. So, in spite of the absence of net longitudinal electric current in the X-direction, there exist two currents equal in value and flowing in opposite directions: acoustoelectric current j_{ae} and compensating electric current $\sigma_{xx}^{(0)}E_{ae}$ from compensating AE field, of course both modified by magnetic field. These currents consist, generally speaking, of electrons of different average energy, having different average mobilities, and so the Hall effects on these two currents do not compensate each other.

The difference of Hall voltages produced by these two currents gives acousto-magnetoelectric (AME) effect.

In the collisionless case $ql \gg 1$, to calculate AME (as well as AE) effect it is necessary to use kinetic equation approach.

Calculations similar to those for the electron absorption coefficient, described above, give for that case (non degenerate semiconductor, weak magnetic field):

$$E_{AME} = \frac{\alpha_e W}{eN_0 v_s}\frac{\mu H}{c}\left[\frac{\Gamma\left(\frac{5}{2}\right)}{\Gamma\left(\frac{5}{2}+\nu\right)}\right]^2 \left\{\Gamma\left(1+2\nu\right) - \frac{\Gamma\left(\frac{5}{2}+2\nu\right)}{\Gamma\left(\frac{5}{2}+\nu\right)}\Gamma\left(1+\nu\right)\right\} \tag{137}$$

where Γ is the known Γ-funcion and $\tau(\epsilon) \sim \epsilon^\nu$.

One may see that in this case AME effect can be used to determine the electron scattering mechanism.

(c) - Longitudinal acoustothermal (AT) effect - As it was already pointed out, two currents flowing in an open-circued specimen in opposite directions (AE current and electric current due to compensating electric field) consist of electrons of different average energies. This means that in these conditions in the specimen there exists a net heat flux produced by electrons interacting with the AW. If the ends of the specimen are thermoisolated, then a compensating longitudinal temperature gradient appears, which we call longitudinal acousto-thermal (AT) effect. Calculation of this effect can be done in a way similar to AE and AME effects. For example, for the hydrodynamic case and for degenerate electron gas one obtains

$$\nabla T = \frac{\alpha_e W}{k_B N_0 v_s}\frac{2}{\pi^2}\frac{\varsigma_0}{T}\left(1+\frac{x_L}{x_e}\right)^{-1} \tag{138}$$

and for collisionless case for nondegenerate gas

$$\nabla T = -\frac{\alpha_e W}{k_B N_0 v_s}\frac{9\sqrt{\pi}}{8}\frac{\Gamma\left(1+\nu\right)}{\Gamma\left(\frac{7}{2}+\nu\right)}\left(1+\frac{x_L}{x_e}\right)^{-1} \tag{139}$$

where $\varkappa_L$, $\varkappa_e$ are the lattice and electron thermoconductivities, k_B is the Boltzmann constant and ζ_0 is the Fermi energy.

One may imagine that due to existence of *acoustic wave dragging force* to electrons in semiconductor (or in any other conducting material) there will appear the whole lot of kinetic phenomena similar to those which are produced by electric field, temperature gradient etc. All these effects have been calculated. They have same very interesting features. For example, Nernst-Ettinghausen and Peltier effects due to AE current are $\left(\frac{\zeta_0}{kT}\right)^2$ times bigger than for the same value of ordinary electric current. So for degenerate electron gas they may be very large. Also, the acoustoelectric Peltier coefficient, Π_{ae}, does not go to zero when $T \to 0$ as it is the case for ordinary electric current. This means that Peltier effect on AE current may produce much deeper cooling than ordinary Peltier effect.

2.3 Acoustoelectric effect on SAW.

A surface acoustic wave (SAW) propagating along the surface of a conducting body (semiconductor) will drag electrons, as one may imagine, along that surface (see Fig.12a). It will lead to the appearance of a circular AE current (and hence to a d.c. magnetic momentum induced by SAW in the specimen). Then, due to the presence of an alternating transverse electric field of SAW in piezoelectric material, comparable with longitudinal one ($E_{z\sim} = \frac{\varkappa}{iq} E_{x\sim}$, $\varkappa$ is the decay coefficient of SAW in Z-direction, see Fig.12a), a transverse AE effect appears. Of course, there will be also the ordinary longitudinal AE effect.

In order to calculate these effects one must first find the *driving* currents

$$\begin{cases} j_{ae}^{\parallel} &= \overline{e\mu n_{\sim} E_{x\sim}} \\ j_{ae}^{\perp} &= \overline{e\mu n_{\sim} E_{z\sim}} \end{cases} \tag{140}$$

They are caculated in a similar manner as we did it above. Then one must solve the problem of d.c. current distribution inside specimen using proper boundary conditions. For example, in hydrodynamic case after such a calculation we find for transverse AE voltage:

$$V_{ae}^{\perp} = \frac{\alpha_e W}{eN_0 v_s} d \frac{F}{\omega' \tau_M} \left\{ 1 + \frac{1}{2\left(1 + 2\sqrt{1 + \omega'^2 \tau_M^2}\right)} - \right.$$
$$\left. -2xr_D \frac{1 + \sqrt{1 + \omega'^2 \tau_M^2}}{1 + 2\sqrt{1 + \omega'^2 \tau_M^2}} \right\} \quad (141)$$

where $\omega' = \omega - qv_d$ and F is the factor of the order of 1. One may see, that the order of magnitude of the transverse AE effect is comparable with that ofthe longitudinal one, and can change sign depending on frequency, parameters of material and drift velocity. The physical origin of transverse AE effect is the bunching of electrons by the longitudinal alternating electric field of SAW and the presence of the transverse alternating electric field of SAW shifted in phase with respect to electron bunches, so that transverse electric current averaged over the period of the wave is not equal to zero.

2.4 Convolution of SAW signals.

If there are two SAWs propagating in opposite directions (see Fig.12b), the acoustoelectric current will be produced by both waves jointly:

$$\begin{aligned} j_{ae}^{\perp} &= \overline{e\mu (n_1 + n_2)(E_1 + E_2)} \\ &= e\mu \left(\overline{n_1 E_1} + \overline{n_2 E_2} + \overline{n_1 E_2} + \overline{n_2 E_1} \right) \\ &= j_{ae}^{\perp(1)} + j_{ae}^{\perp(2)} + j_{ae}^{conv} \end{aligned} \quad (142)$$

Here the mixing term acoustoelectric convolution current is equal to

$$j_{ae}^{conv} = e\mu \left(\overline{n_1 E_2} + \overline{n_2 E_1} \right) \quad (143)$$

where $n_{1,2}$ and $E_{1,2}$ are the alternating electron concentration and alternating electric field in both waves.

$$j_{ae}^{conv} = \frac{e\mu}{2} \mathcal{R} \left(n_1 E_2^* + n_2 E_1^* \right) \quad (144)$$

Let us put

$$\begin{cases} n_1 = n_{10}e^{i\omega t - iqx} \quad , \quad E_1 = E_{10}e^{i\omega t - iqx} \\ n_2 = n_{20}e^{-i\omega t - iqx} \quad , \quad E_2 = E_{20}e^{-i\omega t - iqx} \end{cases} \tag{145}$$

Here $n_1, E_1 \sim f$ and $n_2, E_2 \sim g$ where f and g are the amplitudes of two SAWs propagating in opposite directions (see Fig.12b). Then one has

$$j_{ae}^{conv} = \frac{e\mu}{2}\Re\left[n_{10}E_{20}^{*}e^{2i\omega t} + n_{20}E_{10}^{*}e^{-2i\omega t}\right] \tag{146}$$

So the convolution current is uniform along the surface and changes in time with double frequency. Calculating the amplitudes $n_{10}, E_{10}, n_{20}, E_{20}$ in a way similar to the one followed above, we obtain for the amplitude of the convolution voltage

$$V_{\perp}^{conv} = \frac{K^2q^3Cd}{eN_0}F\left(qr_D, \omega\tau_M\right)\frac{1}{T}\int_0^T f(\tau)g(t-2\tau)d\tau \tag{147}$$

where $F(qr_D, \omega\tau_M)$ is the same factor of the order of 1 and T is the transit time of SAW propagation in the specimen.

Thus, one sees that the transverse acoustoelectric voltage on double frequency is indeed the convolution of two SAW signals $f(t)$ and $g(t)$, which justifies the term we already used for j_{ae}^{conv} and $V_{\perp}^{conv}$.

Let us estimate the loss at the operation of acoustoelectric field:

$$E_{\perp}^{conv} \simeq \frac{K^2q^3C}{eN_0}fg \tag{148}$$

$$E_1 = \beta q f \tag{149}$$

then:

$$\frac{E_{\perp}^{conv}}{E_1} \simeq \frac{\beta q^2}{eN_0}g$$

So the convolution loss (in dB) is given by the expression

$$\Gamma = 20\log\frac{E_{\perp}^{conv}}{E_1} = 20\log\left(\frac{\beta q^2 g}{eN_0}\right) \quad dB \tag{150}$$

Thus, this loss depends on the amplitude of the other wave g. Usually the intensity of the second wave is taken equal to 1 mW; then the loss is then expressed in the units called dBm.

The amplitude of SAW, u, is connected with the intensity W by the approximate formula

$$qu \simeq \sqrt{\frac{W}{cv_s}} \qquad (u = f, g)$$

So by putting $W_2 = 1\ mW$ we have

$$qg = \sqrt{\frac{1mW}{cv_s}}$$

or typical case $c = 10^{12} dyn/cm^2$, $v_s = 3 \cdot 10^5 cm/s$; so

$$qg_{(W=1mW)} \simeq 10^{-7} \tag{151}$$

Let us take the frequency of SAW equal to $200MHz$, so that $q \sim 10^3 cm^{-1}$, $N_0 = 10^{14} cm^{-3}$ and $\beta = 3 \cdot 10^7 V/cm$. The estimate following formula (150) gives

$$\Gamma \simeq -60dBm \tag{152}$$

which is in reasonable agreement with the experiment.

It is necessary to note that the loss in acoustoelectric convolvers is much less than in convolvers based on nonlinearity of elastic properties of a material. Acoustoelectronic properties of different types are used in practice for signal processing.

3. Nonlinear Theory of AW Propagation, Absorption and Amplification in Semiconductors.

When the amplitude of the acoustic wave in semiconductor is large enough, the nonlinear effects become noticeable. The main nonlinear effects are:

1. Dependence of absorption (amplification) coefficients of AW and AW velocity on its amplitude.

2. Nonohmic behaviour of the current-voltage curve (CVC) of the specimen.

3. Generation of harmonics.

4. Current oscillations due to formation and motion of acoustoelectric domains.

5. Parametric amplification of subharmonics.

On one hand they restrict the dynamic range of various acoustoelectronic devices, and on the other hand they may give rise to creation of new devices based on acoustoelectronic nonlinearity. We shall give here a brief sketch of the analytical nonlinear theory of acoustoelectronic phenomena, published by the authors in IEEE Transactions in 1970.

Let us consider a piezoelectric semiconductor oriented in such a way that propagation of the AW can be described as a one-dimensional problem. We shall consider the hydrodynamic case and neglect electron trapping and heating. All the theory is made under the assumption that the amplitude and the velocity of AW change little on the wavelength and during the period of the wave.

The whole problem is divided in two parts:

1) *Quick part:*

(a) Solving the problem of propagation of the AW on the small space-time interval when the amplitude and the velocity of the wave do not change nociteably.

(b) Deriving the average equation describing the small variations of the wave amplitude in space and time (the continuity equation for the sound flux).

2) *Slow part:*

(a) Solving the equation for the amplitude of the wave together with the Poisson equation and continuity equation for the current.

(b) Taking into account proper boundary conditions and finding the distribution inside specimen of AW intensity, electric field and current.

So the main aim of the whole calculation is to find the law of variation in space and time for intensity, velocity and shape of AW and distribution of average electric field in specimen and its CVC.

Basic equations for the *quick* problem are the following:

$$\begin{cases} \rho\dfrac{\partial^2 u}{\partial t^2} - c\dfrac{\partial^2 u}{\partial x^2} = -\beta\dfrac{\partial^2 \phi}{\partial x^2} + \chi\dfrac{\partial^3 u}{\partial x^2 \partial t} \\ \dfrac{\partial^2 \phi}{\partial x^2} + \dfrac{4\pi\beta}{\epsilon}\dfrac{\partial^2 u}{\partial x^2} = -\dfrac{4\pi e}{\epsilon}(N - N_0) \\ \dfrac{\partial N}{\partial t} + \dfrac{1}{e}\dfrac{\partial j}{\partial x} = 0 \\ j = eN\mu\left(\epsilon - \dfrac{\partial \phi}{\partial x}\right) - eD\dfrac{\partial N}{\partial x} \end{cases} \tag{153}$$

where χ is the viscosity coefficient of the crystal.

The results of solving the first part of the quick problem (finding the nonlinear electronic, α_e, and lattice, α_L, AW absorption coefficients and the shape of the wave) are the following:

$$\alpha_e = -\eta q \frac{p - \gamma}{a^2 w} \tag{154}$$

$$\alpha_L = xq\left(\frac{\gamma}{p} + \frac{3\sqrt{3}}{2\pi}\sqrt{w}\frac{p - \gamma}{p + \gamma\sqrt{w}}\right) \tag{155}$$

Where

$$\begin{cases} \eta = \dfrac{4\pi\beta^2}{\epsilon c} \equiv K^2 \qquad p = \dfrac{\mu\epsilon - v_s}{Dq} \qquad \gamma = \dfrac{v_d - v_s}{Dq} \\ x = \dfrac{\chi\omega}{c} \qquad a = qr_D \qquad w = \dfrac{W}{W_c} \\ W_c = q^2 r_D^2 \dfrac{kTN_0 v_s}{\eta} \end{cases} \tag{156}$$

The shape of the wave is nonsinusoidal due to effective generation of harmonics, as it is shown on Fig.13.

Let us consider now the slow problem to find the stationary amplification of AW of very large amplitude in finite crystal.

The basic equations are:

$$\left\{\begin{array}{l} \frac{\partial W}{\partial t} + v_s \frac{\partial W}{\partial x} = -\left[\alpha_e(\epsilon, W) + \alpha_L(\epsilon, W)\right] v_s W \\ I = eN_0\mu\epsilon - eD\frac{\partial N_0}{\partial x} + \frac{\mu}{v_s}\alpha_e(\epsilon, W)W \\ \frac{\partial \epsilon}{\partial x} = \frac{4\pi e}{\epsilon}(N_0 - N_i) \\ \frac{\partial N_0}{\partial t} + \frac{1}{e}\frac{\partial I}{\partial x} = 0 \\ \alpha_e = -\eta q \frac{p-\gamma}{a^2 w} \\ \alpha_L = xq\left(\frac{\gamma}{p} + \frac{3\sqrt{3}}{2\pi}\sqrt{w}\frac{p-\gamma}{p+\gamma\sqrt{w}}\right) \end{array}\right. \tag{157}$$

Here ϵ is the d.c. electric field inside the specimen.

The boundary conditions are

$$W_L(x)\big|_{x=0} = W_0 \tag{158}$$

$$\int_0^L \epsilon(x)dx = V \tag{159}$$

where L is the length of the specimen, V the applied voltage.

The results of the calculation are presented in Fig.14 and Fig.15. One may see that there is good agreement between theoretical and experimentally measured current-voltage curves of the specimen for similar initial AW intensities. Also there is good agreement between theory and experiment for the shape of the wave (Fig.13), which was obtained by Mansfeld et al. and other authors.

In this chapter we have presented the nonlinear theory of AW amplification for the hydrodynamic case, finite size specimen and initially monochromatic acoustic wave. Today many works of different authors exist, developing nonlinear theory of AW propagation and amplification for various cases: infinite specimen (stationary waves), collisionless case (so called *momentum nonlinearity*), nonmonochromatic case (acoustic noise amplification), etc. Because of the shortage of space and time we shall not describe these theories (and many experiments). We only note here that in spite of differences in the physics of various nonlinear cases, the main features of nonlinear acoustoelectronic phenomena described in this chapter remain generally the same.

Conclusions

In these lectures we have tried to present with simple examples the modern state of the theory (and a bit of experiments) of phenomena connected with free carriers (electrons) in semiconductors. Of course, today a lot of papers of various authors, including very distinguished names, are devoted to this topic. It is not possible to mention even a substantial part of them. Since these simple lectures are directed to the listener of the School in physical acoustics, which are in general not working in this particular direction, called acoustoelectronics, we apoligize with those authors whose results are used in preparation of the lectures for not mentioning their names. Just for orientation we give here a short list of references. For a more complete list of references see the review paper by Yu.V. Gulyaev and A.G. Kozorezov in Soviet Scientific Reviews, 1981 and the paper by Yu.V. Gulyaev in IEEE Transations, SU, 1970. We hope that already from these simple lectures the reader can get the image of interesting phisics in the acoustoelectronic phenomena bringing to a better understanding of the properties of condensed matter and also of the mathematics (sometimes quite high level) needed to describe these phenomena. As far as the pratical applications of these phenomena are concerned, they still are not so many: mainly convolvers, correlators, spectrum analyzers, modulators and demodulators and some kinds of storage devices. As it often happens, the fast development of acoustoelectronics in the early sixties was provoked by the experimental discovery by A.R. Hutson, J.H. McFree and D.L. White of the AW amplification via supersonic drift of electrons, and by the later suggestion to use for this amplification the surface acoustic waves in semiconductor-piezoelectric layered structures; the main practical development of acoustoelectronics in general went in the direction of microwave acoustics, mainly using surface acoustic waves in piezoelectric dielectrics. So now, most of the production of acoustoelectronic devices consists of various kinds of effective filters, delay lines, Fourier-trasformers, spectrum analyzers, coding and decoding devices, matched receivers etc. (all using SAW or skimming AW in piezoelectrics). The acoustoelectronic amplifier which attracted the attention of many physicists and engineers into this field is still only an iteresting toy waiting for its application. By I will dare to say that due to complexity of acoustoelectronic phenomena in the narrow sense (AW interaction with free electrons) there are many possibilities to create completely new acoustoelectronic devices: let us hope that they will come!

In conclusion, I would like to express my great gratitude to the organizers of the 1988 Summer School of Physical Acoustics in Erice, Italy, and especially Professor A. Alippi for inviting me there and giving me the possibility to deliver these lectures to the scientific community.

References

[1] V.L.Gurevich, Fiz. Tekh. Poluprov, Sov. Phys. Semicond. 2, 1557, 1968.

[2] V.I.Pustovoit, Usp. Fiz.Nauk, Sov. Phys. Usp. 97, 257, 1969.

[3] Yu.V.Gulyaev, A.G.Kozorezov, Soviet Scientific Review, Physics A, 1981.

[4] A.R.Hutson, J.H.McFee, D.L.White, Phys. Rev. Lett. 7, 237, 1961.

[5] D.L.White, J. Appl. Physics 33, 2547, 1962.

[6] G.Weinreich, Phys. Review 107, 317, 1957.

[7] Yu.V.Gulyaev, IEEE Transations on Sonics and Ultrasonics, April 1970.

[8] Yu.V.Gulyaev, Physics Lett. 29A, 187, 1969.

[9] E.M.Epstein, Yu.V.Gulyaev, Fiz.Tverd.Tela, Soviet Phisics Solid State 9, 376, 1967.

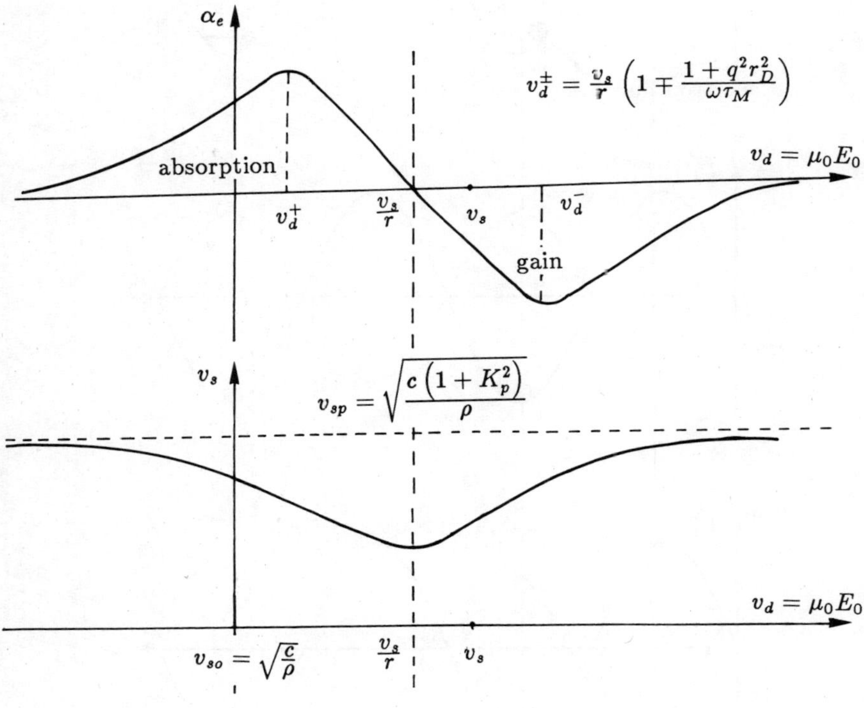

Fig.1: Dependencies of acoustic wave absorption coefficient α_e and its velocity v_s on drift field E_0.

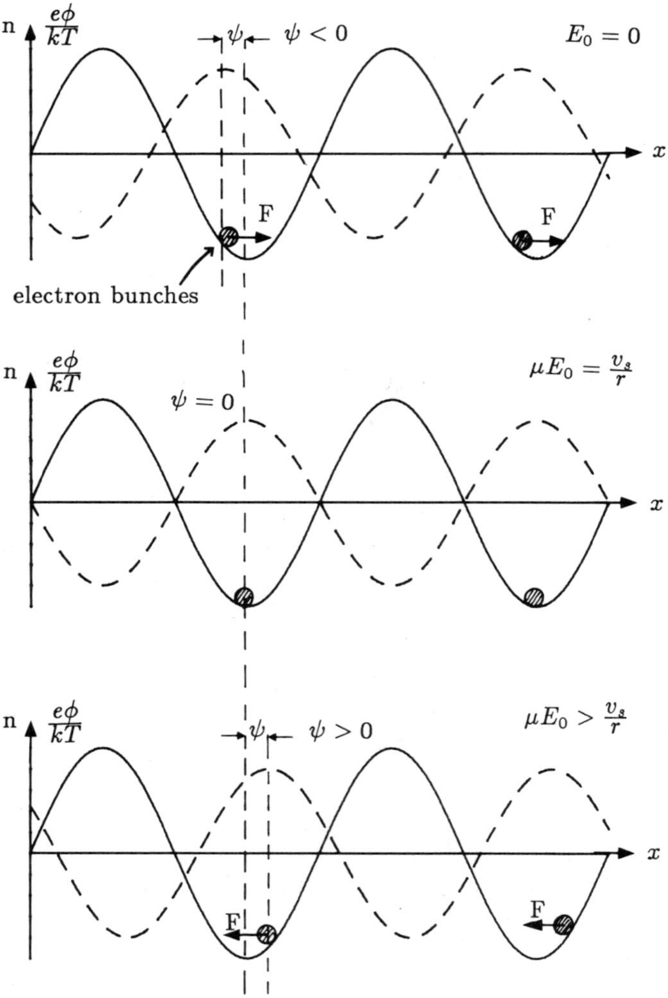

$$\text{sign}\psi = -\text{sign}\left(1 - \frac{\mu E_0 r}{v_s}\right)$$

F - force acting on electron bunch from the acoustic wave.

Fig.2: Physical explanation of electron absorption and amplification of acoustic waves.

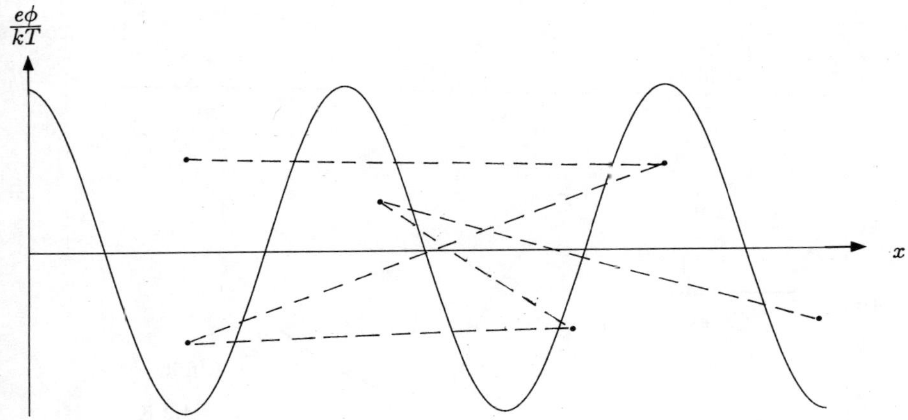

Fig.3: Motion of electron in the potential of acoustic wave in the case $ql >> 1$.

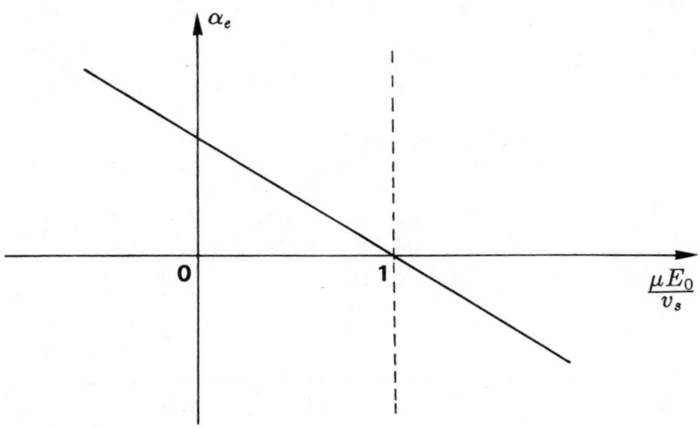

Fig.4: Dependence of electronic absorption (amplification) coefficient, α_e, of AW on drift electric field E_0 in classical collisionless case.

absorption amplification

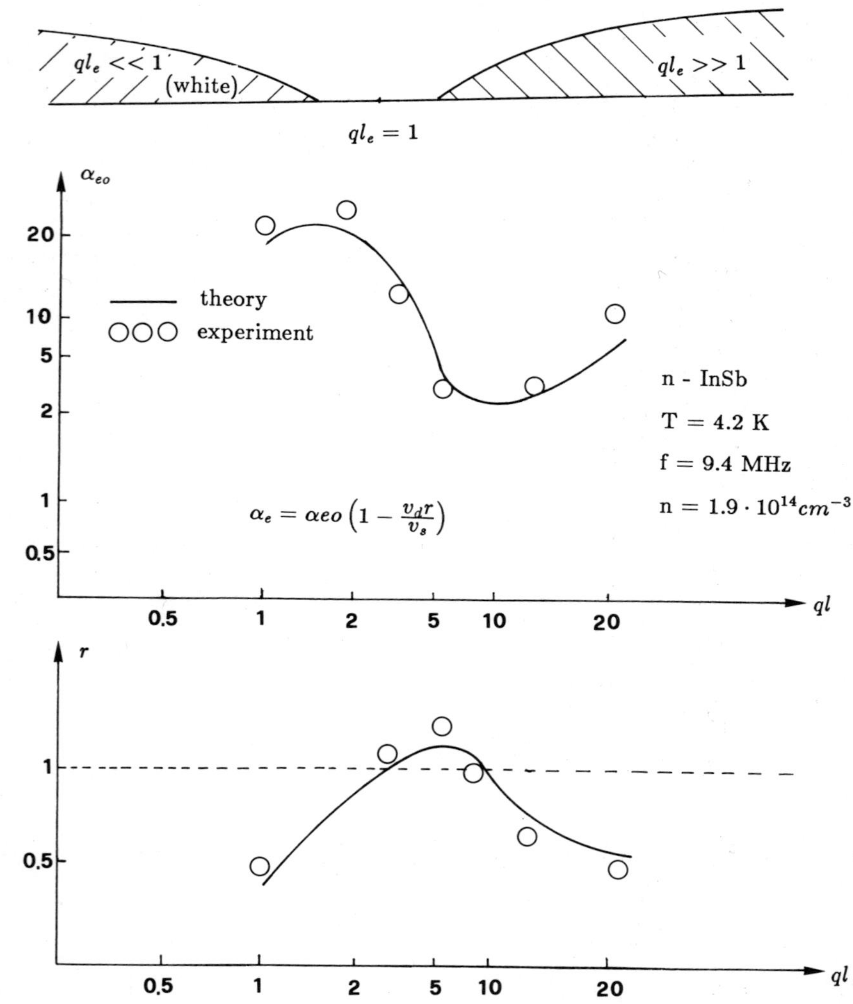

Fig.5: The comparison of the theory of electronic absorption and amplification of acoustic waves at intermediate frequencies, presented above, with the experiment of Iwanov et al.

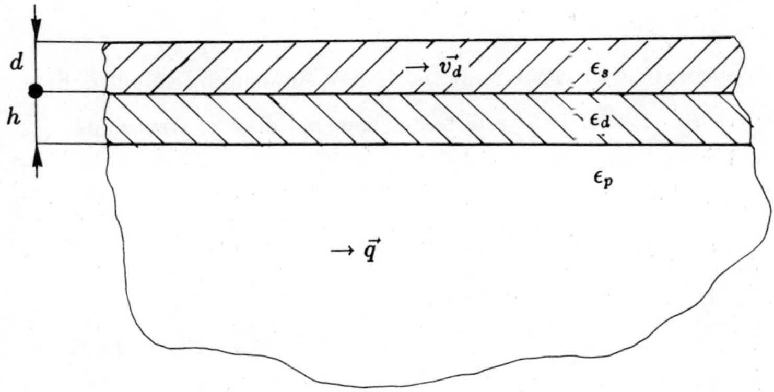

Fig.6: Layered structure semiconductor - dielectric - piezoelectric, optimal for devices, based on interaction of surface acoustic waves (SAW) with electrons. semiconductor dielectric piezoelectric

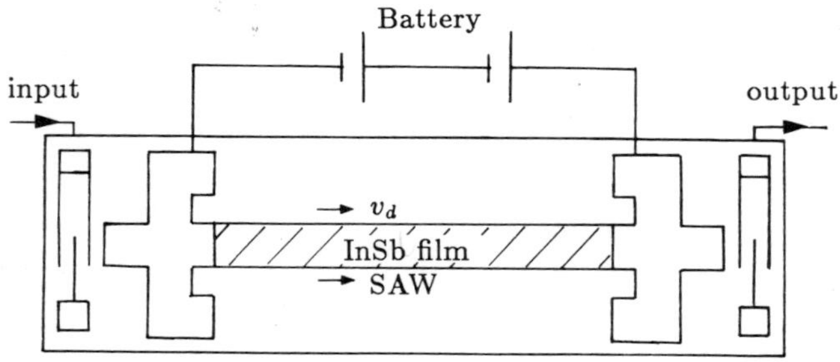

Fig.7: Configuration for CW operation of SAW amplifier based on configuration $LiNbO_3 - SiO_x - InSb$. $InSb$ film has a shape of a narrow waveguide.

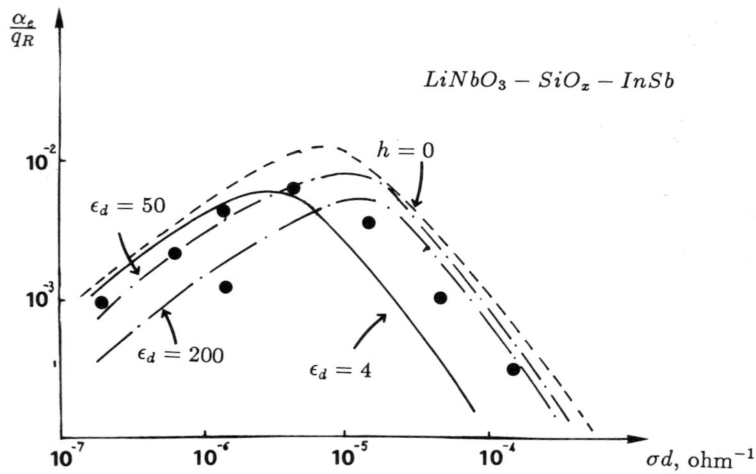

Fig.8: Electronic absorption of SAW versus conductivity for the structure $LiNbO_3 - SiO_x - InSb$, $q_R h = 0.07$, $K^2 = 0.048$, $\epsilon_p = 50$. The experiment (black circles) and theoretical calculation (solid and dashed lines).

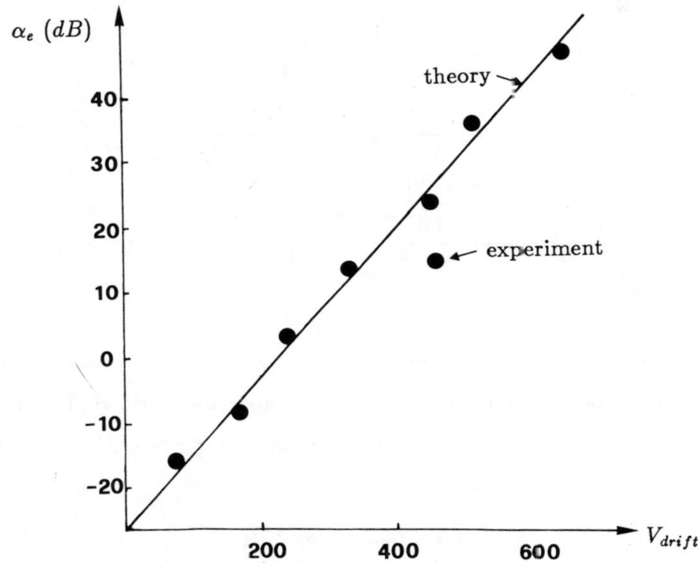

Fig.9: Electronic gain versus drift voltage for $LiNbO_3 - SiO_x - InSb$ structure. Frequency is 200 MHz, mobility of electrons 3000 $cm^2/V \cdot s$, conductivity $4.4 \cdot 10^{-6}$ $ohm^{-1}cm^{-1}$, length of semiconductor 1 cm.

Short-circuited case: Open-circuited case:

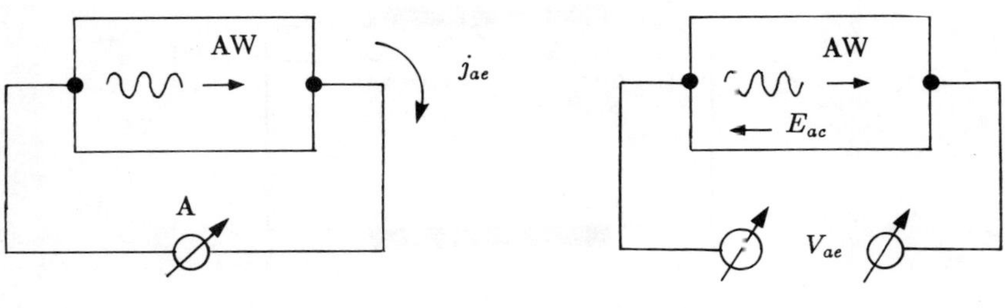

Fig.10: Configuration for observation of AE effect in short-circuited and open-circuited cases.

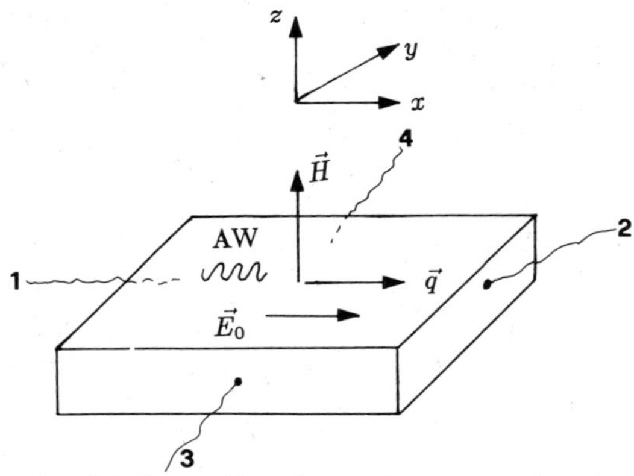

Fig.11: Configuration for observation of acousto-magnetoelectric effect from contacts 3-4 (contacts 1-2 are open-circuited) and acoustoelectric Hall effect from the game 3-4 contacts (contact 1-2 in this case are short-circuited).

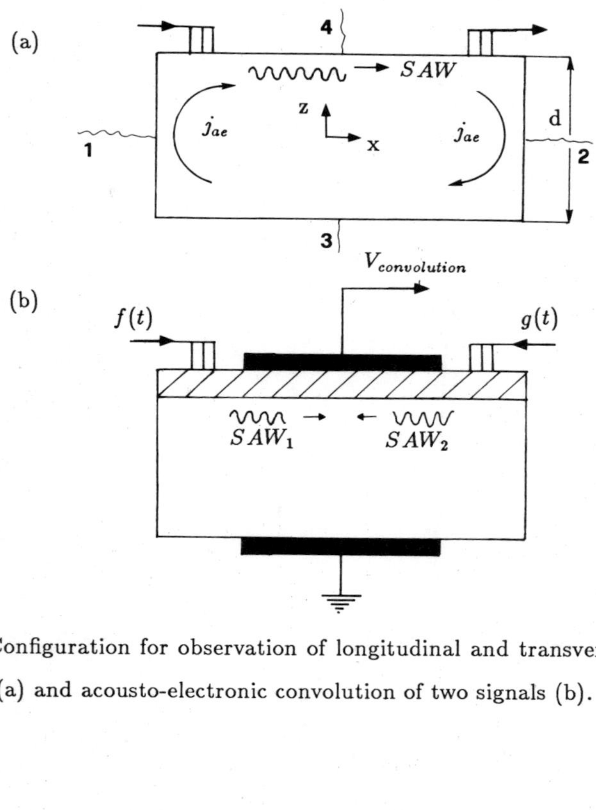

Fig.12: Configuration for observation of longitudinal and transverse AE effects on SAW (a) and acousto-electronic convolution of two signals (b).

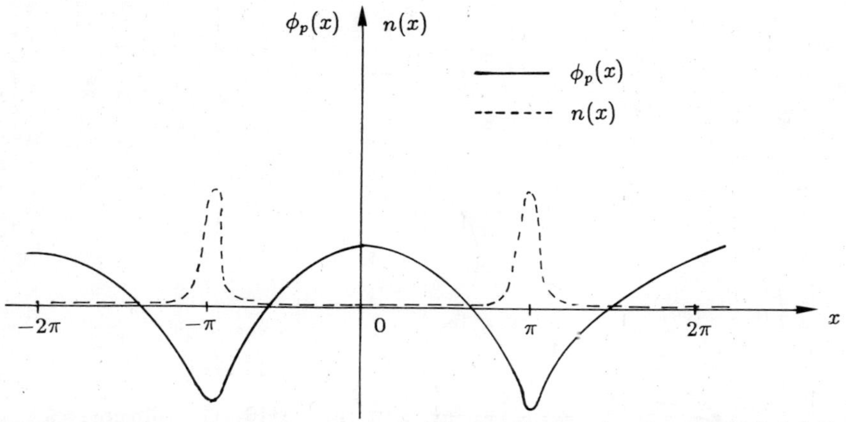

Fig.13: The shape of acoustic wave (coinsiding with piezoelectric potential $\phi_p(x)$) and electron concentration distribution in the wave in strongly nonlinear case.

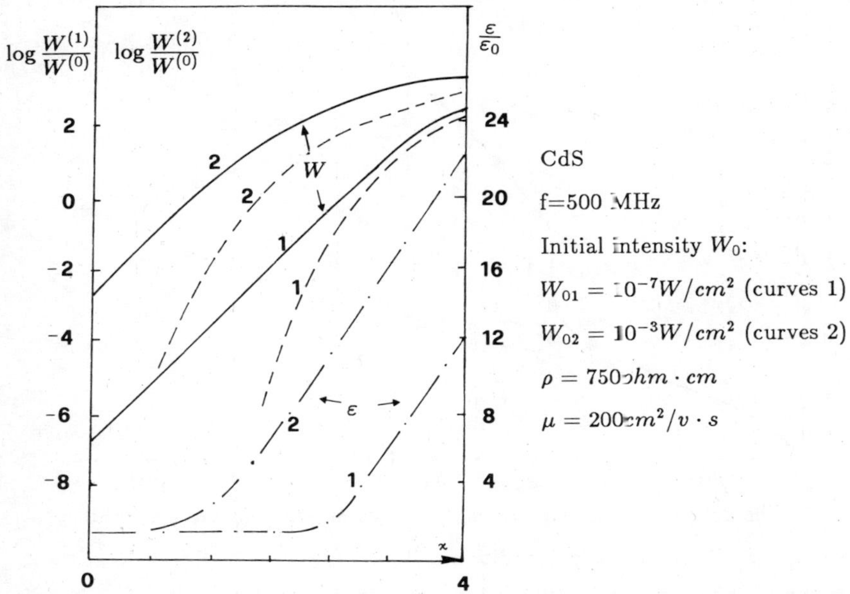

Fig.14: The intensities of first and second harmonics of acoustic wave, $W^{(1)}$ and $W^{(2)}$, for two initial AW intensities, W_{01} and W_{02}, and distributions of electric field along the specimen versus spatial coordinate for Cds sample.

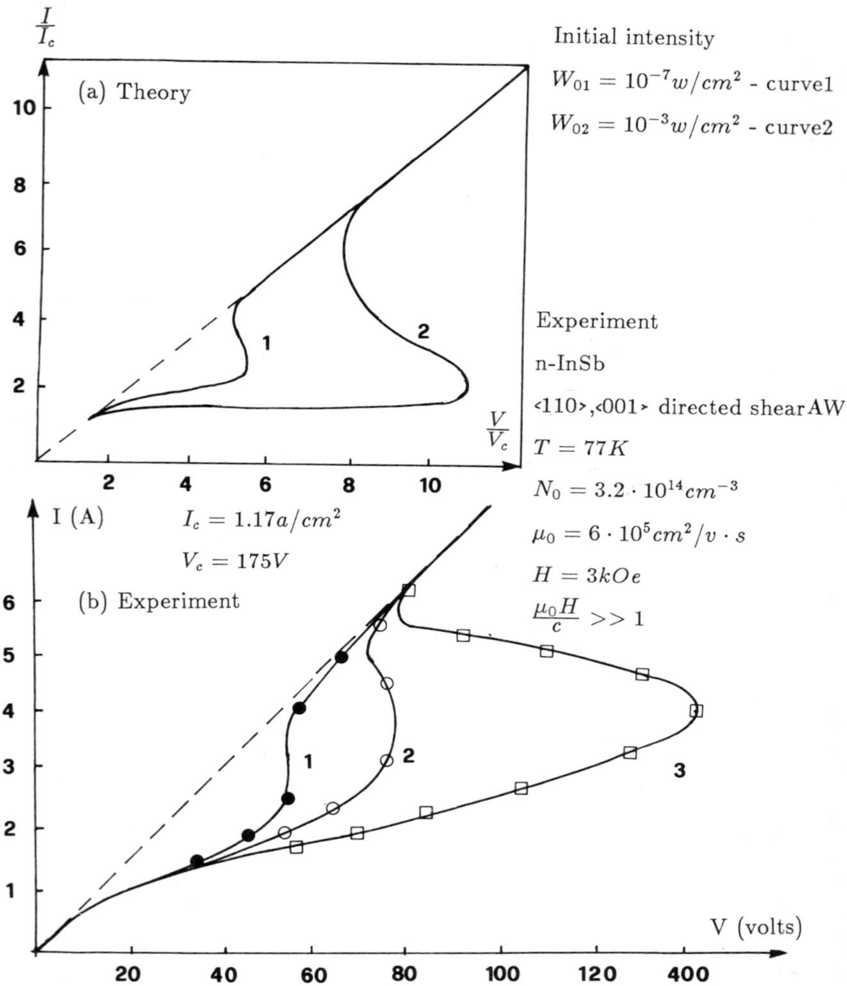

Fig.15: Theoretical (a) and experimental (b) current-voltage curve of InSb specimen under AW amplification conditions.

Initial intensities: $W_{01} = 10^{-7} W/cm^2$ (curves1) $W_{02} = 10^{-5} W/cm^2$ (curves2) and $W_{03} = 10^{-3} W/cm^2$ (curve 3)

(Mansfield et al.)

COMPOSITE CHARACTERIZATION USING GENERALIZED LAMB WAVES

Laszlo Adler
Department of Welding Engineering
The Ohio State University

INTRODUCTION

The objective of this work is to describe several techniques to study wave propagation in thin laminated fiber reinforced composite materials. Although these materials are highly anisotropic and inhomogeneous, they are desirable to use in many applications because of their high strength and stiffness to weight ratios. A comprehensive study of physical and elastic characteristics of fiber reinforced composites and ultrasonic evaluation of their elastic properties was given by Galego Juarez[1] with many references. All the work reported in Ref. 1 deals with bulk waves. In this paper, methods to study composites with generalized Lamb waves will be described. The terminology "generalized Lamb waves" implies that a wave propagates in layers bonded to some materials on either side (special cases would be a free plate: Lamb wave[2], a plate immersed in water: leaky Lamb wave[3], etc.

The structure of this paper is the following: First the theoretical approach to describe generalized Lamb wave propagation in isotropic materials treated by Viktorov[2] for the free plate case and by Pitts, Plona, and Mayer[3] for the plate immersed in water will be briefly described. This treatment was extended by Nayfeh and Chimenti[4,5] for composite plates immersed in water. Next a two-transducer contact method introduced by Rose, Rokhlin, and Adler[6] will be described. This method was developed to demonstrate anisotropy in composite plates and measure phase and group velocity variation with fiber direction. Then the two-transducer ultrasonic immersion technique described first for leaky Rayleigh waves by Breazeale, Adler, and Smith[7] and adapted for Lamb wave in composites by Nayfeh and Chimenti[4] will be introduced. Next a single-transducer immersion method developed by Nagy, Rose, and Adler[8] will be described which can be used to evaluate composite plates. Another method based on backscattering from composite plates will also be described[9]. After correlating some experimental measurements to theoretical predictions for generalized Lamb waves in composites, a few examples will be given to NDE applications to detect delamination and porosity[10].

THEORETICAL CONSIDERATIONS

Lamb Waves

A detailed study of Lamb wave propagation is given by Viktorov[2]. There are elastic waves propagating in a free plate. These waves are dispersive, and at any given value of phase or group velocity, Lamb modes will appear at discrete values of the product of frequency and plate thickness. Figures 1 and 2 are plots of typical dispersion curves for the phase velocity and group velocity behavior. These modes are divided into symmetric and antisymmetric modes as seen in Fig. 3.

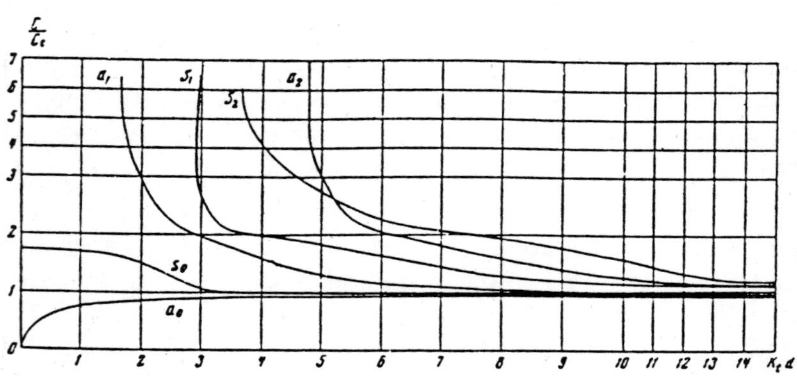

Fig. 1. Dispersion curves for phase velocity (s = symmetric modes, a = antisymmetric modes).

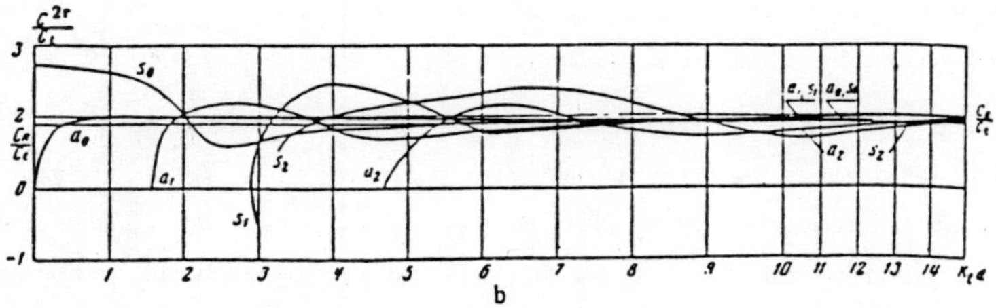

Fig. 2. Dispersion curve for group velocity (s = symmetric modes, a = antisymmetric modes).

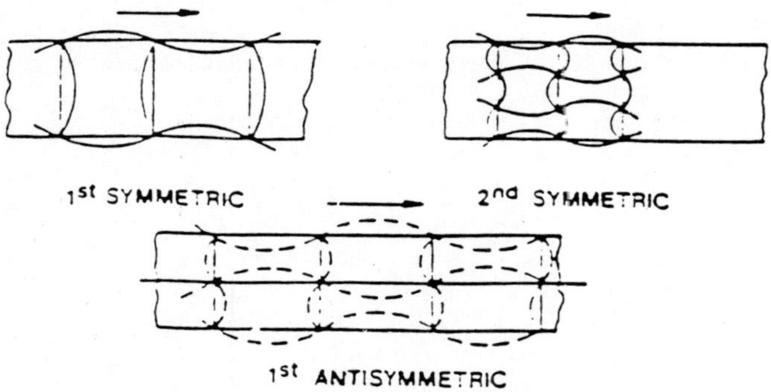

Fig. 3. Formation of symmetric and antisymmetric Lamb modes.

Leaky Lamb Waves in Isotropic Plates

Pitts, Plona, and Mayer[3] considered an extension of the Lamb wave problem for a plate immersed in water. They calculated the reflection and transmission coefficients of liquid-solid layer.

The liquid-solid-liquid reflection coefficient R_{LSL} is given as

$$R_{LSL} = A_r/A_i = N/(f_s f_a), \tag{1}$$

where

$$N = (k_s^2 - 2k_x^2) + 16k_x^4\kappa_s^2\kappa_d^2 - (\rho^2 k_s^8\kappa_d^2/\kappa^2) + 8k_x^2\kappa_s\kappa_d(k_s^2 - 2k_x^2)^2$$
$$\times\ [1 - \cos(d\kappa_d)\cos(d\kappa_s)]\ [\sin(d\kappa_d)\sin(d\kappa_s)]^{-1} \tag{2}$$

$$f_s = (k_s^2 - 2k_x^2)^2\cot(d\kappa_d/2) + 4k_x^2\kappa_s\kappa_d\cot(d\kappa_s/2) - (i\rho k_s^4\kappa_d/\kappa),$$

$$f_a = (k_s^2 - 2k_x^2)^2\tan(d\kappa_d/2) = 4k_x^2\kappa_s\kappa_d\tan(d\kappa_s/2) + (i\rho k_s^4\kappa_d/\kappa), \tag{3}$$

A free wave (Ai = 0) by definition indicates that A_r can be nonzero only when $f_s = 0$ or when $f_a = 0$ which will correspond to the symmetric and antisymmetric leaky Lamb modes. This method, which identifies the poles of the reflection coefficient with the modes of the Lamb modes of the free waves in plate, has been adapted also for the modes of the composite plate by Chimenti and Nayfeh[5].

Leaky Plate Waves in Unidirectional Fiber Composite Laminates

The leaky plate wave problem[4] is shown in Fig. 4. For detailed theoretical treatment, the reader is directed to Ref. 4. In this problem the composite is treated as an anisotropic plate with orthotropic symmetry. An incident plane wave in the X_1 - X_3 plane is reflected and transmitted through a unidirectional composite laminate with arbitrary orientation.

The reflection and transmission coefficients are given in the form

$$R = \frac{AS - Y^2}{(s + iY)(A - iY)} \tag{4}$$

$$T = \frac{iY(S + A)}{(s + iY)(A - iY)} \tag{5}$$

where

$$S = D_{11}G_1\cot(\gamma\alpha_1) - D_{13}G_3\cot(\gamma\alpha_3) + D_{15}G_5\cot(\gamma\alpha_5) \tag{6}$$

$$A = D_{11}G_1\tan(\gamma\alpha_1) - D_{13}G_3\tan(\gamma\alpha_3) + D_{15}G_5\tan(\gamma\alpha_5) \tag{7}$$

$$Y = \frac{\rho_f c^2}{\alpha_f}(W_1G_1 - W_3G_3 + W_5G_5) \tag{8}$$

The D_{ij} and G_i are terms related to the elastic constants, α_i are

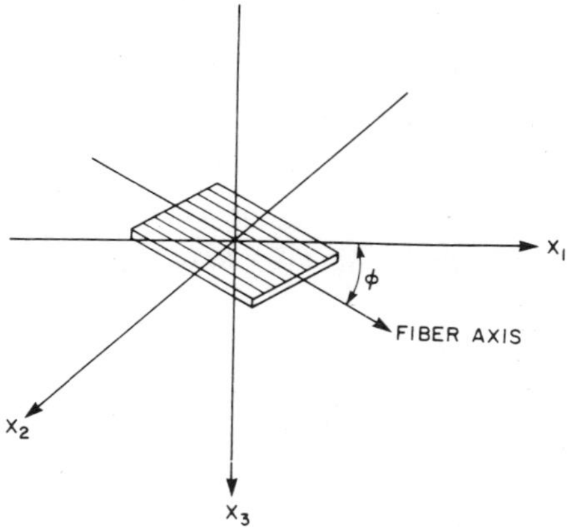

Fig. 4. Coordinate system for leaky plate wave problem. Incident beam strikes plate at angle θ, and ϕ is the angle between fiber axis and plane of incidence.

the x_3 wave vector components with the subscript f denoting the fluid, and the W_i are equivalent to α_i; c is the phase velocity, ρ_f is the fluid density, and $Y = \omega d/2c$. Setting either denominator equal to zero, namely

$$(S + iY)(A - iY) = 0 \tag{9}$$

defines the characteristic equations for such waves where the vanishing of the first term and of the second term corresponds to symmetric and to antisymmetric mode propagation, respectively.

SAMPLES

The type of composite materials used in most of these works were made up of graphite fibers in an epoxy matrix. The average fiber volume fraction was approximately 60% in the AS4/3501-6 prepreg tape, and all samples were made by the Hercules Corporation. Figure 5 illustrates the three samples.

The first sample was an eight-ply unidirectional panel with dimensions of 1.35 mm X 300 mm X 300 mm. The second sample was a nine-ply two-directional panel with dimensions of 1.40 mm X 300 mm X 300 mm and with ply orientation 0/90/0/90/0/90/0/90/0. The third sample was a nine-ply quasi-isotropic panel with dimensions of 1.40 mm X 300 mm X 300 mm and with ply orientation of 0/+45/90/-45/0/-45/90/+45/0.

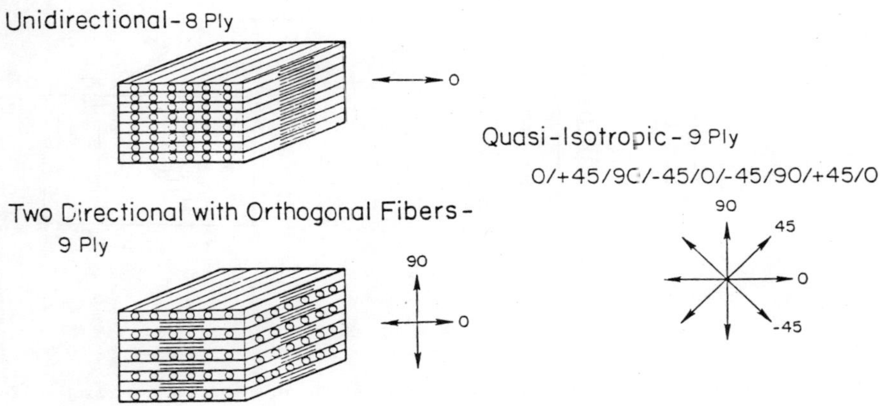

Fig. 5. Graphite/epoxy composite samples.

EXPERIMENTAL CONSIDERATIONS

The Effect of Anisotropy on Phase and Group Velocities

One of the interesting phenomena associated with anisotropic materials is the deviation in direction that can exist between group and phase velocities. Figure 6 demonstrates this effect with bulk waves. The transducer coupled onto the isotropic block causes a wave to propagate (group velocity) in a direction normal to the transducer face. The individual wavefronts (phase velocity) also travel in this direction. When a transducer is coupled to an anisotropic material, the resulting wave may propagate in a direction not normal to the transducer face but along some axis of symmetry in the material. However, the individual wavefronts will travel normal to the transducer face just as they would in an isotropic material. Therefore, the distances used in velocity calculations would be as shown in Fig. 6 and yield different values for group and phase velocities.

Many composite applications involve the use of relatively thin panels and it is, therefore, advantageous to test them in their own plane as opposed to through their thickness. Lamb waves travel in the planes of thin plates with particle displacement out of these planes.

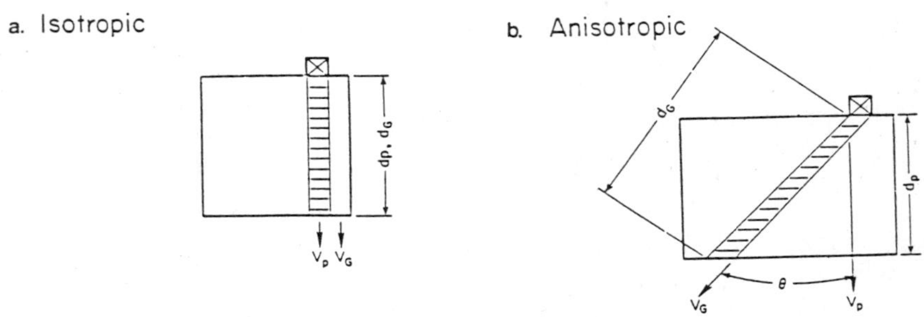

Fig. 6. (a) Wave propagation direction in an isotropic medium and (b) possible direction in an anisotropic medium.

Contract Transducer Technique

One method of studying the anisotropy of a given material is to plot the variation in velocity of ultrasonic waves at various propagation directions in a sample. Using a discrete frequency tone burst at 0.6 MHz, it is possible to isolate a single Lamb mode and measure its group and phase velocities. Variable angle contact transducers were used with the equipment shown in Fig. 7. The Wavetek generator produced a constant frequency signal which was gated to give a tone burst and, once amplified, sent to the transmitting transducer. This generated a Lamb wave in the composite plate which was picked up by the receiving transducer, channeled through the receiver, and displayed on the oscilloscope.

Figure 8 illustrates the received signal. These constant frequency waveforms have varied amplitudes which come together to form a signal envelope with a peak. The movement of the entire signal envelope in time on the oscilloscope screen (Point 1 on Fig. 8) is related to the separation of the transducers in the direction of propagation on the composite plate to calculate group velocity. The movement of a

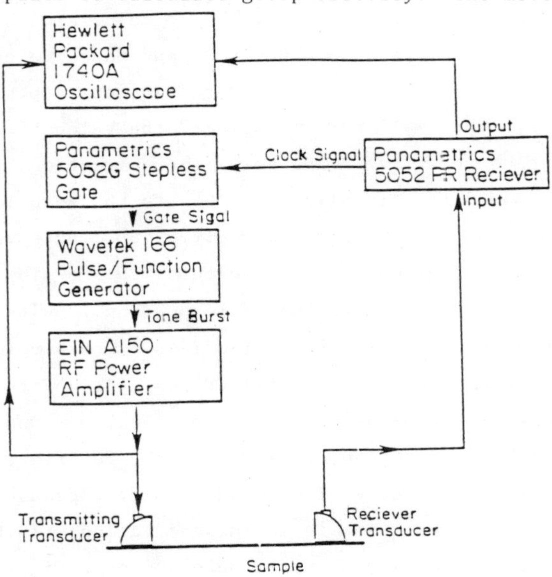

Fig. 7. Equipment set up for contact technique.

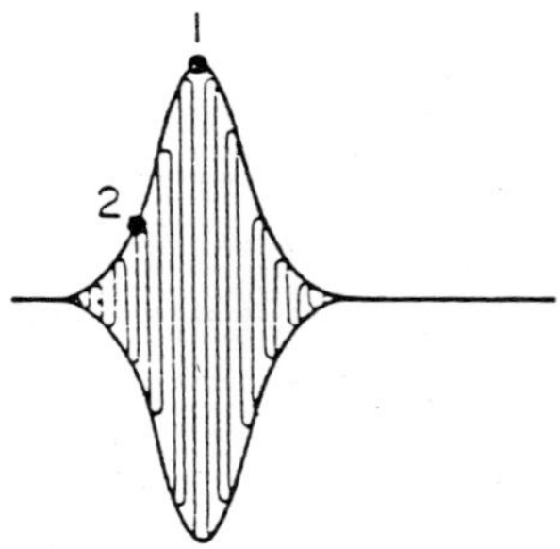

1 - Peak of Imaginary Signal Envelope Used for Group Velocity Measurement.

2 - Point on Single Waveform Used for Phase Velocity Measurement.

Fig. 8. Illustration of received signal.

single waveform in time on the oscilloscope (Point 2 on Fig. 8) is related to the separation of the transducers in a direction normal to the transmitter face, on the composite plate, to calculate phase velocity.

Figure 9 is a polar plot of the group and phase velocities for the unidirectional plate. The zero fiber direction corresponds with the horizontal axis. The small arrows indicate the velocity directions at the various angles relative to the zero fiber direction. Note that even when the transmitter is aligned at 85° to the zero fiber direction the wave propagates close to the fiber direction. Both velocities rapidly drop off as a direction perpendicular to the fibers is approached. Figure 10 displays the corresponding results for the plate with fibers in both the 0° and 90° directions. Phase velocity drops to a minimum in between the two fiber directions. At 45° to the two fiber directions there are three values of group velocity. One wave turns toward the 0° fibers, a second turns toward the 90° fibers, and a third propagates along the 45° direction between them. Because there is one extra 0° ply, the fastest wave is in this direction with the second fastest in the 90° direction as would be expected. Figure 11 illustrates the results in the quasi-isotropic. The velocities are more

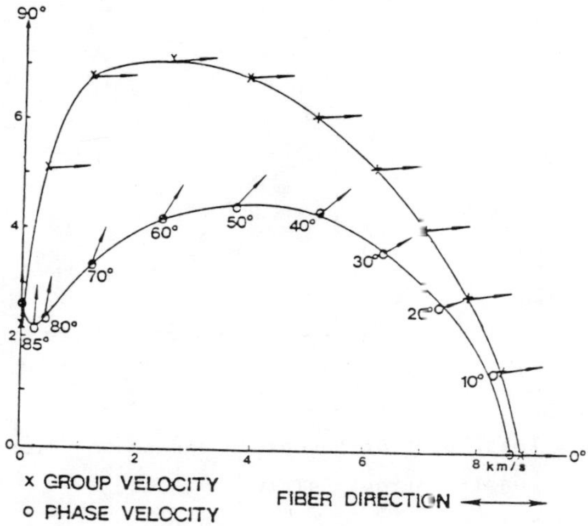

Fig. 9. Polar plot of the group and phase velocities for the unidirectional plate.

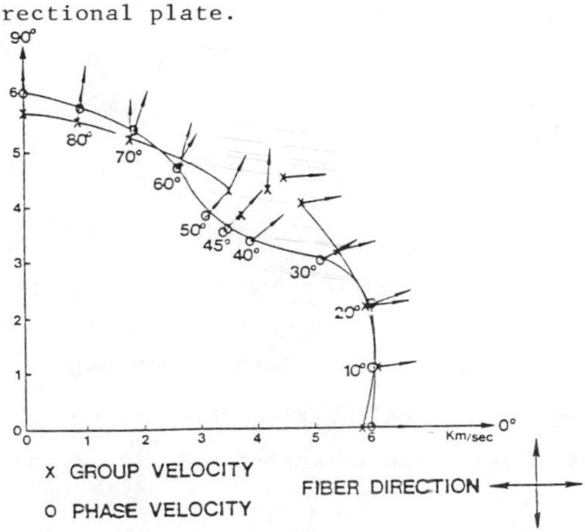

Fig. 10. Polar plot of the group and phase velocities for the two-directional plate.

constant and phase and group velocities tend to be in the same directions.

Figures 12 and 13 show the results for the unidirectional panel plotted alongside theoretically calculated group and phase velocity curves for bulk waves in a unidirectional graphite/epoxy panel.

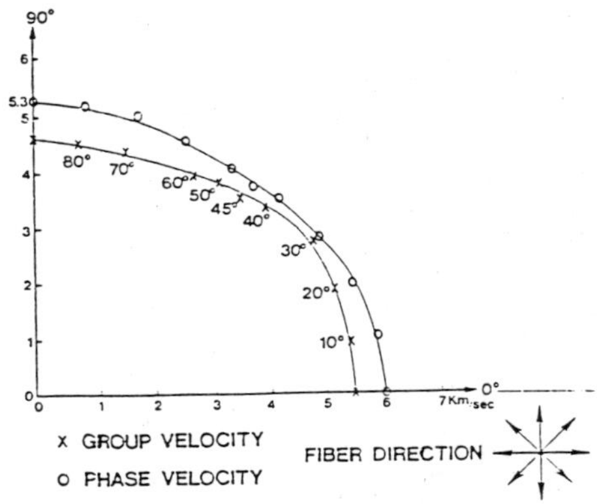

Fig. 11. Polar plot of the group and phase velocities for the quasi-isotropic plate.

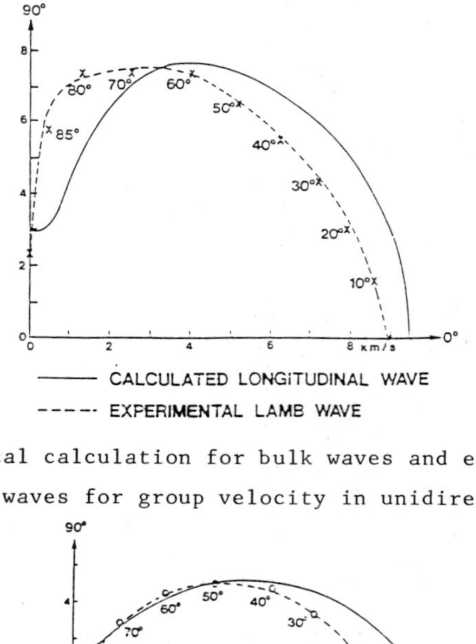

Fig. 12. Theoretical calculation for bulk waves and experimental data for Lamb waves for group velocity in unidirectional plate.

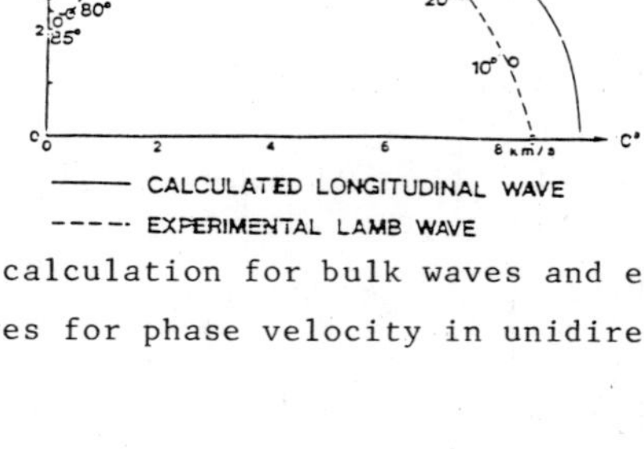

Fig. 13. Theoretical calculation for bulk waves and experimental data for Lamb waves for phase velocity in unidirectional plate.

Although the waves travel at different velocities, they do share the same general shape with a significant decrease in velocity near a direction perpendicular to the fibers.

Two-Transducer Immersion Method

The work presented here is based on and is an extension[11] of the work of Nayfeh and Chimenti[4] on leaky Lamb waves in fibrous composites. Nayfeh and Chimenti used a tone burst method to induce Lamb waves in an immersed composite plate and a broadband receiver to collect data. At frequencies where the phase matching condition for Lamb waves is not satisfied, a specular reflection will occur. However, at frequencies where Lamb modes are excited, there is a distortion of the reflected beam due to the destructive interference between the specular portion and the leaky Lamb portion. With the receiver positioned over this null, or interference zone, a definite decrease in reflected signal will appear at each frequency where the Lamb modes are present. This work[11] has two major differences from that described above. First, a broadband pulse was used instead of a swept frequency tone burst. Second, while Nayfeh and Chimenti[4] kept polar angle constant and varied phase velocity (incident angle), this work involves varying polar angle as well.

Frequency times plate thickness values where Lamb modes occurred were plotted against polar angle. The frequency shift of the modes as polar angle is varied can be used as an indicator of anisotropy. The more anisotropy, the more prevalent the frequency shift.

Experimental setup. Figure 14 shows the equipment as configured for this experiment. The pulser excites the 5 MHz transmitting transducer with a broadband signal, which in turn insonifies the composite plate. The pulser also emits a trigger signal which is fed to an oscilloscope and the delay unit. The received signal from the 5 MHz receiving transducer is preamplified and sent to the oscilloscope. This signal is also sent to the LeCroy transient recorder which digitizes it and sends it on to the computer where it is stored for subsequent data manipulation. The computer is also used to control the amount and rate of movement of the turntable during the experiment.

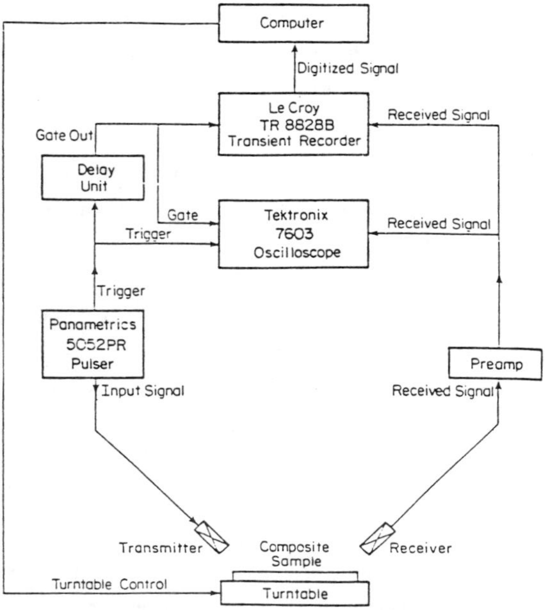

Fig. 14. Two-transducer immersion system experimental setup.

All three samples were studied using this experimental technique.

Procedure. With the equipment setup as discussed, the 5 MHz transducers were set at 10° from the normal and the receiving transducer positioned over the null zone of the signal. The unidirectional composite plate is positioned so that the Lamb wave propagates parallel with the fibers. Data is collected and the computer turns the turntable by 3.6°. Data is collected at this orientation to the fibers and the process is repeated until the sample has been turned by 90°.

The transducer incident angle is increased to 15° and data is once again taken from 0° to 90° polar angles at 3.6° increments. This process is carried out for incident angles of 10, 15, 20, 25, 30, and 35°.

This experimental technique is then repeated for both the two-directional and the quasi-isotropic panels.

Results. The data acquisition procedure resulted in 468 digitized and stored time signals. These were transformed into the frequency domain using a fast Fourier transform algorithm. Figure 15 is an example of one of these signals in the frequency domain. In this case the incident angle is 10°, with propagation parallel with the fibers in the unidirectional plate.

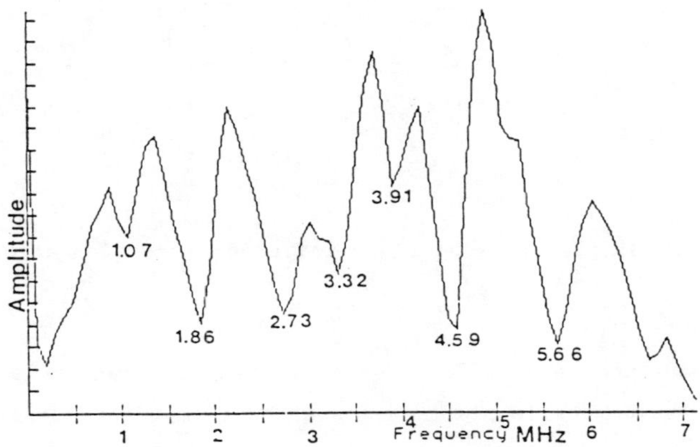

Fig. 15. Sample spectrum using a broadband signal and 5 MHz transducers at a 10° incident angle on the unidirectional plate. Propagation was parallel with the fibers.

Single Transducer and Reflector Immersion Techniuqe

Another method of studying the anisotropy of a given material which was developed by Nagy, Rose, and Adler[8] is to look at the different values of frequency times plate thickness where Lamb modes appear, at constant phase velocity, as the direction of propagation is varied in a sample. When insonifying an immersed plate at a given incident angle with a broadband signal, leaky Lamb waves within that frequency range will be excited. The leaky Lamb signal from the back face can be detected and found to contain peaks at frequencies where Lamb modes occurred in its frequency range. A single transducer and reflector method has been used to detect these back face signals and is

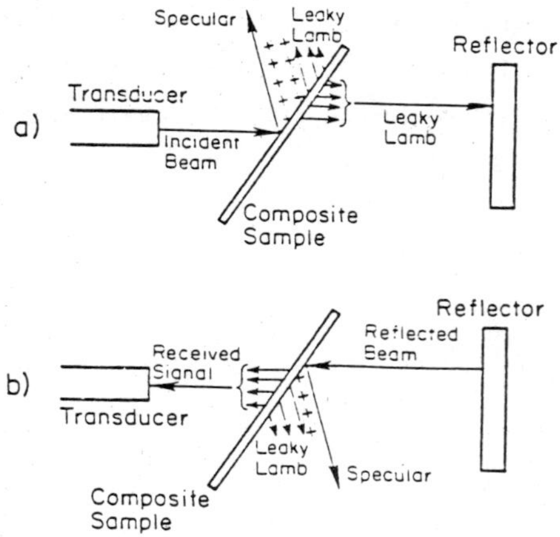

Fig. 16. Single transducer and reflector immersion technique.

illustrated in Fig. 16.

The transducer and reflector are aligned so that their faces are parallel. The composite plate is suspended from a goniometer between them so that its face is also parallel to the reflector face. The plate can then be rotated using the goniometer to obtain any incident angle desired. In this way it is not necessary to move the transducer or reflector once they have been positioned. The plate is insonified at a given incident angle using a broadband ultrasonic signal as displayed in Fig. 17(a). This causes specular and leaky Lamb signals on

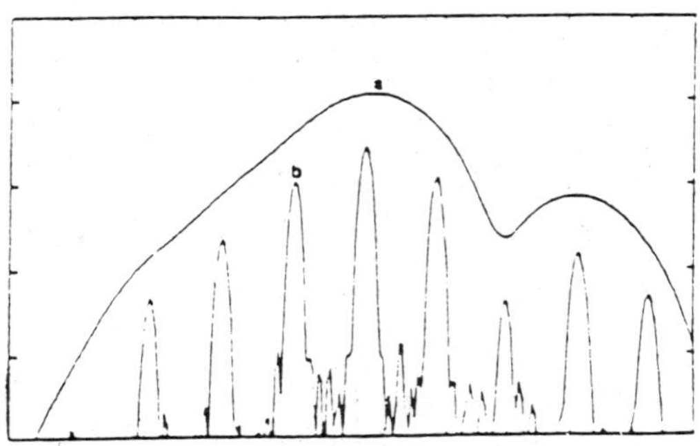

Fig. 17. Example of frequency response of (a) input signal and (b) received signal with peaks indicating Lamb modes.

the front face and leaky Lamb signals on the back face. The back face leaky Lamb signals hit the reflector at normal incidence and are reflected back along their own path where they reinsonify the plate. The back face leaky signals are now on the same side of the plate as the transducer and are thus received. When this received signal is put into the frequency domain it will have peaks, such as those seen in Fig. 17(b), at frequencies where Lamb waves occurred. By plotting these frequency times plate thickness values versus polar angle, the frequency dependence can be observed as propagation direction relative to the fibers is varied. Figure 18 is a plot at a 15° incident angle for the unidirectional plate, of the frequency times plate thickness values where Lamb modes occurred as propagation direction (polar angle) is varied from 0° to 90°. The greatest number of Lamb modes seem to be excited between polar angles of 50° to 70° indicating some

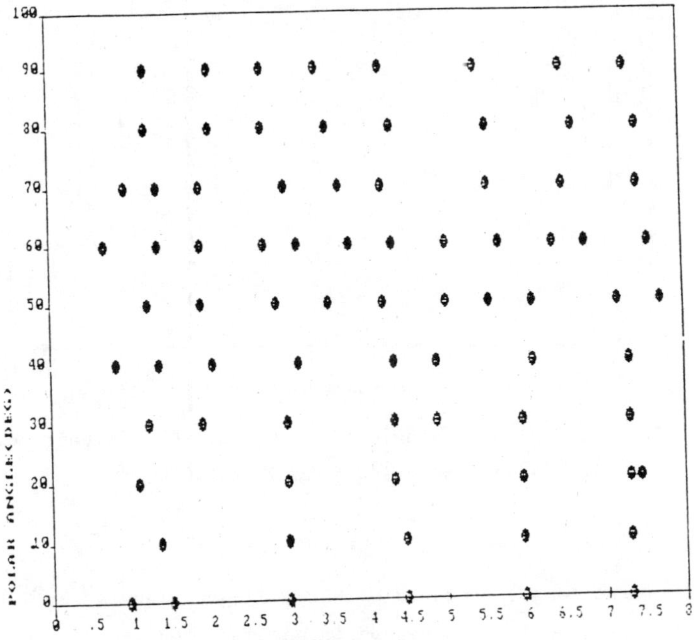

Fig. 18. Frequency dependence of Lamb modes at 15° incident angle as polar angle is varied from 0° to 90°.

sort of mode splitting in this region. In general, there seems to be a decrease in the frequency times plate thickness values of given Lamb modes as propagation is moved further away from the fiber direction. To illustrate this point, a polar plot is shown in Fig. 19 representing the frequency dependence of one mode at an incident angle of 10° in the unidirectional sample. There is more than a 10% drop in the frequency at which the mode was excited as propagation is varied from parallel with, to perpendicular to, the fibers.

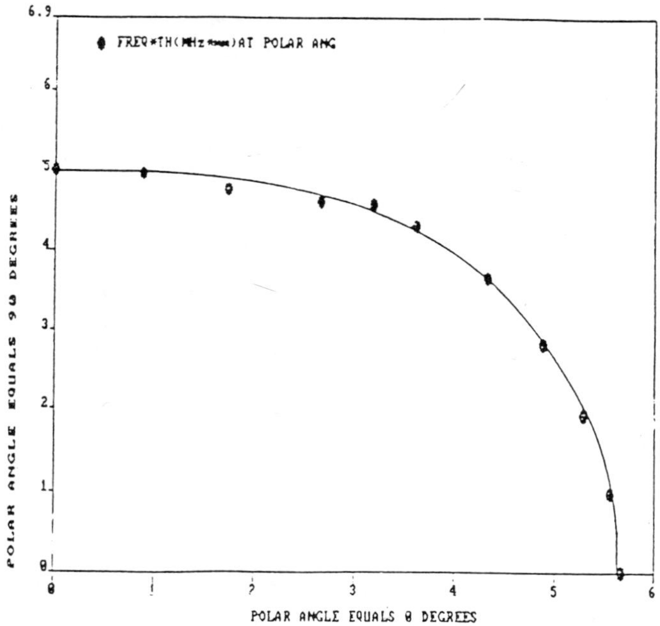

Fig. 19. Polar plot of single Lamb mode at 10° incident angle as polar angle is varied from 0° to 90°.

Backscattered Leaky Lamb Wave Technique

This method was introduced by Nagy, Jungman, and Adler[9] to study resonances in composite plates and is based on the backscattering technique of deBilly, Adler, and Quentin[12].

The schematic diagram of the experimental setup is shown in Fig. 20. The fiber-reinforced composite plate was immersed in water and

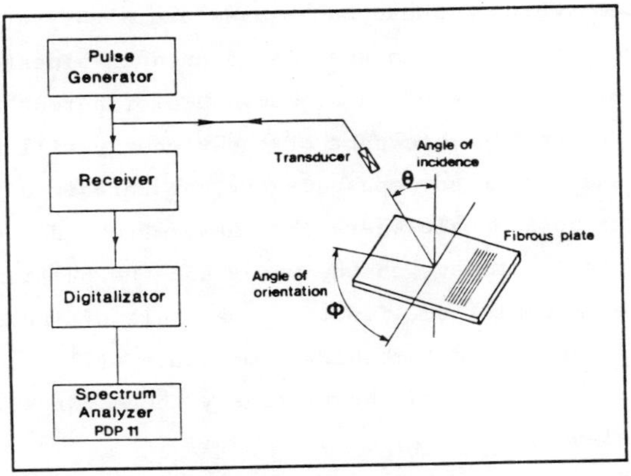

Fig. 20. Schematic diagram of the experimental setup.

rotated into its own plane to change the azimuthal angle. A broadband ultrasonic transducer was used to interrogate the plate at different angles of incidence. The received backscattered signal was digitalized and fed into a PDP 11 minicomputer for spectrum analysis.

Figure 21 illustrates the double-mode conversion at Lamb resonances in the thin plate. Those frequency components of the incident broadband pulse, which are phase-matched to one of the possible Lamb modes in the plate at this particular angle of incidence, will generate

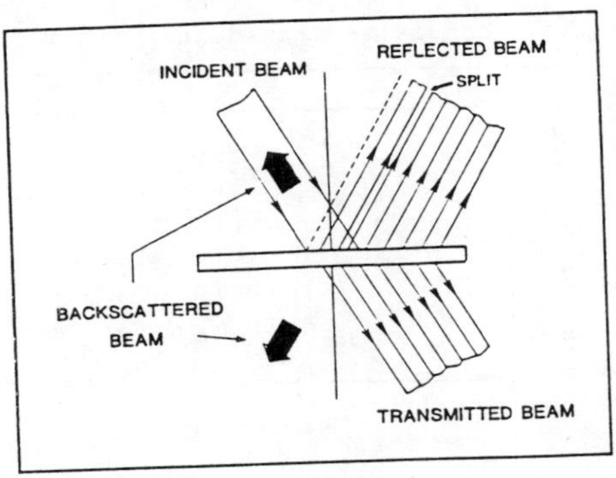

Fig. 21. Schematic diagram of leaky Lamb wave generation in an immersed plate.

strong forward Lamb waves. These Lamb waves are highly attenuated by radiating their energy back into the liquid on both sides of the plate as they propagate. In spite of this, because of inherent material inhomogeneities, a considerable part of their energy will be scattered in the plane of the plate and, consequently, be reradiated into the liquid at the same polar angle where the phase-matching condition between the compressional wave in the water and the particular Lamb mode in the plate is satisfied, i.e., at the angle of incidence. Obviously, the backscattered Lamb wave components will reradiate half of their energy directly toward the receiver, thereby resulting in a strong detected signal.

Figure 22 shows an example of a backscattered ultrasonic spectrum from an eight-ply unidirectional graphite-fiber-reinforced epoxy sample 1.35 mm thick. As expected, the spectrum is strongly modulated by the random interference among incoherent scattering inhomogeneities. A major difference between this broadband technique and the often-used narrow-band technique is that the latter inherently involves some spatial averaging because of its very poor time resolution. To learn more about the physical nature of the acoustical wave being scattered, it is necessary to recover the characteristic power spectrum of the backscattered signal. This can be achieved by averaging many uncorrelated power spectra taken at different spatial positions. In this way, in spite of use of a phase-sensitive transducer, the resultant spectrum

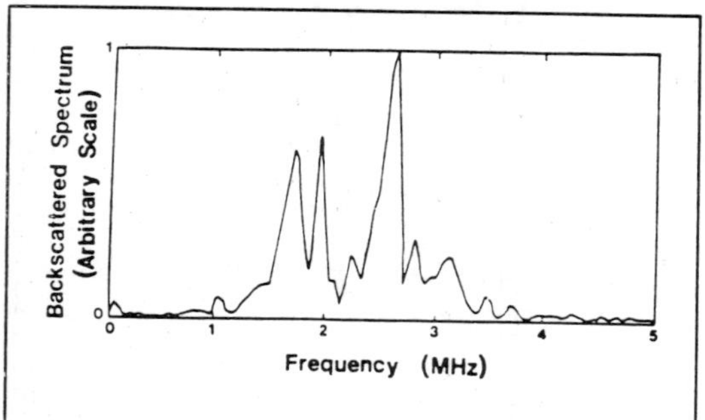

Fig. 22. Frequency spectrum of the backscattered signal from a unidirectional composite plate at 24° angle of incidence and perpendicular fiber orientation.

will be proportional to the power spectrum of the scattered wave (as well as to the backscattering characteristic of the average inhomogeneity).

The characteristic comblike power spectrum of a broadband Lamb wave can be recognized easily in the average backscattered spectrum. Figure 23 shows the same spectrum as Fig. 22 after spatial averaging over 20 positions. The indicated Lamb resonances were measured by an independent, very precise technique based on the detection of the characteristic split beam in the reflected wave, a split caused by destructive interference between the specularly reflected and double-mode-converted components.

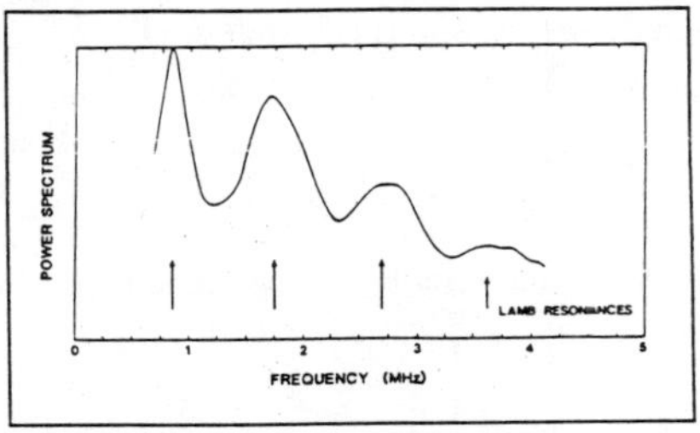

Fig. 23. Average power spectrum of the backscattered signal from a unidirectional composite plate at 24° angle of incidence and perpendicular fiber orientation (deconvolved by the transducer response).

Figure 24 shows the averaged power spectrum of the backscattered signal at a different angle of incidence and different fiber orientation. The peaks of the spectrum obviously correspond to the Lamb resonances. The average discrepancy in more than 20 different cases between the Lamb resonance frequencies as measured by the backscattering and split-beam techniques was about 3%. The only exceptions were the lowest modes, where the relative difference was sometimes as high as 10 - 15%. This is probably a result of distortions caused by smoothening of the spectra. In most cases, averaging of 20 - 30 spatial positions was sufficient to identify the sought Lamb resonances, but further smoothening was applied to reduce the remaining interference ripple on the resultant spectrum. Naturally, the necessity for

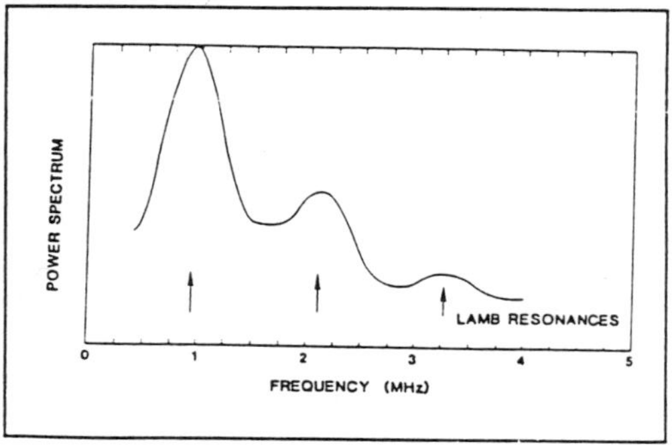

Fig. 24. Averaged power spectrum of the backscattered signal from a unidirectional composite plate at 20° angle of incidence and parallel fiber orientation (deconvolved by the transducer response).

this additional smoothening can be eliminated by more extensive spatial averaging.

COMPARISON OF EXPERIMENTAL RESULTS TO THEORETICAL PREDICTIONS

The results presented here were developed by Nayfeh and Chimenti[4] who presented an extensive evaluation of their theoretical development using the two-transducer immersion method.

In comparisons between the results of many measurements at several azimuthal angles and the model calculation, they concentrated their attention on the reflection coefficient. Since the basis of the experimental data is an amplitude signal in the reflected field of the incident beam, they calculated the plane-wave reflection coefficient for the plate and investigated it for the same type of behavior as one observes in the measurements. In addition to direct comparisons of amplitude spectra, they have also expressed their results as dispersion-like curves, where the results correspond to functions conditioned by the reflection coefficient. Both of these aspects of their studies are presented below.

Figure 25 shows the measured and predicted spectra from 1 - 8 MHz for an incident angle $\theta = 12°$ and an azimuthal angle $\phi = 30°$. The two curves have been vertically scaled, but in no other way adjusted. The

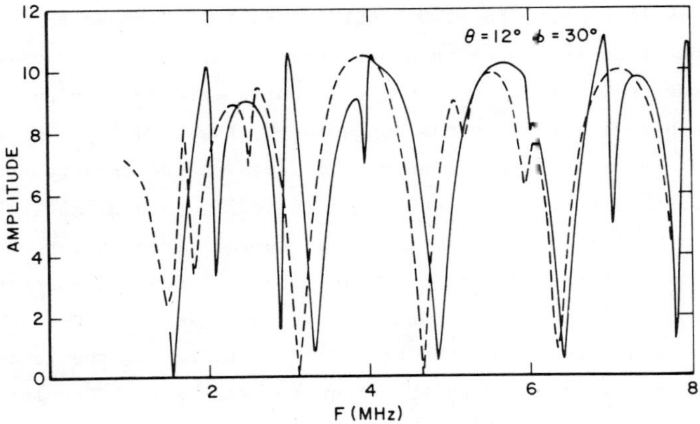

Fig. 25. Reflection spectrum for θ = 12° and ϕ = 30°. Theory is solid curve and experiment is dashed curve. Data has been deconvolved.

solid curve is obtained from the reflection coefficient by integrating over the angular spectrum of the finite acoustic beam. The dashed curve is the experimental data deconvolved to remove transducer response. Positions of the deep minima in the two curves are nearly coincident. The positions of the minima are well modeled by the theory whereas the minor differences in the lineshape details may be attributed to wavefront distortion and sound absorption, neither of which is considered in the model.

Taking the results of many dozens of experimental spectra, such as those shown in Fig. 25, and recording the minima as a function of the incident angle (expressed through Snell's law as a phase velocity), yielded a dispersion-like plot of the ultrasonic reflection behavior. In previous work they demonstrated that the hypothesis of Cremer concerning the coincidence of reflection minima with the excitation of Lamb wave modes is not well satisfied in all regions of plate wave dispersion in graphite-epoxy composites. Therefore, although they present these data in the manner of a velocity dispersion curve, it must be stressed that it is the reflection properties which are being reported.

FLAW DETECTION IN COMPOSITES WITH SINGLE MODE LEAKY LAMB WAVE

Flaw detection in composite plates presents difficulties with the use of conventional normal incidence echoscopic methods. An oblique incidence Lamb wave technique was suggested by Nagy, Adler, Mih, and Sheppard[10]. The inspection technique is based on the one-transducer method[8] described earlier and shown schematically in Fig. 16. The transmitted signal is greatly amplified at certain frequencies where dispersive Lamb modes are phase matched to the incident wave at that particular angle of incidence. One can obtain a Lamb wave C-scan of the composite plate by moving the transducer parallel to the reflector and using a tone burst tuned to the frequency of the most sensitive mode at that particular angle of incidence.

Figure 26 demonstrates the substantial improvement offered by the Lamb wave inspection technique to detect porosity in composite plates. At 22° angle of incidence corresponding to approximately 4000 m/s phase velocity, the strongest mode is generated at 6 MHz. This mode is attenuated by as much as 12 dB, roughly four times more than the

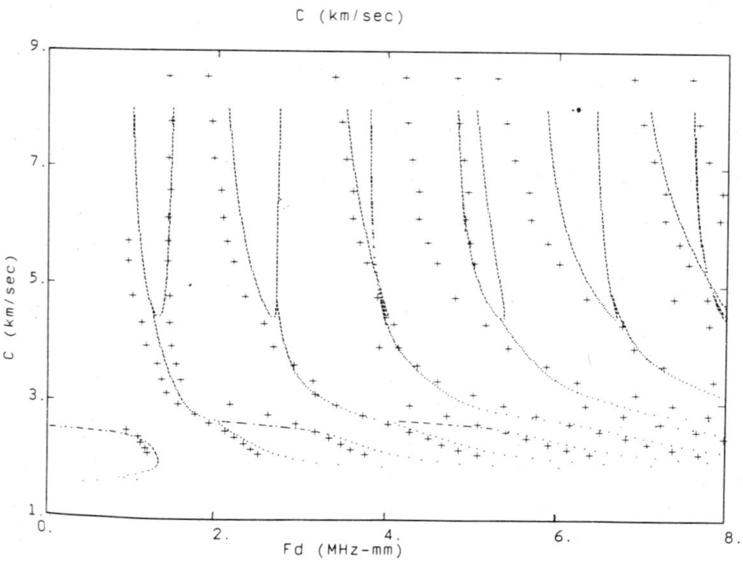

Fig. 26. Reflection results expressed as a phase velocity dispersion plot for ϕ = 90°.

normally incident longitudinal wave, by the same weak porosity. This enhancement can be taken advantage of in order to push the detection threshold well below 0.1%.

Another fairly common defect in composite plates which is difficult to detect at normal incidence is a transverse crack. Such cracks usually run parallel to the reinforcing fibers, and they tend to grow in transverse directions through ply boundaries separating plies of the same fiber orientation, but they stop at boundaries between plies of different orientation. Consequently, the strongest cracks run in certain directions which have the most consecutive plies of the same fiber orientation. Such cracks can reach four or six plies, or approximately 0.5 - 0.8 mm, and they produce roughly 3 dB drop in the C-scan intensity. Unfortunately, smaller cracks of two-ply height have been repeatedly missed by conventional normal incidence C-scan inspection. The detection threshold is mainly limited by the inherent inhomogeneity of the composite plate which causes about 1 dB uncertainty in the transmitted signal. As an example, Fig. 27 shows a cluster of one- and two-ply cracks in a fiber reinforced composite plate.

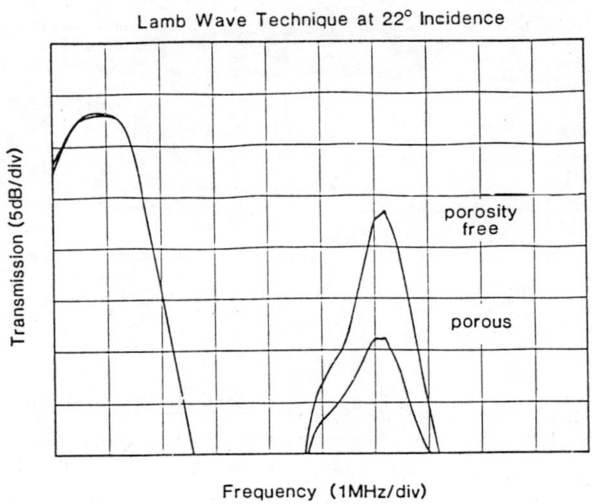

Fig. 27. Transmission spectra through porosity-free and porous parts of a composite plate (sample #ACL 6804).

The ultrasonic contrast produced by such clusters of transverse cracks at normal incidence is in the order of 1 - 3 dB, i.e. barely detectable. Figure 28 demonstrates this low detectability through examples of transmission spectra at defect-free and defective parts. The scatter of the defect-free spectra is approximately 3 dB at 5 MHz, while the average contrast produced by these multiple cracks is roughly 3 - 5 dB. Figure 29 shows the ultrasonic transmission spectra of the same defect-free and defective parts at 20° angle of incidence. The principal Lamb mode of 4 MHz is attenuated by as much as 20 dB, approximately five times more than at normal incidence.

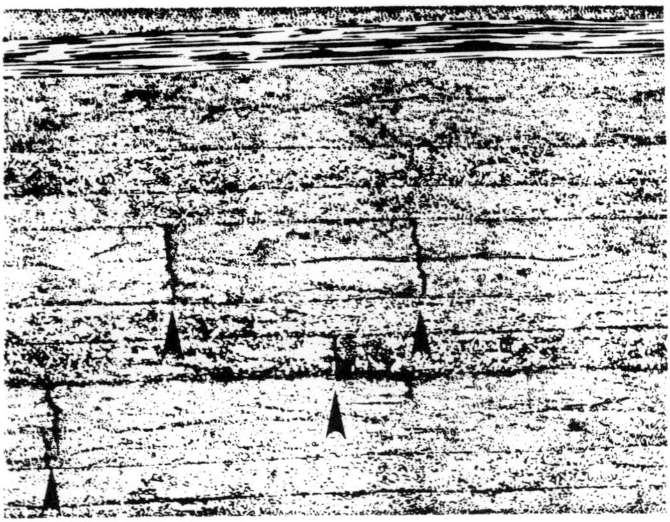

Fig. 28. One- and two-ply transverse cracks in a composite plate (sample #AMD 1530, x 50 magnification).

REFERENCES

1. Galego-Juarez, F. A., Ultrasonic Methods in Evaluation of Inhomogeneous Materials, A. Alippi and W. Mayer, eds. (Martimus Nijhoff Publishers, NATO, SAD, 1987).

2. Viktorov, I. A., Rayleigh and Lamb Waves, (Plenum Press, New York, 1967).

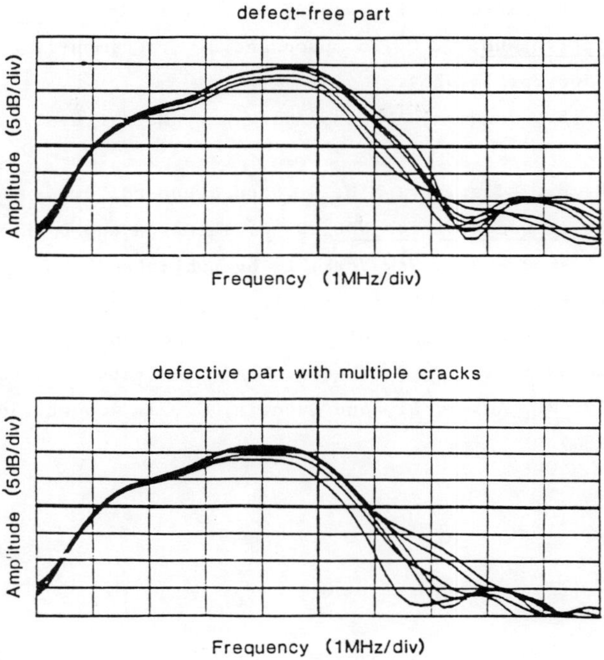

Fig. 29. Ultrasonic transmission spectra at normal incidence through defect-free and defective parts (sample #AMD 1642).

3. Pitts, L. E., Plona, T. J., and Mayer, W. G., IEEE Son. Ultrason. SU 24, 101 (1977).
4. Nayfeh, A. H., and Chimenti, D. E., in Review of Progress in Quantitative NDE, D. O. Thompson and D. E. Chimenti, eds., (Plenum Press, New York, 1987).
5. Nayfeh, A. H., and Chimenti, D. E., J. Acoust. Soc. Am. 83, 1736 (1988).
6. Rose, W. R., Rokhlin, S. I., and Adler, L., in Review of Progress in Quantitative NDE, D. O. Thompson and D. E. Chimenti, eds., (Plenum Press, New York, 1986).
7. Breazeale, M. A., Adler, L., and Smith, F., J. Acoust. Soc. Am. 56, 866 (1974).

8. Nagy, P. B., Rose, W. R., and Adler, L., in Review of Progress in Quantitative NDE, D. O. Thompson and D. E. Chimenti, eds., (Plenum Press, New York, 1986).

9. Nagy, P. B., Jungman, A., and Adler, L., Mater. Eval. 46, 97 (1988).

10. Nagy, P. B., Adler, L., Mih, D., and Sheppard, D., in Review of Progress in Quantitative NDE, D. O. Thompson and D. E. Chimenti, eds., (Plenum Press, New York, to be published).

11. Rose, W. R., "Polar Characteristics of the Frequency Dependence and the Group and Phase Velocities of Lamb Modes in Graphite/ Epoxy Composites," M. S. Thesis, The Ohio State University (1986).

12. de Billy, M., Adler, L., and Quentin, G., J. Acoust. Soc. Am. 75, 998 (1984).

THREE LECTURES ON ACOUSTO-OPTICS

Adrian Korpel

Department of Electrical and Computer Engineering
The University of Iowa
Iowa City, IA 52242, USA

ABSTRACT

The paper summarizes the lectures by giving an outline of the fundamental principles on which acousto-optics is based, describing general applications which are inherent in these principles and discussing how some of these general applications are relevant to acousto-optic signal processing.

1. FUNDAMENTALS

Acousto-optics, or the interaction of coherent light and ultrasound, may be explained by:

a) A rigorous treatment by means of Maxwell's equations applied to a medium in which, phenomenologically speaking, the propagating sound induces variations of the refractive index. Although this is the most rigorous approach, it will not be used in these lectures, being too cumbersome and providing but little intuitive insight.

b) Modeling the sound field as a rectangular sound column that acts as a phase grating on the incident light. The latter is usually treated as a plane wave, although in practical cases it is ,of course, always in the shape of a finite beam. To the extent, however, that the angular spread of the incident light beam is much smaller than that of the sound beam, this model gives reasonably accurate results , provided the sound beam does not show excessive curvature within the region of interaction.

c) Treating the sound and light fields as compositions of plane waves, and considering their individual interaction. For weak scattering this approach is excellently suited to to explaining the interaction of fields of arbitrary shape , the bandwidth of light deflectors and modulators, the phenomenon of Bragg diffraction imaging, the action of beam steering phased arrays etc. For strong interaction this approach does not always lead to easily derivable results, but it is physically transparent and explicit through the use of Feynman diagrams and path integrals, and facilitates the choice of appropriate approximations through the selection of plausible pathways of scattering.

d) The application of an eikonal theory that provides a way of ray tracing of diffracted rays and a calculation of their amplitude. This method is particularly suitable for configurations that rely on imaging of diffracted light, and for interactions of curved wavefronts of sound and light.

The series of lectures summarized in this paper are be based on a recent book[1] written by the present author. Although this book treats all aspects a)-d) of acousto-optics, I will in this series of lectures concentrate mainly on items b) and c), and follow a heuristic treatment wherever possible so as to bring out clearly the underlying physics of the phenomenon. Ref. 1 itself contains an extensive list of references , of which I have listed a few that are easily accessible, reasonably up-to-date and of a not too specialized review-tutorial character.

1.1 The Sound Field as a Thin Phase Grating

In this model the sound field is considered to be a thin sound column with perfectly straight wave fronts. Although in reality, of course, the sound field will spread and acquire phase curvature, the model is remarkably accurate in predicting the basic phenomenon, provided in the region of interaction the two fields look like beams with straight crests, and many sound wavelengths are contained within the width of the light beam. In such a model then, the sound beam acts essentially as a thin phase grating and, like in all gratings, it splits the incident light beam up into a multiplicity of beams, all of which are separated by the same angle ≈ λ/Λ (assumed <<1), where λ is the wavelength of light and Λ that of sound [2,3]. The relative amplitude of the nth beam is given by $J_n(v)$ where v is called the Raman-Nath parameter, and physically denotes the peak phase shift of the light induced by the sound. In addition , because the sound grating is moving, the nth beam is Doppler shifted in frequency by $n\Omega$ where Ω denotes the frequency of the sound. A schematic picture of the process is shown in Fig.1. The analysis of such a thin sound grating is straight forward. The "rays" perpendicular to the incident plane wave front of the light beam are delayed in phase by the sound-induced refractive index variations. It is, however, assumed that, because the sound column is so thin, the rays are neither diffracted nor refracted.

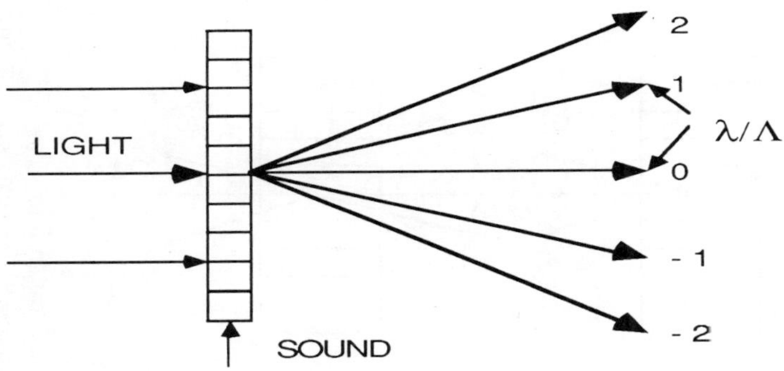

Fig.1 Sound field acting as a thin phase grating.

The result of this simple calculation is that the outgoing beam is transversely corrugated in phase and splits up into many beams as indicated in Fig.1. The region in which the sound field behaves as a thin phase grating is called the Raman Nath region and characterized by the criteria:

$$Q \ll 1 \text{ and } Qv \ll 1 \tag{1}$$

where $Q = K^2L/k$, $K=2\pi/\Lambda$, $k=2\pi/\lambda$ and L is the width of the sound field.

1.2 The Sound Field as a Thick Phase Grating

In this particular case it is no longer justified to ignore refraction and diffraction of the rays used to analyze the configuration of Fig.1. In fact it is now easier to abandon the model of rays altogether and invoke the concept of constructive interference of plane waves reflected off the crests of the sound wave. This method of analysis is very similar to that of the reflection of X rays off atomic planes in a crystal. It turns out that diffraction only occurs for two specific critical angles of incidence of the light: the so-called positive and negative Bragg angles: $\pm\lambda/2\Lambda$. Furthermore, in each case only one additional order is generated, either upshifted in frequency (Fig.2a) or downshifted in frequency (Fig.2b).

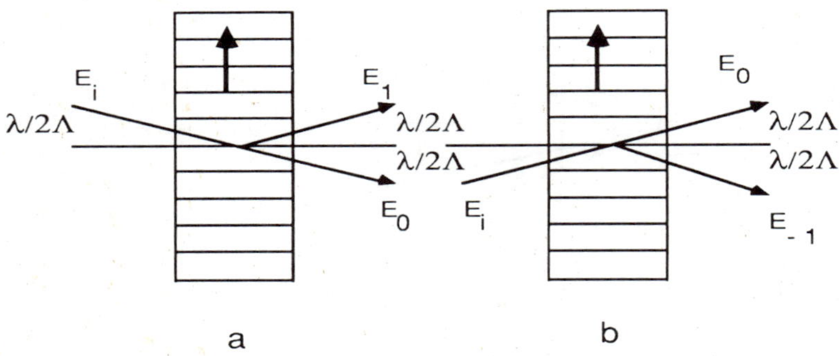

Fig.2 Sound field acting as a thick phase grating with a) upshifted interaction, b) downshifted interaction.

In this so called Bragg mode of operation the efficiency of diffraction is theoretically 100% for $v=\pi$:

$$E_{1,(-1)} = E_i \sin(v/2) \tag{2}$$

$$E_0 = E_i \cos(v/2) \tag{3}$$

The criteria for Bragg diffraction are as follows:

$$Q>>1 \text{ and } Q/v>>1 \tag{4}$$

1.3 The Sound Field as a Plane Wave Spectrum

It is a well known fact that each field satisfying a Helmholtz type wave equation may be decomposed into a spectrum of plane waves, at distances greater than a few wavelengths from a real source. The plane waves of light are characterized by their wave vector **k** ($k = 2\pi/\lambda$) and those of sound by **K** ($K = 2\pi/\Lambda$). The interaction of individual plane waves of incident light $\mathbf{k}_i$, scattered light $\mathbf{k}_1$ or $\mathbf{k}_{-1}$ and sound **K** can be illustrated by wave vector diagrams such as shown in Fig 3a and 3b for upshifted and downshifted interaction respectively.

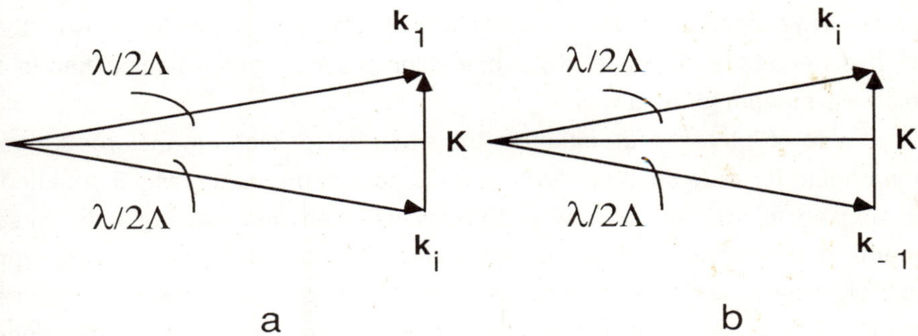

Fig.3 Wave vector diagrams for a) upshifted interaction, b) downshifted interaction.

In terms of classical physics, the diagrams of Fig.3 symbolize phase synchronous interaction. In the context of quantum mechanics, they represent conservation of momentum in photon-phonon collisions. The latter may be seen readily by multiplying each wave vector with $h/2\pi$ where h denotes Planck's constant. In that way each wave vector diagram becomes a momentum diagram. It is also of interest that the quantum mechanical conservation of energy in this collision process translates classically into a Doppler shift. Note that the concept of the Bragg angle $\lambda/2\Lambda$ required for interaction is immediately obvious from the condition that the wave vector triangle must be closed.

If the interaction between sound and light waves were really in the form of plane waves, then the Bragg angle condition should be exactly satisfied. For instance the slightest change in the frequency of the sound wave , i.e a change in the length of the **K** vector, would nullify the diffraction effect completely. In practice this is not the case and there exists a certain tolerance about the Bragg angle. This due to the fact that, although the light beam usually acts as a plan wave because of its small wave length, the sound beam exhibits a much larger spread of wave vectors, inversely proportional to its width. Hence, if the sound frequency now changes , the Bragg angle condition will simply be satisfied by some other **K** in the total distribution of sound wave vectors. It will be clear that, if this distribution is known, the relative frequency response of the diffraction process can be calculated in a relatively straight forward way [4].

It is also of interest that, although the wave vector diagram picture is most appropriate for wide sound beams, there is no a priori reason why it could not be applied to the very narrow sound beams exemplified in Fig.1. In these beams the distribution of sound wave vectors is so wide that incident light scattered by a particular **K** may be rescattered by another **K** in the same distribution with large efficiency. In other words: the appearance of the many orders in Raman-Nath diffraction is due to efficient rescattering of light by a narrow sound beam having a wide angular spectrum of plane waves. In modern treatments of acousto-optics [1] this concept of rescattering is used in the sense of calculating, by means of suitable path integrals, what the total contribution is to a given order, taking into account the most likely scattering paths through successive intermediate orders .

Finally, in the context of acousto-optic diffraction in crystals, anisotropic effects may be taken into account by incorporating in Fig.3 the known dependence of the length of the wave vectors $\mathbf{k}_{1,(-1)}$ and **K** on their direction of propagation.

2. APPLICATIONS

In what was discussed so far, particularly with reference to the Bragg diffraction configuration illustrated in Fig. 2, we already suspect a number of potential applications of acousto-optics inherent in the particular properties of the phenomenon itself:

2.1 Beam Deflector

The angle of deflection is inversely proportional to Λ, i.e proportional to the acoustic frequency. By electronically changing the latter we should be able to deflect the beam rapidly, thereby in fact constructing an *inertialess beam deflector.*

2.2 Spectrum Analyzer

By the same token, if various sound frequencies are present in the sound beam simultaneously, each one should deflect its own light beam in a direction proportional to its frequency. We should therefore be able, by using appropriate optics, to construct a *parallel addressed frequency analyzer.*

Note that the only real difference between a beam deflector and a spectrum analyzer is that in the former successive frequencies are applied sequentially, while in the latter they are applied simultaneously.This is schematically indicated in Figs. 4a and b.

2.3 Light Beam Amplitude Modulator

The amount of diffracted light depends on the strength of the sound beam, or more precisely, the amplitude of the high frequency sound. Therefore , by modulating the electronic carrier that generates the sound, we should be able to devise a *light beam modulator* .

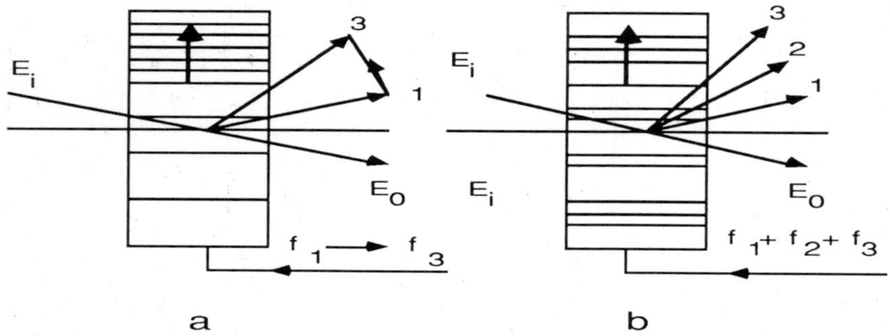

Fig.4 Bragg cell used as a) beam deflector, b) spectrum analyzer

2.4 Spatial Light Amplitude Modulator

If the sound cell is long enough or, in more general terms, if the modulation signal is fast enough compared to the transit time through the sound cell, then the locally diffracted light is proportional to the modulation signal at that location. Hence the outgoing wavefront carries a (moving) picture of the modulation signal. With some ingenuity we should be able to use this for *image display* or *parallel optical processing of electrical signals.*

It will be clear that the only difference between a light beam modulator and a spatial light modulator is the width of the incident beam of light. This is illustrated schematically in Figs. 5a and b where the thickness of the sound wave crest symbolizes the local amplitude of the carrier and the thickness of the light rays the locally diffracted light amplitude.

2.5 Frequency or Phase Shifter

The diffracted light is up- or downshifted by the sound frequency. Thus an acousto-optic Bragg cell is inherently an *electronic frequency shifter* or , because the phase of the sound is transferred to that of the diffracted light, an *electronic phase shifter.* In fact, the two properties are inherently connected as frequency is just the time derivative of phase. By tailoring the local phase of the carrier, the Bragg cell may be used as a spatial phase filter by the same principle as when it is used as a spatial amplitude filter.

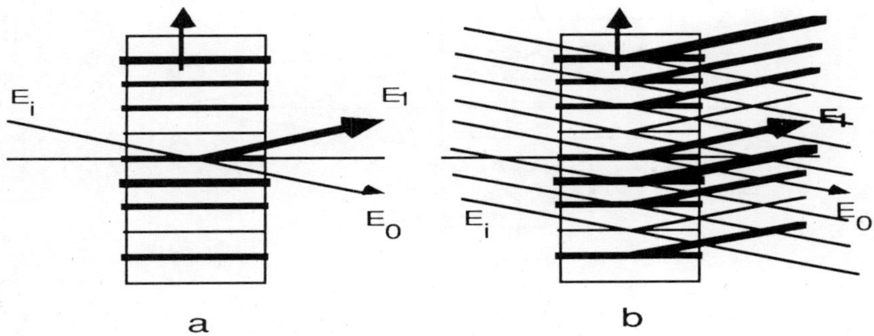

Fig.5 Bragg cell used as a) beam amplitude modulator, b) spatial amplitude modulator.

2.6 Sound Field Visualizer

With reference to Fig.3, it is clear that there exists a one-to-one correspondence between the propagation vectors of the diffracted light and those of the sound. Given enough spread in the incident light, the distribution of diffracted wave vectors (i.e the distribution of diffracted plane waves) ought to mirror the distribution of the plane waves of sound. But the latter constitute by definition the field of the sound itself. Thus we must conclude that the diffracted light somehow carries an image of the sound field. By appropriate optics this image can indeed be made visible in a process called Bragg diffraction imaging. Hence acousto-optic diffraction is capable of *sound field visualization.* [5]. A typical configuration is shown in Fig.6. The areas O_- and O_+

are images of the sound field region O, demagnified by a factor λ/Λ. Appropriate astigmatic and anamorphic optics is necessary to magnify and re-scale the images.

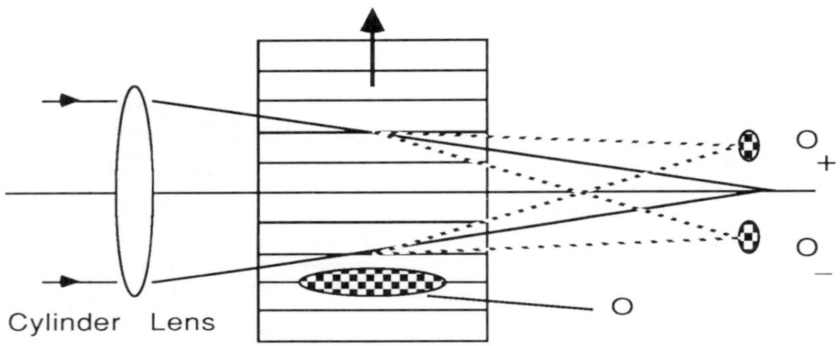

Fig.6 Configuration for sound visualization

3. SIGNAL PROCESSING

In general, acousto-optics lends itself well to acousto-optic signal processing [6] because, as we have already seen, the soundcell can be thought of as a spatial light modulator (Fig.5b) or as a spectrum analyzer (Fig.4). Both concepts may be used for signal processing. In the first case we speak of image plane processing, in the second case of Fourier plane processing.

3.1 Spectral Analysis

A particularly simple case of Fourier plane processing is shown in Fig 7 where the lens L concentrates the diffracted beams of Fig.4b onto an array of photo diodes. A sequential scanning of these diodes then will display on an oscilloscope a plot of the frequency spectrum in the sound cell. Strictly speaking, of course, there is not much signal processing going on other than displaying the frequency spectrum. It should be realized, however, that this kind of spectrum analyzer operates on a substantia portion of the signal in parallel, and is thus in principle less likely to miss transient occurrences, than a conventional electronic spectrum analyzer.

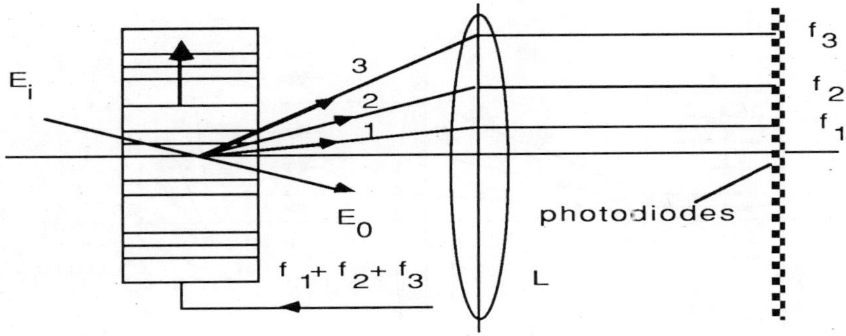

Fig.7 Configuration for incoherent spectral analysis.

3.2 Spectral Shaping

In a refinement of the technique just discussed, a strong reference signal at the light frequency is added to the array of photodiodes and the latter are modified so as to deliver parallel r.f outputs which are summed. This is shown in Fig 8, where the dashed line indicates a reference field. Each frequency component at, say, $\omega + \Omega_n$ at the nth diode, is mixed with the reference signal at ω to give an out put at Ω_n, i.e its proper Fourier frequency. The phase and amplitude of this output depends obviously on the phase and amplitude of the reference field at the location of the nth diode. For instance if the reference field has a uniform phase and amplitude distribution across the array of photo diodes, then summing all the r.f outputs is equivalent to performing an inverse Fourier transform and will results in an output which is the exact duplicate of the original signal in the soundcell (apart from a time delay due to transit time effects in the soundcell).However, by phase and/or amplitude shaping of the reference beam , it is possible to shape the frequency spectrum before summing the outputs, thereby, in fact creating a frequency filter by optical means. Recent developments have extended these concepts to two dimensions such that, for instance, the horizontal dimension represents a coarse frequency representation, the vertical dimension a fine one [6].

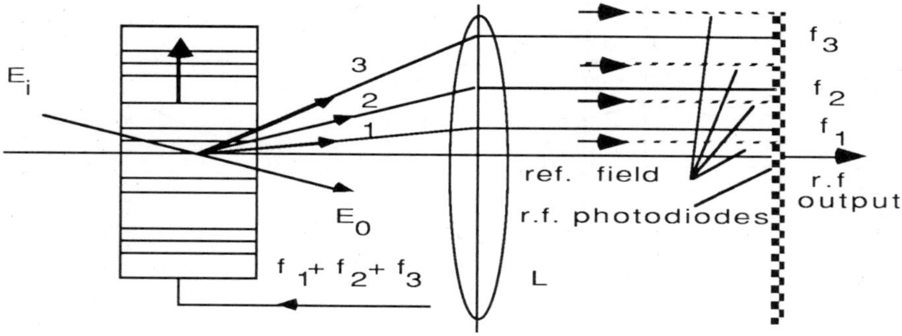

Fig.8 Spectral shaping with coherent reference field

3.3 Correlation / Convolution

Correlation and convolution are good examples of image plane processing [7]. The simplest case is illustrated in Fig.9 where a "moving" image g(x) of Fig. 5b is projected onto a correlation mask m(x) by an optical system, in this case a simple lens L.

A photodiode behind the mask m(x) collects all the light, and it is simple to show that the electrical output i(t) is of the form:

$$i(t) \propto \int_{-\infty}^{\infty} |g(Vt \pm x)m(x)|^2 dx \tag{5}$$

In expression (5) the + or - sign depends on the direction of the movement of g(x) past m(x) and on the original direction of the sound in the Bragg cell. By proper choice of configuration both convolution and correlation are possible.

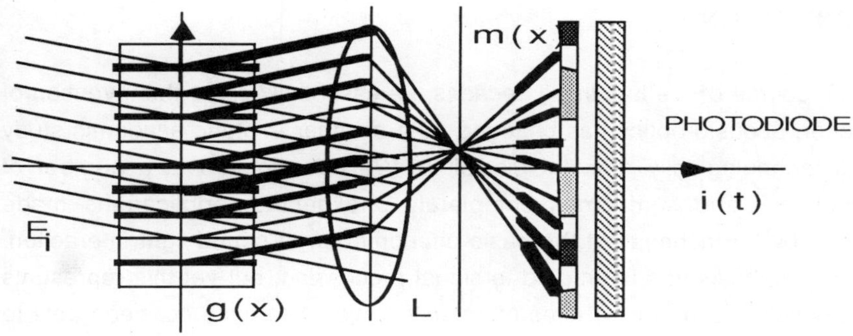

Fig.9 Correlation / convolution in the image plane

However, it will be noticed that (5) does not really represent a real correlation/convolution because of the squaring of the argument of the integral What is necessary, is to avoid this by integrating over the amplitude rather than the intensity as done in Fig.9. This is , in fact, possible by a) imaging g(x,t) onto m(x) by a double lens (telescope) system so as not to introduce phase curvature, b) form the Fourier transform of the transmitted field g(Vt±x)m(x) in the back focal plane of a third lens, and c) collect all the light in the center of this focal plane by using a pinholed photodiode. It may be shown readily that the photodiode output i(t) in this case is given by:

$$i(t) \propto \left| \int_{-\infty}^{\infty} g(Vt \pm x) m(x)^2 dx \right|^2 \tag{6}$$

It will be seen that expression (6) , although considerably more useful than (5), still cannot give negative outputs, nor handle complex functions.Such complex signals may be processed by applying reference signals instead of masks, and heterodyning in much the same way as was done in Fig.8 for Fourier transform processing . In fact, it may be shown that there really is no essential difference between Fourier transform plane and image plane processing[1] , and that, by the insertion of proper optics, one system may be converted into another.

4. CONCLUSION

In the course of the last three decades, i.e essentially since the invention of the laser, acousto-optics has changed from a rather esoteric academic study into a technologically relevant discipline. We have looked at a number of interesting, and sometimes completely unexpected, applications made possible by the richness of the basic phenomenon of sound light interaction. Special emphasis has been paid to signal processing, but yet this represents only a small fraction of the uses of acousto-optics. We have not been able to touch at all upon other applications such as acoustic surface wave monitoring, bulk wave probing, sound wave generation, phased array beam steering, optical filtering etc. Nor has the time been sufficient to obtain more than a fleeting impression of the fascinating formalisms and algorithms that have been developed in order to analyze the phenomenon itself[1]. It is hoped that the interested reader, guided by the references quoted in this article, will pursue his or her own studies, in order to gain an intuitive understanding of this field, so rich and rewarding in potential applications.

REFERENCES

1. Korpel, A., "Acousto-optics", Marcel Dekker Inc., New York, 1988
2. Adler, R., IEEE Spectrum 4, 42 (1967)
3. Klein, W.R. and Cook, B.D. , IEEE Trans. SU-14 , 123 (1967)
4. Korpel, A., Proc. IEEE 69 , 448 (1981)
5. Korpel, A., IEEE Spectrum 5 , 45 (1968)
6. See special issue Proc. IEEE 69 , Jan. 1981
7. Rhodes, W.T., Proc. IEEE 69 , 65 (1981)

RESONANCE SPECTROSCOPY

Gérard QUENTIN

Groupe de Physique des Solides de l'E.N.S. - Tour 23 -

Université Paris 7 - 2 Place Jussieu -

75251 Paris Cedex 05 - FRANCE -

ABSTRACT

Resonance phenomena are observed in all domains of Physics (optics, Electromagnetism, Thermodynamics, NuclearPhysics, Acoustics...) We shall focuss these lectures on the resonances observed in the field of Acoustics and specifically on their observation in the frequency domain (spectroscopy)

1- RESONANCES IN ACOUSTICS

1-1- Resonance of Bulk Waves

That is the most simple case. When two surfaces of a medium with parallel tangent planes exists we can observe resonances of bulk acoustic waves in the medium. The first simple case to which we shall devote the begining of these lectures is the infinite plane plate (Fig. 1-1). More generally if a plane wave is incident upon a scatterer and if the wave vector hits the surfaces of the scatterer at points where the plane tangent to the surface are parallel, one can observe resonances (Fig 1-2).

The scatterer can be a cylinder or a sphere or a more complicated object immersed inside a fluid. It can also be can inclusion or a defect inside a solid for example.

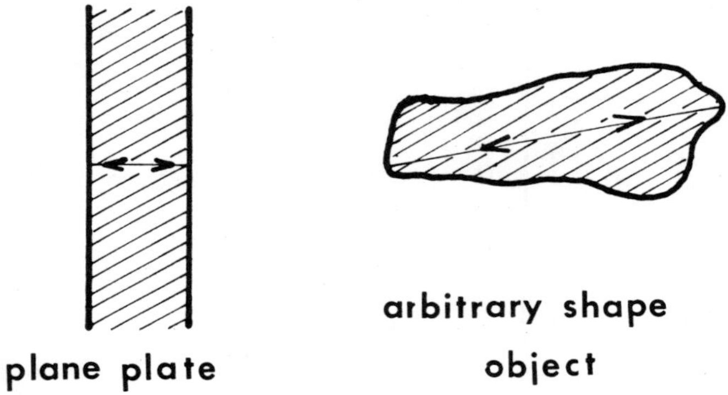

Fig.1Resonances of Bulk waves.

1-2 Resonance of Surface Waves

One of the most simple is observed in the scattering of an acoustic wave by an elastic cylinder or a sphere immersed in a fluid. On such scatterer many different kind of surface waves may propagate (Franz, Stoneley, Rayleigh, Whispering Gallery) and if the insonification is due to a plane wave we can observe resonances due to surface waves circumnavigating around the cylinder. Fig. 2-1 shows the case of backscattering from a cylinder. The resonance occur when the surface waves'path around the shell is an integer number of wavelengths. Fig. 2-2 shows that this phenomenon can also accur for a scatterer of arbitrary geometry.

2- COMPRESSIONAL WAVE RESONANCE OF AN ELASTIC PLANE PLATE

2-1 Harmonic Plane wave insonification

We shall study first a very simple case. Consider a plane plate with width e, density ρ_2 and longitudinal wave velocity C_2 immersed on the left in a fluid medium with density ρ_1 and sound velocity C_1 and on the right in a fluid with density ρ_3 and sound velocity C_3 (Fig.3). We shall suppose that the incident plane wave has its wave vector perpendicular to the plane and is of the form

$$A_{inc} = A_o \, e^{j(\omega t - k_1 x)} \qquad (1)$$

The problem is one-dimensional

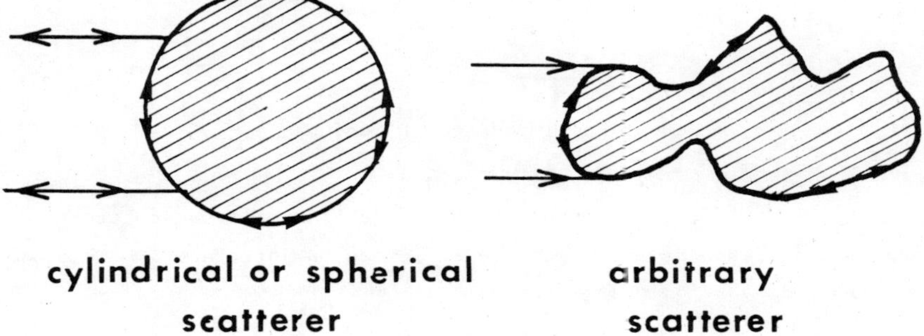

Fig. 2 Resonance of surface waves

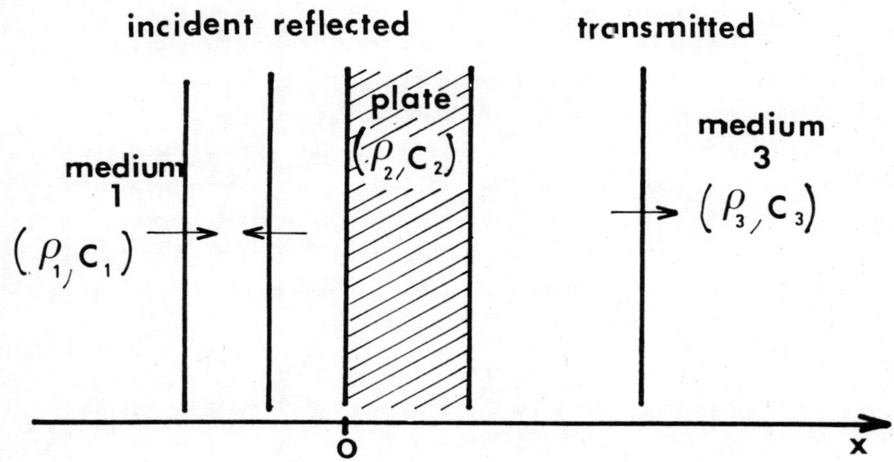

Fig. 3 Geometry of the problem for an incicent plane wave incident normally upon a plane elastic plate.

We shall take the origin of coordinates at the incident point ($x = 0$) and study the various components of the reflected wave A_r inside fluid 1 and the transmitted wave A_t inside fluid 3. For simplification reason we shall suppose that there is no attenuation inside both fluid 1 and 3.

The reflexion and transmission coefficients at the interfaces takes the following form with obvious index notations.

$$R_{i\to j} = \frac{Z_j - Z_i}{Z_j + Z_i} \qquad (2)$$

$$T_{i\to j} = \frac{2Z_j}{Z_j + Z_i} \qquad (3)$$

Z_i = acoustical impedance of medium $Z_i = \rho_i \; C_i$

$i = 1,2,3.$

The first component of the reflected amplitude returning in medium 1 corresponds simply to the reflexion on interface 1/2. This componant has the following expression at abcissa d<o

$$A_r^o = R_{1\to 2} \; A_o \; e^{\,j(\omega t + k_1 d)}$$

that is

$$A_r^o(d) = R_{1\to 2} \; A_{inc}(0) \; e^{\,jk_1 d} \qquad (4)$$

It will be very important for the following developpments to notice that this first component bear no information about the plate except throug R_{12} it depends of the acoustical impedance of medium 2 This reflected wave has only seen an infinite interface between medium 1 and medium 2. We shall denote later this kind of response a rigid response. At the opposite the following reflected waves have crossed the plate and have been reflected at the interface 2-3 and for orders different from 1 at the interface 2-1. Their expression are the following

$$A_r^1(d) = T_{1\to 2} \; R_{2\to 3} \; T_{2\to 1} \; e^{-2jk_2 e} \; A_{inc}(0) \; e^{jk_1 d}$$

$$A_r^2(d) = T_{1\to 2} \; R_{2\to 3} \; R_{2\to 1} \; R_{2\to 3} \; T_{2\to 1} \; e^{-4jk_2 e} \; A_{inc}(0) \; e^{jk_1 d}$$

$$A_r^n(d) = T_{1\to 2} \; R_{2\to 3} \; T_{2\to 1} \; (R_{2\to 1} \; R_{2\to 3})^{n-1} \; e^{-2jk_2 en} \; A_{inc}(0)$$

wich can be rewritten

$$A_r^n(d) = \frac{T_{1\to 2} \; T_{2\to 1}}{R_{2\to 1}} \; A_{inc}(0) \; e^{jk_1 d} \times (R_{2\to 1} \; R_{2\to 3} \; e^{-2jk_2 e})^n \qquad (5)$$

The total reflected wave is the sum of all components and takes the form

$$A_r(d) = A_{inc}(0)e^{jk_1d}[R_{1\to2} + \frac{T_{1\to2}\,T_{2\to1}}{R_{2\to1}} \sum_{n=1}^{\infty} (R_{2\to1}\,R_{2\to3}\,e^{-2jk_2e})^n]$$

or

$$A_r(d) = A_r^o(0)\,[1 + \frac{T_{1\to2}\,T_{2\to1}}{R_{1\to2}\,R_{2\to1}} \sum_{n=1}^{\infty} (R_{2\to1}\,R_{2\to3}\,e^{-2jk_2e})^n] \tag{6}$$

Resonance occur when the various terms of the sum add in phase. We shall consider now the phase of these terms. The wave vector is complexe in the more general case where the wave can be attenuated inside the elastic medium 2 :

$$k_2 = k' - j\alpha \quad k' = \alpha\varepsilon \mathcal{R} \tag{7}$$

We shall consider three cases, two simple ones and then the general situation.

$\underline{1}$ $(R_{2\to1}\,R_{2\to3}) \in \mathcal{R}^+$

In this case constructive interference between partial waves is observed for

$$e^{-2jk'e} = 1$$

$$2k'e = m\,2\pi$$

$$e = m\frac{2\pi}{2k'} = m\frac{\lambda}{2}; \text{ m integer} \tag{8}$$

This is for example the case of an ordinary metallic plate in a classical fluid.

$\underline{2}$ $(R_{2\to1}\,R_{2\to3}) \in \mathcal{R}^-$

In this case constructive interference between partial waves is observed when

$$e^{-2jk'e} = -1$$

$$e = (m + 1/2)\frac{\lambda}{2} \qquad \text{m interger} \tag{9}$$

$\underline{3}$ $(R_{2\to1}\,R_{2\to3})$ complex

In this case the resonance condition is much more complicated. Nevertheless this situation occurs for example is medium 2 is ground because the ground impedance is complex at sonic frequencies and its real and imaginary parts have the same order of magnitude. We shell not deal with the situation and we shall only consider in the following the simple case 1

We shall use simplified notations in order to write formula (6)

$$A_r = A_r^o \left(1 + \beta \sum_{n=1}^{\infty} \gamma^n\right) \tag{10}$$

Where

$$A_r = A_r(d) \qquad A_r^o = A_r^o(d)$$

$$\beta = \frac{T_{1\to 2}\ T_{2\to 1}}{R_{1\to 2}\ R_{2\to 1}}$$

$$\gamma = R_{2\to 1}\ R_{2\to 3}\ e^{-2jk'e}\ e^{-2\alpha e}$$

We shall suppose first that the incident wave is at the resonance frequency and that the reflexion coefficiants $R_{2\to 1}$ and $R_{2\to 3}$ are such that.

$$\gamma = R_{2\to 1}\ R_{2\to 3}\ e^{-2\alpha e} \in \mathcal{R}^+ \tag{11}$$

Consider now the factor β, it can be written as a function if the acoustique impedances of medium 1 and 2.

$$\beta = -\frac{4\ Z_1\ Z_2}{(Z_1+Z_2)^2} < 0 \tag{12}$$

In addition in the case that we consider

$$\gamma < 1$$

The sommation of the geometrical series is obtious and we obtain from (10)

$$A_r = A_r^o \left(1 - |\beta| \frac{\gamma}{1-\gamma}\right) \tag{13}$$

The total amplitude reflected from a plate at the resonance frequency is smaller than the amplitude reflected from the first interface A_r^o. If we vary the frequency minima in the spectrum occur at every resonant frequencies

Similar developments can be used to calculate the total transmitted signal. In most real ultrasonic experiments continuous harmonic waves cannot be used for obvious experimental considerations. We shall now consider two situations very often encountered in practice :

- narrow band exitation by a train of sinusoïds.
- wide band exitation.

2-2 Narrow-band Plane Wave exitation

We shall suppose now that the insonfying signal is a train of sinusoïds

$$A_{inc} = A_o e^{j(\omega t - k_1 x)}\ \Pi_T(t) \tag{14}$$

Where $\Pi_T(t)$ is the window function

$$\Pi_T(t) = 0 \text{ if } |t| > \frac{T}{2}$$
$$\Pi_T(t) = 1 \text{ if } |t| < \frac{T}{2} \qquad (15)$$

The mathematical formulation of the various partial reflected waves is as follows

$$A_r^o(d) = R_{1\to2}\, A_o\, e^{j(\omega t + k_1 d)}\, \Pi_T\left(t - \frac{c}{C_1}\right) \qquad (16)$$

$$A_r^n(d) = A_r^o(d)\left[1 + \frac{T_{1\to2}T_{2\to1}}{R_{1\to2}R_{2\to1}} \sum_{n=1}^{\infty} \left(R_{2\to1}\, R_{2\to3}\, e^{-j2k_2 e}\right)^n \times \Pi_T\left(t - \frac{2ne}{C_2}\right)\right] \qquad (17)$$

We have made here an important simplification. We have supposed that the wave packet is not deformed inside medium 2 that is this medium is non dispersive.

It is not easy to see what happens from mathematical formulas. The best way to understand is to look for a geometricla interpretation which is given on Fig. 4. We have considered a resonance with the following assumptions.

$$\beta = \frac{T_{1\to2}\ \bar{T}_{2\to1}}{R_{1\to2}\ R_{2\to1}} = -0.1$$

$$\gamma = R_{2\to1}\ R_{2\to3}\ e^{-2\alpha e} = 0.8$$

The length of the incident wave packet is supposed to be 20 times the time of a round trip inside the plate.

It is rather easy to see what happen. The wave packet reflected on the front (left) interface arrives first with an amplitude A_r^o. The second following wave packet has an amplitude $\beta\gamma A_o = 0.08\ A_r^o$ but it is in phase opposition with the first one. So the amplitude decrease to $0.92\ A_r^o$. The third wave packet has an amplitude $\beta\gamma 2\, A_o = 0.064\ A_r^o$ and is in opposition of phase with the first one and the amplitude decrease to $0.856\ A_r^o$ and so on. Finally the amplitude goes assymptoty cally towards.

$$A_r^o\left(1 - 0.1\,\frac{0.8}{1-0.8}\right) = 0.6\ A_r^o$$

At a time T after the beginning of the reflected signal, the whole

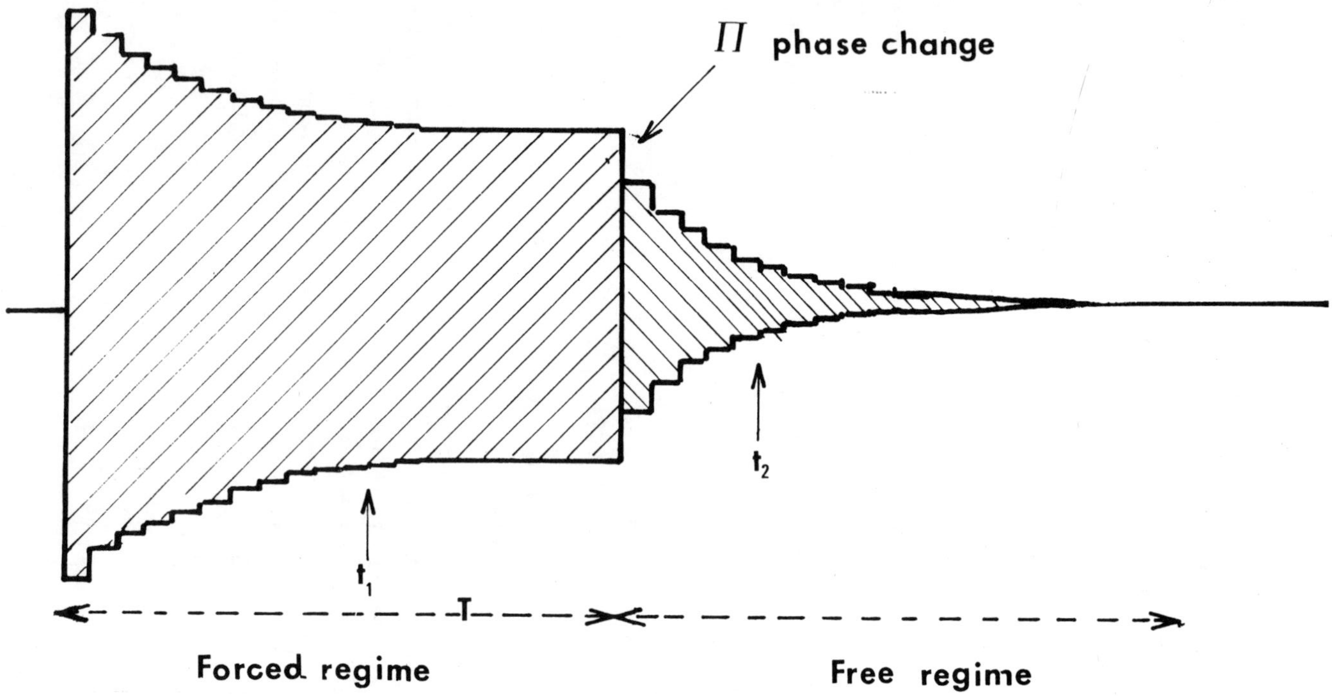

Fig. 4 - Resonance of a long train of sinusoïds

signal signal reflected by the front face has arrived and the succursive arriving signals have been transmitted and multiply reflected inside the plate. We shall first notice that there is a π phase change in the signal. The amplitude goes abruptly from $\simeq 0.6\, A_r^c$ if T is sufficiently long to $0.4\, A_r^o$ and then decreases toward zero. This regime will be considered as a "free" vibration regime because no energy is entered inside the plate. Similar signals had been observed firstly by the team of Professor Ripode in France for a different situation (scattering from cylinders) and we shall discuss these observation later on. For plane plates signals similar to the one displayed on Figure 4 have been observed later by the same team and by M. de Billy in my team.

We shall now explain how to study resonances from such signals Two possibility are offered. Either you study versus frequency the amplitude of the signal in the "forced" vibration regime. We have omitted to notice that far from a resonance frequency the signal has an almost constant amplitude with a length T and is almost zero after this time (Fig.5)

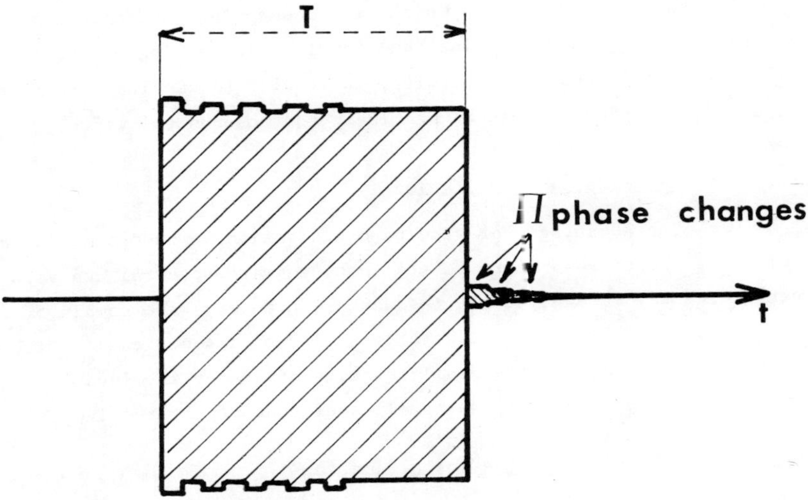

Fig. 5 - Reflection of a wave train near antiresonance

As a consequence if the frequency of the sinusoïds inside the train is scanned continuously and if we record versus frequency the amplitude at time t_1 for exemple within the "forced" regime we shall observe a "spectrum" with minima at each resonance frequency (form function). This typical spectrum displayed on Fig. 6 a). At the opposited if the amplitude is measured at time t_2 during the "free" regime the spectrum observed will have maxima at each resonance frequency. (fig. 6 b). If we want only to measure the resonances frequencies there is no problem with this experimental method in the case considered because the minima in the form function or the maxima in the resonance spectrum occur exactly at these frequencies. A little problem arises if we want to measure the resonance width. The incident signal is of the form

$$e^{j\omega t} \times \Pi_T(t)$$

in the time domain. This means that in the frequency domain there will be a convolution of the reponse of the scatterer by the Fourier transform of the window function $\Pi_T(t)$ that is by the cardinal sine function. This function has a natural width which will add to the resonance's width. The width of the cardinal sine function is a decreasing function of T. If experimental conditions permits, it is wise to use long trains of sinusoïds in order to measure precisely the resonances' width.

2-3 Wide band excitation

The advantages of wide band excitation an numerous. In particular the use of very short pulses combined with spectral analysis is a very quick way to analyse the resonances in a very wide domain of frequencies. The only disadvantage ofthis kind of insonification with respect to narrow band is the small quantity of energy involved at a given frequency and consequently the small amplitude of the received signal.

2-3-1 δ impulse insonification

We shall first consider the case where the free media are non dispensive and media 1 and 3 not attenuating. The insonifying pulse is supposed to be

$$A_{inc}(x,t) = A_o\delta(t-\frac{x}{c_1})$$

If we take the origin of coordinates at the interface 1/2 the first reflected pulse A_r^o at the form.

$$A_r^o\ (d,t) = R_{1\to 2}\ A_o\delta(t-\frac{d}{C1})$$

and the successive seradiated in medium (1) can be written

$$A_r^n\ (d,t) = A_o T_{1\to 2}\ R_{2\to 3} T_{2\to 1}(R_{2\to 1}\ R_{2\to 3})^{n-1}\ e^{-2\alpha n e}\ \delta(t-\frac{d}{c_1} - \frac{2ne}{c2})$$

where we have supposed that the attenuation α in medium 2 is constant for all frequencies. This can be written, in same namer as previously :

$$A_r^n\ (d,t) = A_o R_{1\to 2}\ \frac{T_{1\to 2}\ T_{2\to 1}}{R_{1\to 2}\ R_{2\to 1}}\left[R_{2\to 1}\ R_{2\to 3} e^{-2\alpha e}\right]^n \delta\left(t-\frac{d}{c_1} - \frac{2ne}{c_2}\right)$$

The series of reflected impulses can be separated into two parts. The first reflected pulse A_r^o is only the result of specular reflection on the front interface. The following ones A_r^n are a series of impulse functions regularly separated in time and with decreasing amplitude. They can be represented as the product of the Dirac series

$$Ш_{(\frac{2e}{c2})}(t) = \sum_{n=1}^{\infty} \delta\ (t-\frac{d}{C_1} - \frac{2ne}{C_2})$$

by a decrasing function

$$g\ (t) = [R_{2\to 1}\ R_{2\to 3} e^{-2\alpha e}]^{(\frac{-C_2 d}{2eC_1} + \frac{C_2 t}{2e})} A_o\ R_{1-2}\ \frac{T_{1\to 2}\ T_{2\to 1}}{R_{1\to 2}\ R_{2\to 1}}$$

The total reflected signal can be written :

$$A_r(d,t) = A_r^o(d,t) + g\ (t)\ Ш_{\Delta t}(t)$$

where

$$\Delta t = \frac{2e}{C_2}$$

is the round-trip tim of the signal inside the plate.

Such signal has been plotted on Fig. 7 a.)

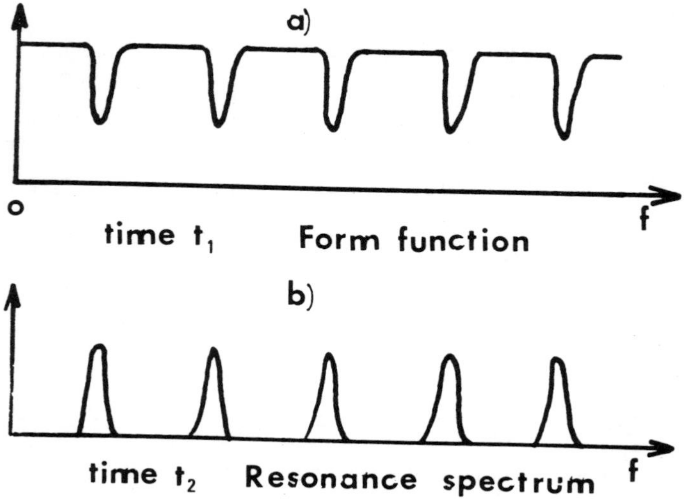

Fig. 6. Spectra obtained using narrow band insonification

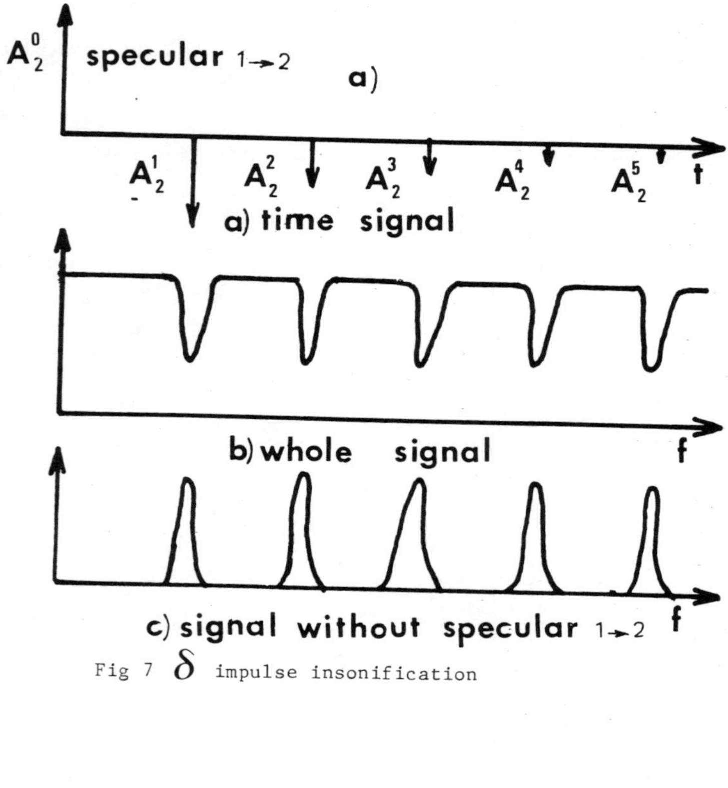

Fig 7 δ impulse insonification

Now we shall study the same signal in the frequency domain for the same experimental situationas previously used for narrow band insonification (plane metallic plane inside a fluid). The Fourier transform of A_r^o (d,t) is simply a constant with a value proportionnal to the reflexion coefficient $R_{1\rightarrow 2}$. The Fourier transform of the series $Ш_{\Delta t}(t)$ of δ impulses is simply a series $Ш_{1/\Delta t}(f)$ of δ impulses

$$Ш_{1/\Delta t}(f) = \sum_{n=1}^{\infty} \delta\,(f - \frac{n}{\Delta t})$$

occuring at frequencies

$$f = \frac{n}{\Delta t} = \frac{2ne}{C_2}$$

That is at the resonances frequencies of the plate. We shall denote G(f) the Fourier transform of function g(t). It has a bell shape around zero frequency and we have plotted on Fig. 7 a) the spectra of the whole signal plotted (form function) and also the spectrum of the partial signal $A_r(d,t) - A_r^o(d,t)$ where the speculary reflected signal has been substracted.(resonance spectrum). We have taken into account the fact that in the conditions under study $A_r^o(t)$ and g(t) have opposite signs and used the convolution theorem : $\mathcal{F}(g(t)\, Ш_{\Delta t}(t)) = G(f) * Ш_{1/\Delta t}(f)$

Comparing Fig. 7 and Fig. 5 shows that the same information about resonances can be deduced from wide band or from narrow band experiments.

2-3-2 <u>Wide band excitation</u>

Physical systems are notable to emit δ impulses. Ultrasonic spectroscopy uses very short pulses consequently wide band spectra. If we replace the δ impulse by such a signal h(t) with Fourier transform H(f) the following change will occur. In the time domain each $\delta(t-\tau)$ impulse will be replaced by $h(t-\tau)$ that is $\delta(t-\tau) * h(t)$. The time domain signal A_r' will consequently be obtained simply by a convolution operation :

$$A_r'(t) = A_r(t) * h(t)$$

The Fourier transform of a convolution being the ordinary product the signal in the frequency domain will be obtained by the product of the signal obtained with a δ impulse insonification by H(f). That is the frequency domain signal is modulated by the spectrum of the incident signal. This fact is explained on Fig 8. This property is valid whateven the shape of the time domain signal is a pure sinusiïdal function H(f) si a δ impulse. That means that the spectrum obtained by varying the frequency under cintinuous wave insonification is exactly the same as obtained with δ impulse insonification.

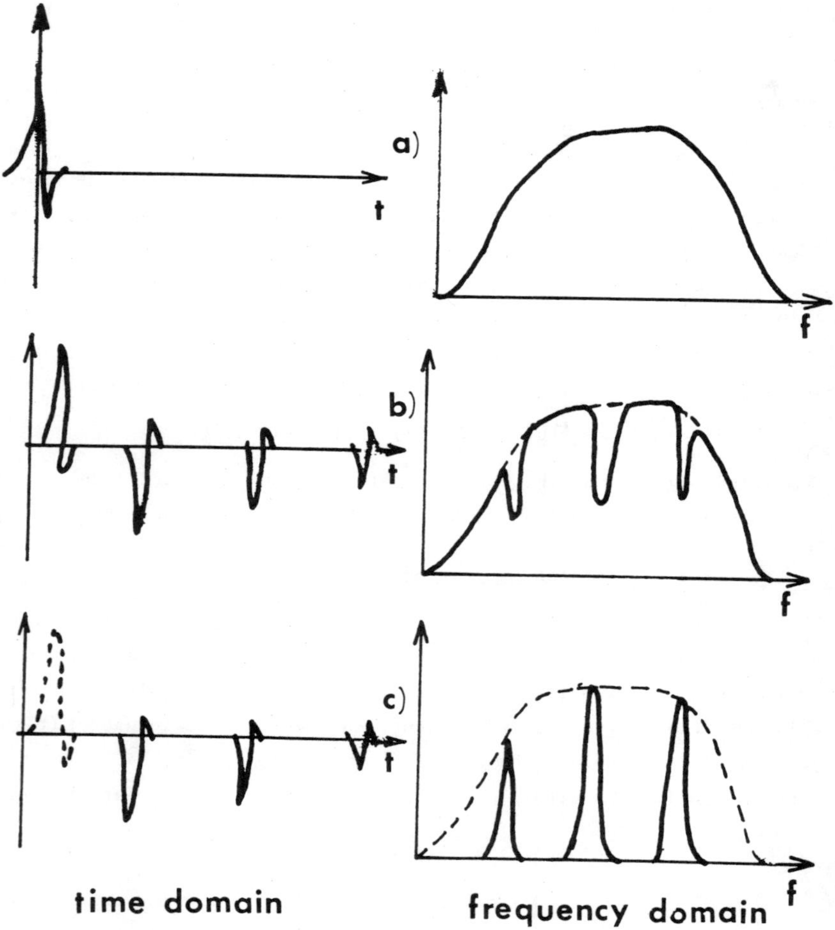

Fig 8 - Wide band insonification

a) incident signal b) whole reflected signal analysis

c) analysis without specular 1→2

It is, in principle, easy from results obtained with a real signal to return to the time spectra by dividind the spectra obtained (in absolute value or energy spectrum) by that of the incident signal. In the kinds of experimental situation two limitations occur. The first one concerns the resonance spectrum and is related to the possibility of separating the specularly reflected echo from the other echoes. The second one is due to the fact that in order to obtain the signal in the frequency domain one must use analog or digital processing and that this processing must be suitable adapted not to deform the spectra (especially concerning the widening of the spectral rays).

2 - SURFACE WAVE ON AXISYMETRICAL SCATTERERS

We shall only consider the case of infinite elastic cylinders and mention only the differences for other forms like spheres for exemple.

The situation the more common is that of a cylinder immersed inside a liquid. (Fig 9).

Various kind of surface waves can circumnavigate around the circonference of the cylinder : Rayleigh waves, Franz waves, Scholte-Stoneley waves and whispering gallery waves. There is a good agreement between the theorical prediction of the resonance of the waves and the

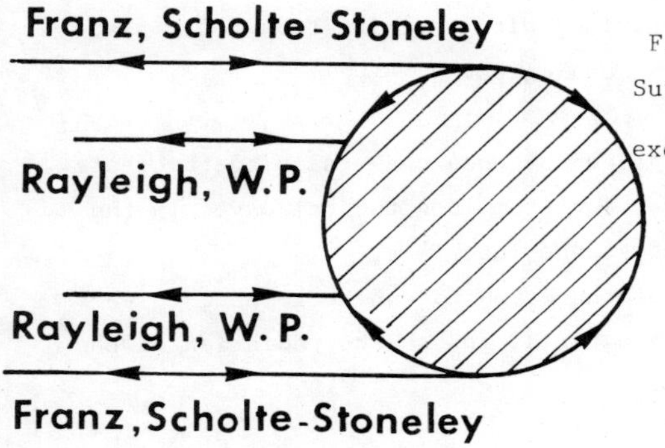

Fig 9
Surface waves
excitation on
axysymetrical objects
(W.l. = Whispering Gallery

measurements made using either narrow band excitation (Ripoche and al) or wich band excitation (Quentin et al).

In a cylinder for circuferential waves the resonances occur when the circumference is equal to an integer number of wave length.

$$2\pi a = n\ \lambda$$

a radius of the cylinder
λ wavelength of the surface wave
n = integer

For a sphere n must be replaced by n + 1/2.

The situation can be represented like in Fig 9 where we have used a ray representation. The first ultrasonic echo received is due to specular reflection at the interface of the cylinder at point A and like for the plane plate bears only informations about the properties of the interface. If we suppose that there exist only one surface wave generated at a given angle of incidence corresponding to points of incidence B or C, the second echo corresponds to a signal having circumnavigated the cylinder from B to C or from C to B (these two echoes adds because they arrive at the same time). The third and the following echoes have in addition accomplished an integer number n of roundtrips at the surface of the cylinder. The resonance condition is obvious and has been written before. If we try to make a comparison with the situation for a plane plate we shall make the following analogies.

$R_{1\to2}$ and $T_{2\to1}$ are replaced by mode conversion coefficients corresponding to conversion from longitudinal waves in the liquid to surface wave at the interface.
α is the attenuation of the wave and can be related to the attenuation inside the material and to the radiation of the surface wave.

This analogy leads for the scattered signal to write expressions rather analogous to those explained for the plane plate.

Consequently the conclusions obtained about the experimental conditions to be used to obtain the resonant spectra of the target are valid. Nevertheless the situation is not totally equivalent (especially distinction must be made between rigid, soft or medium scatterers.Many results obtained in France by our team in Paris and by the team of Professor Ripoche in le Havre are presented during the lecture. They can be found in the book "La Diffusion Acoustique" - N. Gespa Edit - CEDOCAR-Paris.

Surface Acoustic Wave Sensor

E. Verona

C.N.R., Istituto di Acustica "O.M.Corbino"
Via Cassia 1216, I-00189 Roma, Italy

Abstract

Surface acoustic wave devices have been succesfully exploited in recent years in the field of sensors. Because of the peculiar properties of surface wave propagation, these devices are suitable for the implementation of sensors of both physical and chemical variables. In this paper, a review of the most important surface acoustic wave sensors is reported and their operation is described. As the operation of surface acoustic wave sensors strongly depends on the measurand as it can involve both linear and nonlinear properties of the acoustic propagation medium, the treatment has been divided by the variable sensed. Finally the most common methods used for the signal output readout have been reported, as well as techniques used to compensate drifts in the output signal level due to temperature fluctuations.

Introduction

In recent years there has been a growing interest in using *Surface Acoustic Wave* (SAW) devices for the detection of various physical and chemical parameters, including temperature [1,2], mechanical acceleration [3-7], force [8] and pressure [9-11], d.c. and a.c. high voltages [12-14] and electric fields [15], magnetic fields [16], and concentrations of gasses or vapours in the environment [17-31].

The use of SAW's, in the field of sensor, is a very attractive technique since the mechanisms of interaction between the measurand and acoustic propagation, which can contribute to the sensor response, include many different linear and nonlinear properties of the propagation medium, such as mass density, elastic stiffnes and, in piezoelectric materials, also electric and dielectric properties. In addition, as the acoustic energy is confined to a thin near-surface region of the substrate, the propagation is highly sensitive to surface pertubations and the whole propagation path can contribute to the response of the device. Fabrication techniques of SAW sensors include thin film deposition and photolitographic processes which are in common with the technology of planar integrated circuits. The use of thin piezoelectric ZnO films make possible a future monolithic integration of the electroacoustic sensitive structures with the associated signal conditioning circuitry on the same semiconductor substrate, so to obtain small sized devices, robust in construction at a low price [32-33].

As to the changes produced by the measurand on the acoustic propagation, even if there are exemples of sensor based on the attenuation of the acoustic wave, most of them, however, are based on the changes of the phase velocity. In both cases the device is configured as a delay line, and changes in signal attenuation or in time delay are measured, respectively. In sensors, based on phase velocity changes, the readout can be obtained simply and accurately by monitoring the frequency of oscillation of the delay line when connected to a feedback amplifier to configure a SAW delay line oscillator the fractional change in oscillation frequency equals the fractional change in phase velocity.

This techique is very attractive since frequency can be measured with a high accuracy, only simple and inexpensive electronic circuitry is required, and finally the output signal is suitable for digital data acquisition systems.

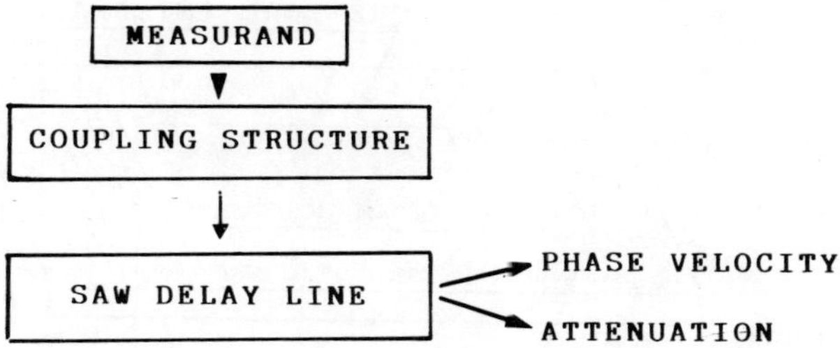

Figure 1: Schematic representation of a SAW sensor

1 SAW Sensors

The principle of operation of a SAW sensor is schematizated in Fig.1. It consists of a SAW delay line, where the measurand interacts through a proper coupling structure with the wave, along its propagation path, causing thus changes in the phase velocity and/or attenuation.

Most sensors are based on the propagation of the Rayleigh wave along the surface of a substrate, which, depending on the sensor, may be free or covered by one or more thin films. There are however examples of SAW sensors which are based on the propagation of Lamb modes in thin plates, of Rayleigh modes in layered structures and, more recently, of surface transverse waves [34-38]. To the interaction between acoustic propagation and measurand may contribute one or more different physical properties of the propagation material, depending on the nature of the measurand itself. As a consequence of this, the construction as well as the operation of SAW sensor may differ quite considerably. It is convenient in the following description to classifly the sensor by the type of variable sensed. Chemical sensor, Electrical, machanical, magnetic and temperature sensor will be briefly discussed here.

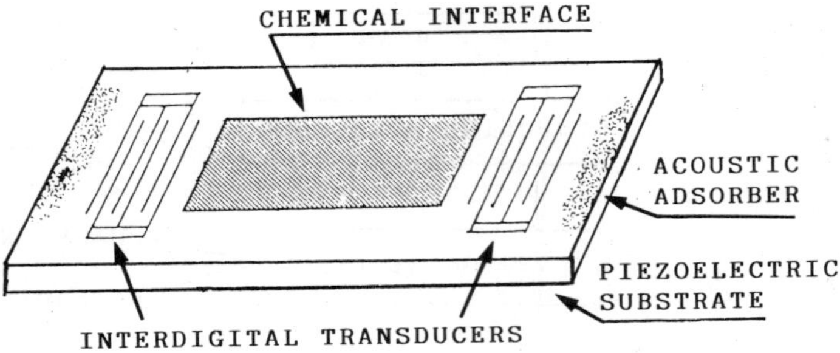

Figure 2: Basic structure of a SAW gas sensor

1.1 - Chemical Sensors

A large number of papers relating with SAW chemical sensors appeared in the last few years. Most of these papers describe the use of SAW devices as sensors of concentrations of gases or vapours. The structure of a SAW gas sensor is shown in Fig.2. It consists of a SAW delay line, whose propagation path is coated by a thin selectively sorbent film (chemical interface). Gas absorption by the film causes changes in its physical properties which, in turn, affect the acoustic propagation.

The physical properties of the interface which contribute to the response of the device are most frequently the mass density and the elastic stiffness. Changes in the thickness of the film can also enter in the response, as well as changes in the electrical resistivity can affect both phase velocity and attenuation when the the substrate is piezoelectric [39-40].

Requirements for the chemical interface are that it should be selectively sorbent to the gas, the process of absorption should be reversible and its properties should be stable in time. In addition the material used should be suitable for deposition in the form of a thin film without introducing high losses in the acoustic propagation.

The gas sensed, the chemical interface material used, as well as data on the substrate material and acoustic propagation conditions are sumarized in Tab.1 for most of the SAW gas sensors which have been reported in the leterature. The first work relating to SAW gas sensors appeared in 1979 and was due to Wohltyen and Dessy [17]. They investigated a SAW gas chromatograph detector by using both $STx - SiO_2$ and a $YZ - LiNbO_3$ SAW delay lines, coated by a gas chromatographic partitioning liquid. In the following years many SAW sensors have been investigated for the detection of concentration of different gases including SO_2 , H_2O , H_2 , NH_3 and NO_2 or of organic vapours. Both organic and inorganic thin layers have been used in these devices as a chemical interface.

Phase velocity is commonly the readout of SAW gas sensors. A tipical time response of a gas sensor is shown in Fig.3. Upon exposure to the gas, the phase velocity changes, reaching a different steady state value in a few minutes. When the gas is switched off, desorption take place and the velocity changes again reaching the initial value. For a given sensor, the steady state velocity change as well the time response during both the absorption and desorption processes depend by the gas concentration and by temperature. Fig.4 shows the response of a SAW Hydrogen sensor vs. Hydrogen concentration in nitrogen. Rise and fall time as measured for the different gas concentration tested are reported in the figure. The response of SAW gas sensors depends also on the thickness of the chemical interface and on the substrate material. Generally, an increase in the thickness of the interface give rise to a higher response but increases rise and fall times. As an example, the steady state fractional velocity change vs. chemical interface thickness, normalized to the acoustic wavelength, is shown in Fig.5 for three different hydrogen sensors implemented on a $YZ - LiNbO_3$, on a $STX - SiO_2$ and on a $(001)(100)Si/SiO_2/ZnO$ substrate.

A theoretical design of an optimized gas sensor would be based on the investigation of the acoustic propagation in the layered structure shown in Fig.2,

Table I

Measurand	Chemical Interface	Substrate Propagation	Reference	Detection
gas chromatograph detector	gas chromatographic partitioning liquid	$LiNbO_3 - SiO_2$	[17]	attenuation or velocity
SiO_2	triethanolamine	$yz - LiNbO_3$	[18,22]	velocity
H_2S	WO_3	$yz - LiNbO_3$	[22,25,29]	"
vceton methanol	photoresist Shipley AZ 13505	$Si/SiO_2/ZnO$ (Lamb wave)	[34]	"
organic vapours	PVC	$Si/SiO_2/ZnO$ (Lamb wave)	[26]	"
H_2	Pd	$yz - LiNbO_3$	[19,20,21]	"
H_2	Pd	$STx - SiO_2$	[24]	"
H_2	Pd	$Si/SiO_2/ZnO$ (Rayleigh wave)	[33]	"
NH_3	Pt	$STx - SiO_2$	[31]	"
NO_2	metal free phtalocyamine	$STx - SiO_2$	[23]	"
NO_2	lead phtalocyamine	$yz - LiNbO_3$	[28]	"
organic vapours	isothermal measurement	$STX - SiO_2$	[27]	"

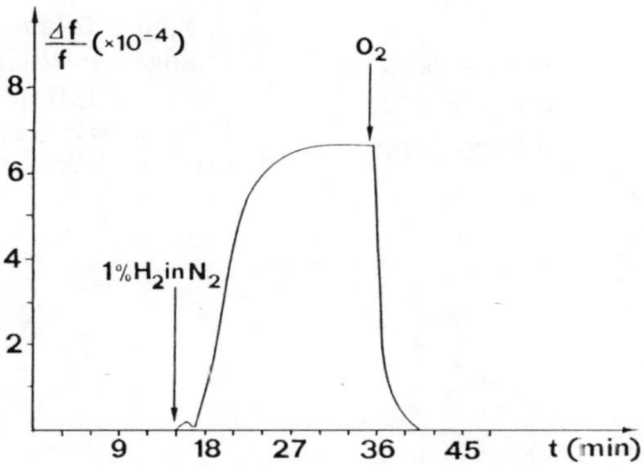

Figure 3: Time response of a SAW H_2 sensor upon gas absorption and desorption

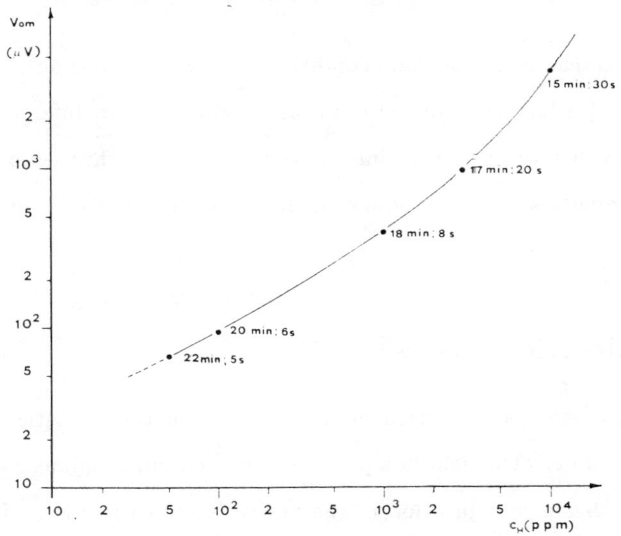

Figure 4: Fractional velocity change vs. H_2 concentration in N_2 for a SAW H_2 sensor (see [20]). Rise and fall time values are reported in correspondance of the experimental points.

once the changes in the physical properties of the interface material as well as the chemisorption processes involved are completly known. Unfortunately these data are not available at the moment for most of the materials investigated, and an

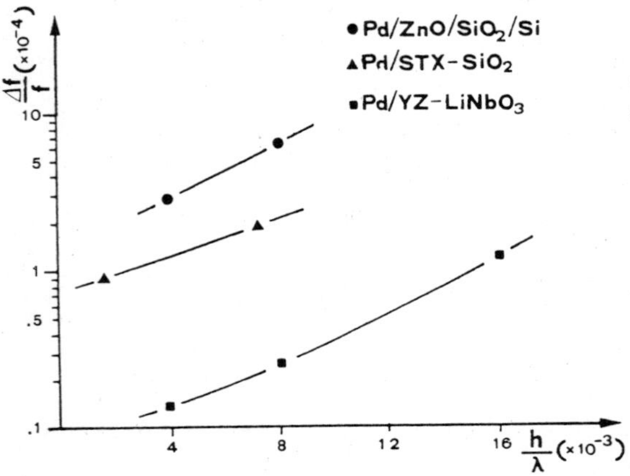

Figure 5: Fractional velocity change vs. thickness of the Pd chemical interface, normalized to the acoustic wavelength, as meas ured on different SAW H_2 sensor (see [33]).

experimental search for optimum conditions has to be carrier out. In this respect, an analysis of a large number of data, as obtained under different experimental conditions, will give important informations about the changes produced in the physical properties of the interface by gas entry and the related chemisorption processes.

1.2 - Electric Sensors

SAW electric sensors are based on the changes in the acoustic phase velocity produced by an electric field in a piezoelectric medium (nonlinear electroacoustic effect). The basic configuration of the device is shown in Fig.6. It consists of a SAW delay line implemented on a highly piezoelectric substrate. The acoustic propagation path is coated by a thin metal layer connected to ground. Ground is also the average potential of the interdigital transducers. The voltage V to be measured is applied to a thin film metal electrode deposited on the back surface of the substrate of thickness h, giving rise to a uniform electric field $E = V/h$ inside the piezoelectric plate. Due to the piezoelectric coupling, both an a electrical and a mechanical strain field are generated in the acoustic propagation

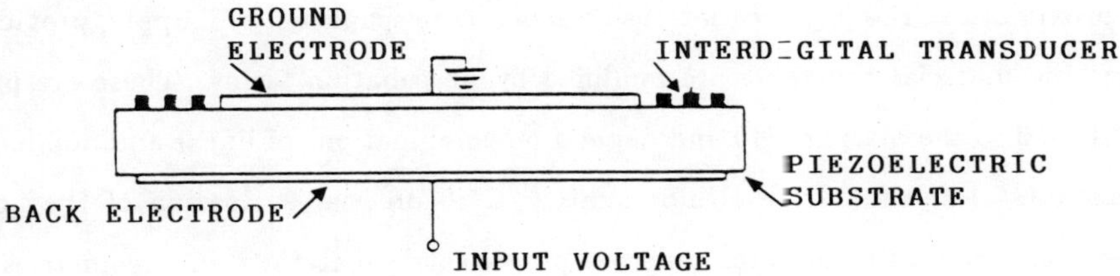

Figure 6: Basic structure of a SAW electric sensor

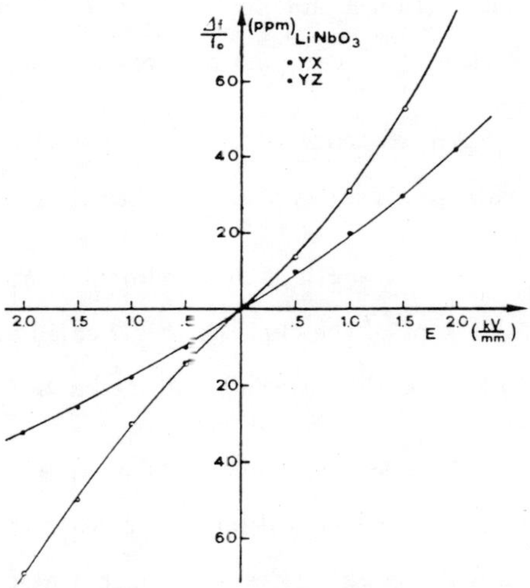

Figure 7: Fractional velocity change vs. applied electric field for SAW propagating along the z-direction and the x-direction of a y-cut $LiNbO_3$ plate (see [14]).

medium. These, in turn, give rise to changes in the phase velocity of the acoustic wave, which are mainly due to the nonlinear properties of the medium. The relative change in phase velocity vs. applied electric field for SAW's propagating along two different directions of a y-cut $LiNbO_3$ substrate, is shown in Fig.7. Requirements for an ideal SAW electric sensor are that the relative change in phase velocity vs. electric field is as high and as linear as possible.

Theoretically, the phase velocity in presence of an electric field, could be evaluated on using a simplified model which consider small amplitude acoustic waves superimposed to a uniform electromechanical biasing field [41]. On using this approximation, the linear equations of acoustic propagation still apply, provided that the material constants are modified by pertubation terms. These are proportional to the biasing field and depend on combinations of linear and nonlinear constants of the material. Unfortunately, a theoretical evaluation of the nonlinear electroelastic effect for different piezoelectric materials and geometries of interaction is not possible, at the moment, because of an incomplete knowledge of the nonlinear constants for most piezoelectric materials.

An experimental search for a substrate showing optimum characteristics for voltage sensor applications has therefore to be performed.

The strength of the nonlinear electroacoustic interaction, which correspond to the slope of the curves of Fig.7 calculated at the origin, as measured for defferent piezoelectric materials and interaction geometries, is reported in Tab.II.

Recently, an analysis of a large number of experimental data obtained under different conditions, on measuring the strength of the interaction for Lamb and Rayleigh modes propagating along both the y-direction of a z-cut $LiNbO_3$ plate and the z-direction of a y-cut $LiNbO_3$ plate, allowed the evaluation of the pertubation terms which enter in the nonlinear electroelastic effect for acoustic propagation in any direction of the yz plane of $LiNbO_3$ in presence of a bias electric field laying on the same plane [41].

SAW electric sensors are suitable for measurement of both d.c and a.c. voltages, provided that the time delay of the acoustic line is small if compared to the a.c. period. Because of the high dielectric properties of the piezoelectric material which are used as substrates. The electric input impedance of the device is a capacitance with very low electric loss. Because of this, the device can be succesfully used as an electric field strength meter, provided that it is connected to a proper coupling structure (see Fig.8).

Table II

Substrate material and orientation	Acoustic mode	Strength of the interaction $\frac{\Delta f}{f}\|_{E=1kV/m}$ (ppm)	Reference
yz-$LiNbO_3$	Rayleigh	20	[12, 14, 15]
yx-$LiNbO_3$	Rayleigh	40	[14]
128° rot y-cut $LiNbO_3$	Rayleigh	90	[12]
yz-$LiNbO_3$	Lamb modes	up to 120	[35, 41]
$(001)(110)Bi_{12}GeO_{20}$	Rayleigh	56	[14]
PZT	Rayleigh	53	[43]
zx-$Z_nO/zx - Si$	Rayleigh layer modes	up to 7	[36]

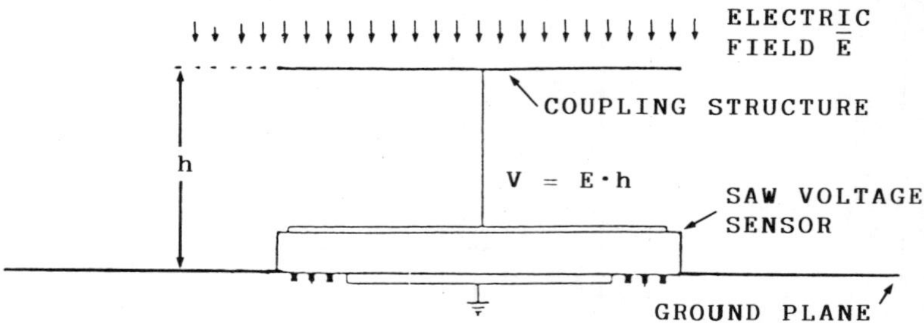

Figure 8: SAW electric field strength sensor.

1.3 - Mechanical Sensors

The principle of operation of SAW mechanical sensors is similar to that of electric sensors. Here, the force to be measured is applied to the acoustic propagation medium, according to one of the two interaction geometries shown in Fig.9, giving rise to a mechanical field which, in turn, affects the acoustic phase velocity. Unlike electric sensors, the use of a piezoelectric substrate is not necessarily needed for mechanical sensors, as an electromechanical coupling of the measurand is not required. Temperature stable substrates such as ST-cut quartz are commonly used for this application. One of the first papers relating with measurements of SAW velocity in quartz versus mechanical strain, appeared in 1975 and was due to Cullen and Reeder [42]. Since then a number of papers have appeared relating with the use of SAW devices as force sensors.

SAW accelerometers can be implemented by loading a force sensor with an inertial mass. The basic structure of the device is shown in Fig.10. SAW accelerometers have been widely investigated in the last fiew years with respect to the choise of the substrate orientation, to the interaction geometry and to the architecture of the device. SAW mechanical sensors can also be used as pressure sensors provided that a proper coupling structure is used to convert the pressure to be measured into a force applied to the device.

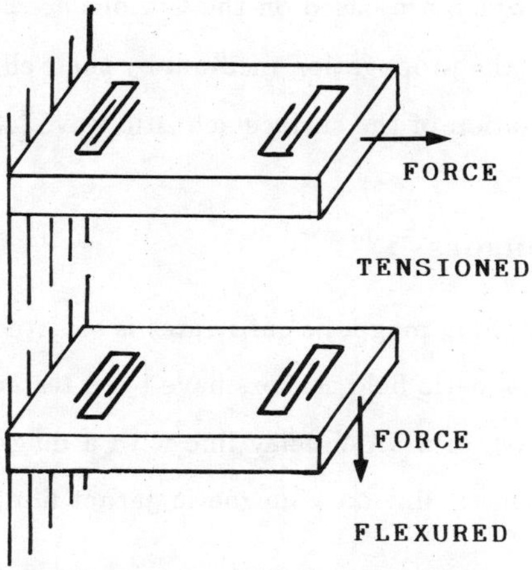

Figure 9: Basic structures of SAW force sensors.

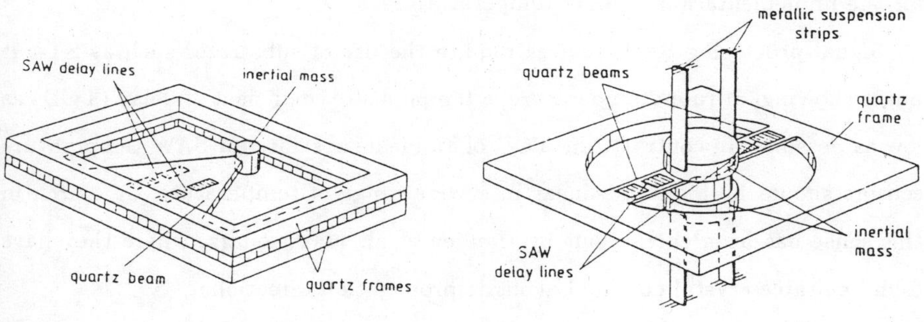

Figure 10: Schematic diagram of SAW accelerometer. (see [3])

Finally, we have to mention a group of SAW devices such as accelerometers and rotation rate sensors, which are based on the Coriolis acceleration produced on the material particles of the propagation medium by their elliptical movement associated with the propagation of the surface acoustic wave [43-45].

1.4 - Magnetic Sensors

The propagation of SAW's along magnetic substrates is sensitive to the presence of magnetic fields. SAW magnetic field sensors have been tested, as obtained by coating the propagation path of a SAW delay line with a magnetostrictive film, or by implementing the acoustic line on a megnetic garnet film.

1.5 Temperature sensors

It is well known that the phase velocity of acoustic waves depends on the temperature of the propagation medium. This effect, which is undesirable in signal processing SAW devices as it affects their performances, can be usefully exploited for the implementation of SAW temperature sensors.

Signal processing SAW devices require the use of substrates such as ST-cut quarz showing, at room temperature, a temperature coefficient of delay (TCD) as low as possible. In contrast, the TCD of an ideal substrate for SAW temperature sensors should be high and linear in a wide range of temperature. A search in this sense has been carried out by Hauden et al. [1] on quarz, where they have found suitable crystal cuts and acoustic propagation directions.

2 Plate mode sensors

The use of acoustic Lamb waves which propagate along thin plates can show several advantages in acoustic sensor applications. In the following we describe two examples of acoustic sensors which are based on the propagation of Lamb modes. The first one is concerned with an electric sensor; here the use of a thin plate as a substrate is suggested by the desire to obtain high electric field intensities at low drive voltages. In addition, the presence of a large number of acoustic modes gives a wide choice of interaction conditions which increase the possibility of improving the overall performances of the device.

As known, the propagation of acoustic Lamb modes, in a plate of thickness h can be described in the $(\omega h,\ Kh)$ plane by the dispersion curves. These, for the specific case of propagation along the z-direction of a y-cut $LiNbO_3$ plate are shown in Fig.11.

Lamb waves in piezoelectric media can be excited and detected by interdigital transducers which are effective in transducing all the acoustic modes corresponding to the intersection in the $(\omega h,\ Kh)$ plane, of the dispersion curves with the straight line given by the specific value of the Kh product which depends on the periodicity of the transducer and on the thickness of the plate.

Each mode can be selectively excited by a proper choice of the operation frequency. The strength of the nonlinear electroelastic effect, as measured in the experimental conditions shown in Fig.11 is reported in Fig.12 for the first acoustic 25 modes. As it can be seen from the figure, the stength of the interaction strongly depends on the mode considered. In particular, the mode at frequency $f = 22.7\ MHz$, show a strength which is about 6 times higher than that of the Rayleigh wave (dashed line).

A second example of Lamb wave sensor is due to White et al [34,37] and consists of the basic structure shown in Fig.13 which can be made sensitive to various different parameters. The structure is obtained by anisotropic chemical etching of a $Si/Si^{p++}/SiO_2$ substrate. A piezoelectric ZnO layer is used for the electromechanical trasduction. The device, as shown in Fig.13, is highly sensitive

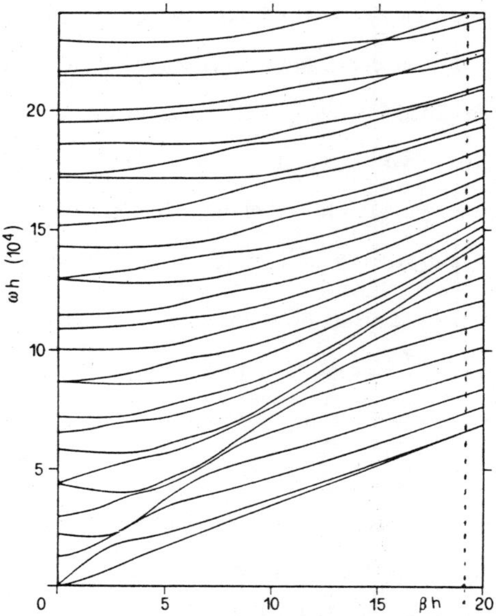

Figure 11: Dispersion curves for Lamb wave propagation along the z-direction of a y-cut $LiNbO_3$ plate.

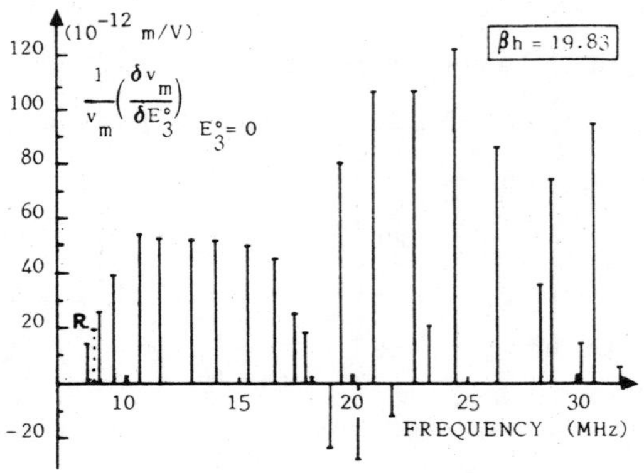

Figure 12: Strength of the nonlinear electroelestic interaction for Lamb waves propagating in the experimental conditions shown in Fig.11 (continuous lines) and for Rayleigh wave propagation (dashed line).

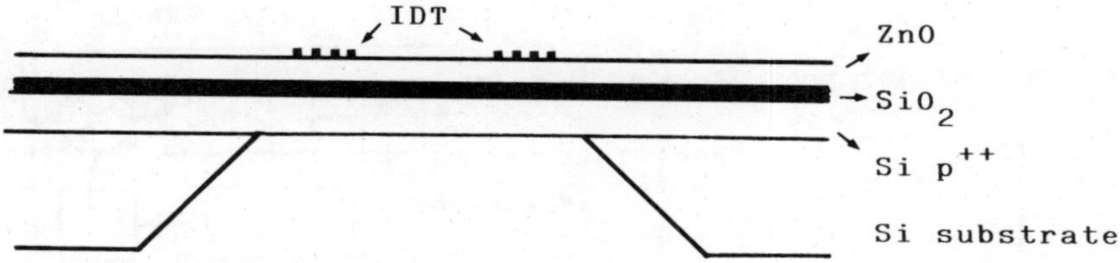

Figure 13: Basic structure of a thin membrane acoustic sensor.

to pressure differences between the two free surfaces (pressure sensor). When coated by a proper chemical interface, it operates as a chemical sensor. For this application, the coating is commonly applied to the lower surface, in order to allow a protection of the delicate trasductor region from the enviroment. A promising feature of this device is due to the extremely low value of the phase velocity of the first antisymmetrical mode (200 m/s or less) which is observed for a specific range of values of the parameter $K\ h$. This can allow the operation of the device in a liquid enviroment without introducing severe propagation losses due to acoustic mode convertion in liquid, so to implement an acoustic sensor of ion concentrations in liquids.

3 Techniques for measuring velocity

Most acoustic wave sensors are based on detecting the change in phase velocity produced by the measurand. In this section we consider the methods which can be used to measure phase velocity changes. Two methods are commonly used: the phase shift and the frequency shift detection technique.

In the phase shift detection technique, shown in Fig.14a, the phase ϕ of the signal at the output of the acoustic delay line is compared with that of the

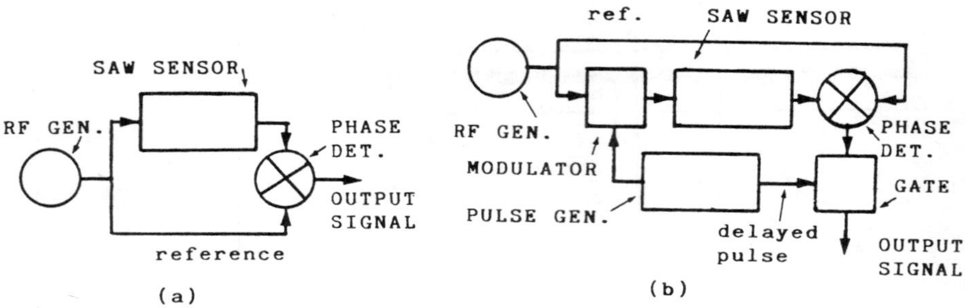

Figure 14: Block diagram of the phase shift detection technique for CW operation (a) and pulsed operation (b).

reference signal. We have:

$$\phi = 2\pi\frac{l}{\lambda} = 2\pi\frac{l}{v_{ph}}\rho \tag{1}$$

l being the length of the acoustic propagation path, v_{ph} and λ the phase velocity and the wevelength at frequency f, respectively. Any change Δv_{ph} in the velocity give rise to a change $\Delta\phi$ in the phase which is given by:

$$\frac{\Delta\phi}{\phi} = -\frac{\Delta v_{ph}}{v_{ph}} \tag{2}$$

The phase shift detection technique needs the use of a highly stable frequency source. It is suitable, however, for laboratory test measurements as the frequency can be freely changed and the measure can be carried out in pulsed operation (Fig.14b) so to cancel the effects due to electric feedthrough between trasducers and multiple transit signals. The frequency shift detection technique, which is commonly preferred in the operation of SAW sensors is shown in Fig.15. Here the acoustic sensor is connected to a reaction amplifier so to configure a SAW delay line oscillator. Oscillation require that both phase and gain loop conditions are satisfied. The phase shift around the loop, which is very close to that of the acoustic lines given by (1), must be an integral multiple of 2π radians. This

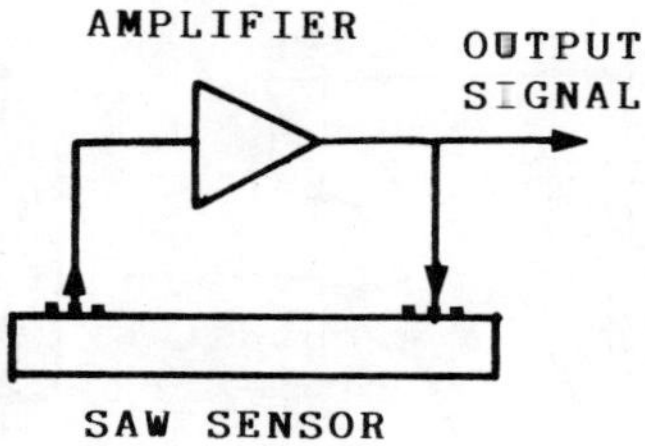

Figure 15: Block diagram of the frequency shift detection technique.

condition gives the comb of frequencies

$$\rho = n\frac{v_{ph}}{l} \qquad (n = 1, 2, 3, \ldots) \tag{3}$$

Oscillation take place at the frequency for which unity gain loop condition is satisfied; this is related to the bandwidth characteristics of both the trasducers and the amplifier. Here the relation between frequency and phase velocity change is:

$$\frac{\Delta\rho}{\rho} = \frac{\Delta v}{v} \tag{4}$$

Frequencies can be measured with a high accuracy (1 Hz or less); this allows at frequency of 100 MHz to detect a relative change in the phase velocity of 10^{-8}. Frequency shift detection technique is very attractive since it allows a high sensitivity in the measurament of the changes in phase velocity; only a simple and unexpensive reaction amplifier is required and finally the output signal (frequency) is readily suitable for digital data acquisition systems.

4 Thermal compensation techniques

An ideal sensor should be at the same time highly sensitive to the measurand and completely intensitive to all the other parameters. This is not true in a real sensor where a more or less pronounced response is observed to changes of many different

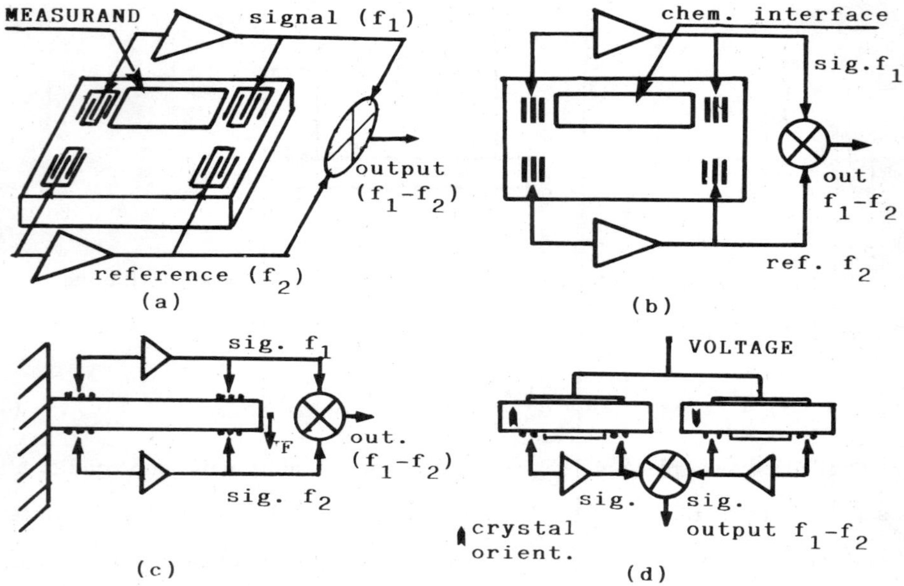

Figure 16: Differential structures for thermal compensation of the drift in the reference value. Principle of operation (a) and structures for chemical (b) mechanical (c) and electrical (d) sensors.

variables. As a consequence of this, the dynamic of the device is limited in the detection of the measurand, at low levels, where the signal has to come out from the noise which is due to the response to fluctuations of the other parameters in the considered conditions of operation. Temperature changes is one of the most common error source in SAW sensors, where it can affect the response, according to two different mechanisms: the former is a drift in the reference level, which is due to the temperature dependence of the acoustic phase velocity, the latter is a change in the calibration of the device which can be due to the dependence from the temperature of the strength of the interaction between measurand and acoustic wave.

Drifts in the reference value, are commonly compensated through use of differential structures, which, for SAW sensors based on measurament of the phase velocity by means of the frequency shift technique, are shown in Fig.16. The mothod relies on the use of two devices which are located very close each other, so to operate at the same temperature, and with the measurand applied only to one of them (Fig.16a). The output signals at frequency f_1 and f_2 are mixed together and the down converted signal at frequency $f_1 - f_2$ is considered. Temperature changes are supposed to affect f_1 and f_2 of the same amount so that their difference is independent from temperature. The measurand, on the contrary, affects only f_1 so that the corresponding shifts can be detected at the output. An exact compensation would require the use of the same acoustic structure for the two devices and their operation under the same enviromental and frequency conditions. These requirements can be only approximately matched in real devices. The worse condition is that of chamical sensors, where the use of the same acoustic structure for the two lines and their exposure to the same enviroment is not possible. Here, differential structures are commonly obtained by implementing two acoustic delay lines on the same substrate and by coating with the chemical interface the propagation path of one only of them (Fig.16b). Conditions are more suitable in mechanical (Fig.16c) and electrical (Fig.16d) sensors, where the response to the measurand can be easily changed in sign. On using these configurations, both the acoustic lines contribute to the overall response, which is two times higher than that of the single sensor. An improvement in the linearity of the response is observed too, which is due to the cancellation of the even terms of the polinomial expansion describing the relation between frequency shift and measurand amplitude.

5 Conclusions

In conclusion, a review on surface acoustic wave sensors which are appeared in the last few years is reported and their operation is described. SAW's seems a promizing technique on which to base sensors because of the many different mechanisms of interaction between the measurand and the acoustic propagation that can contribute to the sensor response. In addition, the confinement of the acoustic energy to a thin near surface region, makes the response of the device rather high and fast.

The operation of SAW sensors has been described as divided by the variable sensed: mechanical, electrical, magnetic, chemical and thermal sensors have been considered. Sensors based on the propagation of surface waves in thin layers or in membranes have been considered too. Finally readout methods for the measurament of the changes in the phase velocity have been reported and differential techniques for the compensation of reference level drift due to temperature are described.

References

[1] D. Hauden, G. Jaillet, R. Coquerel, "Temperature sensor using SAW delay lines", 1981 IEEE Ultrasonics Symposium Proc., 148, 1981

[2] X.Q. Bao, W. Burkhard, V.V. Varadan, V.K. Varadan, "SAW temperature sensors and remote reading system", 1987 IEEE Ultrasonics Symposium Proc., 148, 1987

[3] P. Hartemann and P.L. Meunier, "Surface acoustic wave accelerometers", 1981 IEEE Ultrasonics Symposium, 152, 1981

[4] E.J. Staples, J. Wise and R.E. De Wames, "Pressure and acceleration sen-

sitivity of a SAW interferometer", 1981 IEEE Ultrasonics Symposium, 155, 1981

[5] P.L. Meunier and P. Hartemann, "Cantilver-beamed SAW accelerometers", 1982 IEEE Ultrasonics Symposium, 299, 1982

[6] D. Hauden, F. Bindler, R. Coquerel, "SAW cantilever beam accelerometer sensitivities", 1985 IEEE Ultrasonics Symposium, 486, 1985

[7] D. Hauden and B. Loewenguth, " Acceleration sensitivities of SAW's propagating on a cantilever quartz beam", Appl.Phys.Lett 47, 1271, 1985

[8] J.F. Dias, "Physical sensors using SAW devices", Hewlett-Packard Journal, 18, Dec. 1981

[9] T.M. Reeder, D.E. Cullen, M. Gilden, "SAW oscillator pressure sensors", 1975 IEEE Ultrasonics Symposium Proc., 264, 1975

[10] D.E. Cullen and G.K. Montress, "Progress in the development of SAW resonator pressure transducers", 1980 IEEE Ultrasonics Symposium Proc., 696, 1980

[11] M.R. Risch, "Precision pressure sensor using quartz SAW resonators" Sensors and Actuators, 6, 127, 1984

[12] S.G. Joshi, "A temperature compensated high voltage probe using surface acoustic wave", 1982 IEEE Ultrasonics Symposium Proc., 317, 1982

[13] R. Inaba, Y. Kasahara and K. Wasa, "An electrostatic voltage sensor using SAW's", 1982 IEEE Ultrasonics Symposium, 312, 1982

[14] E. Gatti, E. Palma, E. Verona, "A surface acoustic wave voltage sensor", Sensors and Actuators, 4, 45, 1983

[15] E. Gatti, E. Palma, E. Verona, "A SAW probe for time variable high voltage measurements", 1983 IEEE Ultrasonics Symposium Proc., 331, 1982

[16] S.B. Hanna, "Magnetic field sonsors based on SAW propagation in magnetic films", IEEE Trans. Ultras. Ferroelectr. Freq. Contr. UFFC-34, 191, 1987

[17] H. Wohltjen, R. Dessy, "Surface acoustic wave probe for chemical analysis: part I -Introduction and instrument description, part II -Gas chromatography detector", Analytical Chemistry, 51, 1458, 1979

[18] A. Bryant, D.L. Lee, J.F. Vetelino, "A surface acoustic wave gas detector", 1981 IEEE Ultrasonics Symposium Proc., 171, 1981

[19] A. D'Amico, A. Palma, E. Verona, "Palladium surface acoustic wave interaction for hydrogen detection", Appl.Phys.Lett 41, 300, 1982

[20] A. D'Amico, A. Palma, E. Verona, "Surface acoustic wave hydrogen sensor", Sensors and Actuators, 3, 31, 1982

[21] A. D'Amico, A. Palma, E. Verona, "Hydrogen Sensor using a palladium coated surface acoustic wave delay line", 1982 IEEE Ultrasonics Symposium Proc., 308, 1982

[22] A. Bryant, M. Poirier, G. Riley, D.L. Lee and J.F. Vetelino, "Gas detection using surface acoustic wave delay lines", Sensors and Actuators, 4, 105, 1983

[23] A.W. Barendesz, J.C. Vis, Nieuwenhnizen, E. Nieuwkoop, M.J. Wellekoop, W.J. Ghijsen, A. Venema, "A SAW-chemosensor for NO_2 Gas-concentration measurament", 1985 IEEE Ultrasonics Symposium Proc., 586, 1985

[24] A. D'Amico, M. Gentili, P Verardi, E. Verona, "Gas sensor based on improved SAW devices", 2nd International Meeting on Chemical Sensors, Bordeaux, 743, 1985

[25] J.F. Vetelino, R. Lade, R.S. Falconer, "Hydrogen Sulfide surface acoustic wave gas detector", 1986 IEEE Ultrasonics Symposium Proc., 549, 1986

[26] E.T. Zellers, R.M. White and S.W. Wenzel, "Computer modelling of polimer-coated ZnO/Si surface acoustic wave and Lamb wave chemical sensors", Sensors and Actuators, 14, 35, 1988

[27] S.J. Martin, A.J. Ricco, D.S. Ginley, T.E. Zipperian, Hanna, "Isothermal measuraments and thermal deposition of organic vapours using SAW devices", IEEE Trans. Ultras. Ferroelectr. Freq. Contr. UFFC-34, 142, 1987

[28] A. Venema, E. Nieuwkoop, M.J. Wellekoop, W.J. Ghijsen, A.W. Barendesz, M.S. Nieuwenhnizen, "NO_2 Gas-concentration measurement with a SAW-chemisensor", Ibid, 148, 1987

[29] J.F. Vetelino, R.K. Lade, R.S. Falconer, "Hydrogen Sulfide surface acoustic wave gas detector", Ibid, 156, 1987

[30] H.Wohftien, A.W. Snow, W.R. Barger, D.S. Ballantine, "Trace chemical vapor detection using SAW delay line oscillator", Ibid, 172, 1987

[31] A. D'Amico, A. Petri, P. Verardi, E. Verona, "NH_3 surface acoustic wave gas sensor", 1987 IEEE Ultrasonics Symposium Proc., 633, 1987

II. CONTRIBUTED PAPERS

A. Field reconstruction and imaging

Signal Processing for Acoustic Field Inversion and Imaging in Ultrasonography

G. Jacovitti*, G. Scarano**

*INFOCOM Dpt., via Eudossiana 18, I00184 Rome, Italy

**CNR, Istituto di Acustica, via Cassia 1216, I00189 Rome, Italy

Introduction

Let us consider the general problem of observing structures embedded into a volume by taking measurements on the surface sorrounding this volume.

For this purpose it is necessary to measure some physical entities related to the observed structures. These entities can be the values of the electromagnetic or elastic field propagating throgh the volume. In some cases, the structures constitute the sources of the field, but often they are passive and cause only modification of external fields by reflecting or diffracting incoming waves. In other words, passive structures can be observed by exciting or *illuminating* them with electromagnetic or acoustic radiation, and by measuring the returned waves.

The basic condition for observing structures by this technique is that the medium be sufficiently transparent to the employed radiations. Thus, the frequency (or the wavelength) of the illumination must be chosen in appropriate bands depending on the nature of the medium filling the given volume.

The observation is made throgh a set of sensors or *transducers* placed on the allowed surface of the volume. The sensors translate the measured field into a set of time functions, or signals. This set of signals can be seen as one multidimensional signal, *i.e.* function of both the time and the spatial position of the sensor. This set of signals is entered in a processor which attempts to locate the single points of the observed structures, thus giving an image of their profiles.

The propagation from the structures to the observation surface tends to spread the geometrical information in time and space. Conversely, the processor must compensate the propagation effects by recollapsing the waves by some *inverse propagation* operation.

Ideally, the processor should proceed backward both in time and space up to the time and location of the diffraction events on the structures.

This is not physically possible, but the processor can simulate such a back-propagation on the basis of some mathematical model of the wavefield propagation in the medium. The accuracy of the this operation depends on how well the model fits the actual phenomenon. There are different models for the propagation phenomenon, and different methods for the inversion.

In principle, mathematical physics tells us that, if we know the response of the medium to a point source (the so called *Green's function*) and if we measure the wavefield on the whole surface of a volume sorrounding the structures using infinitely broadband signals, then we can completely reconstruct the wavefield into the volume, and perfectly locate the structures.

In practice this is not possible because:

- we can use only band-limited signals. In other word the *frequency aperture* of the system is limited. Also the *time aperture* is limited.
- we can collect signals on a limited portion of the surface. The *surface aperture* of the system is limited.
- the sensors usually sample the surface field in a discrete set of points into the aperture. We say that the observations are sampled in the space domain. They are also sampled in the time domain.

- we do not know the Green's function exactly.

These considerations lead to the conclusion that the inversion problem admits only **partial** solutions.

Nevertheless, satisfactory technical results are currently obtained in several inversion applicative fields, both in the acoustic and in the electromagnetic enviroments.

In the acoustic application we cite:

- The geophysical prospecting, for observing structures into the earth in the band 10-100 Hz.
- The seismic marine bottom profiling, in the band from 1 KHz to 10 KHz.
- The sonar for underwater object location, in frequencies ranging from 100 Hz to 1 Mhz.
- The acoustical echography, for non-destructive evaluation and for medical diagnosis, in bands from 1 MHz to 10 MHz.
- The acoustic microscopy, in bands from 100MHz to 2GHz

In the electromagnetic applications we recall:

- The RADAR, in frequencies ranging usually from 1 GHz to 100 GHZ.
- The Synthetic Aperture Radar (SAR), for imaging the earth from flying platforms, operating in the same RADAR bands.
- The radiotelescopes, for passive deep space source location, in bands over 30 GHz to 300 Ghz (EHF).
- The optical systems, in the infrared and visible bands.
- The electronic microscope, in the ultraviolet (UV) band.

In all the above applications, the inversion problem is substantially the same, apart from scale factors, so that similar techniques can be employed. In the current technical literature, these techniques are referred as *array processing techniques* because of the nature of the sensor (array) arrangements. However, it has to be noticed that (for instance) in optics continuous apertures (lens) are employed.

The Mathematical Model of Inversion

Let us refer to the simplest model of an ideal homogeneous non-dispersive medium. In this case, the Green's function, *i.e.* the response to a point source, is very simple. Consider a shot in the air. The energy is carried on a spherical wavefront emanating from the source,the radius r of which increases linearly with the time t elapsed from the shot. The proportionality factor is the medium propagation velocity c. Because of the energy invariance principle, if conversion phenomena do not occur, the amplitude of the shot is inversely proportional to r, so that we can say that the response $p(t,r)$ to a shot in the origin of space and time is:

$$p(t,r) = \frac{1}{r} w(t - \frac{r}{c}) \tag{1}$$

where $w(t)$ is the time function of the shot.

Let us consider now a wavefront on the sphere $r = r_0$, and a point at distance r_1. Clearly we must have

$$p(t,r_1) = \frac{r_0}{r_1} p(t - \frac{r_1 - r_0}{c}, r_0) \tag{2}$$

provided that the time function does not change.

On the other hand, if we attempt to calculate $p(t,r_1)$ from the contributions of all the points of the wavefront in r_0, following the Huygens principle we should have:

$$p(t,r_1) = \int_S g(t - \frac{\rho}{c}, r_0) dS \tag{3}$$

where $g(t, r_0)$ does not coincide with $p(t, r_0)$, but rather is related to its derivative. In fact ρ varies along the integration surface, so that $p(t, r_1)$ is a primitive of $g(t, r_0)$ apart from a geometrical weight.

This is a seemingly continuity paradox, when $r_1 \rightarrow r_0$. However, it is explained with the *evanescent wave* the amplitude of which decays as the inverse of the squared distance.

More generally, if the evanescent field is neglected , the Raylegh integral can be used to calculate the field at any point when the field is defined on a surface surroundig it. Let us refer to the plane indefinite surface $z = 0$, and to the half-space $z > 0$. Then the acoustic pressure in the point (x_0, y_0, z_0) is given by:

$$p(t, x_0, y_0, z_0) = \int\int \frac{1}{2\pi c} \frac{\alpha(\frac{z_0}{\rho})}{\rho} \frac{\partial}{\partial t}[p(t - \frac{\rho}{c}, x, y, 0)]dxdy \tag{4}$$

where $\rho = \sqrt{(x - x_0)^2 + (y - y_0)^2 + z_0^2}$ and α is on *obliquity factor*.

This formula can be used in two ways. Given the pressure distribution on the surface at $t = 0$, it foresees the resulting pressure in the considered half-space at $t > 0$. Conversely, returning back in time, it computes the pressure which originated the given distribution. If a shot from a point source did happen at $t = 0$, the Raylegh integral computed for $t = 0$ will give a function of space $p(x_0, y_0, z_0)$ with a sharp peak in the true position of the source.

Let us now interpret Eq.4 in a more explicit way. If we write

$$p(t - \frac{\rho}{c}, x, y, 0) = \delta(t - \frac{\rho}{c}) * p(t, x, y, 0) \tag{5}$$

then the Raylegh integral becomes a convolution (both in space and time) between the surface pressure $p(t, x, y, 0)$ and the time-space point source response

$$h_z(t, x, y) = \frac{1}{2\pi c} \frac{\alpha(\frac{z_0}{r})}{r} \frac{\partial}{\partial t}[\delta(t - \frac{r}{c}) \tag{6}$$

where $r = \sqrt{x^2 + y^2 + z^2}$.

Leaving out the time domain differentiation, this operator is a surface mass the shape of which is, in the (t, x, y) space an hyperboloid with apex in $\frac{r}{c}$.

So, we can interpret the relation between the fields in the plane $z = 0$ and $z = z_0$ as a time-space filtering. We may pass from the plane $z = 0$ to the plane $z = z_0$ and viceversa by the same operator, with time reversal.

We recall that the evanescent field is ignored in this scheme.

Let us refer now to the particular case when all the structures lie in the plane $y = 0$. The same equation holds in two spatial dimensions, and we have:

$$p(x, z_0, t) = p(x, 0, t) \overset{t\,x}{**} h_{z_0}(x, t) \tag{7}$$

This is a very common ideal case, widely referred to in echography, sonar, SAR, etc.

Summarizing, from a measure of $p(x, 0, t)$ we can reconstruct $p(x, z_0, 0)$ if we suppose that the structural point emits a wave at $t = 0$. The reconstruction is made by applying the inverse operator of $h_{z_0}(x, t)$ by means of a computing device.

The same operation can be applied for each depth z_0 by a suite of experiments.

A reference experiment consists of recording the signal $p(x, 0, t)$ when all the points of the structures simultaneously emit a pulse at $t = 0$. For the *explosive source* model, we can apply all the operators to the *time section* $p(x, 0, t)$, and record the values of $p(x, z_0, 0)$. Notice that each operator must be applied only to calculate the output at $t = 0$ and then changed.

The result of this operation is the *depth section* $p(x, z_0, 0)$, which represents an image of the structures.

To fit actual situations, we must take into account that the structures must be excited to emit a wave. This can be done (and is done) in several ways.

One way is to excite the medium with a plane downgoing wave. The *time section* is the same as in the explosive source model, but with the time values doubled (or the medium velocity halved). In fact we must to take into account the forward and the backward travel times.

Let us refer to the physical situation depicted in Fig.1.

Here, an array of piezoelectric transducers spaced by the sampling interval d, is placed at $z = 0$. The transducers are fired simultaneously in *transmission mode*. Then they are switched to *receiving mode* to record the echoed waves. In practical echographic arrangements, the experiment is repeated shifting the transducers by a quantity d sequentially. The inversion operator is applied in real time for each position of the array. In other words, the image is built by scanning the z_0 axis for different successive values of x_0: $x_0 + d$, $x_0 + 2d$, $x_0 + 3d$, etc, just as in a TV image rotated by 90 degrees (linear B-mode scanning technique).

Another method, widely employed in SAR, consists in using one transducer at time, either in transmission and receiving mode. In this case the obtained *time section* differs from the reference explosive source model because not only the time axis must be doubled but also the x axis. In fact we must account for the lateral travel time of the transmitted wave, which doubles the effective distance of the transducer with respect to the observed point. In this case, the excitation differs also because of different directions of exciting waves, which cause different reflection effects on the structures. Other methods make use of different sets of transducers for transmission and receiving modes. The distance between the sets on the x axis is called *offset*. The techniques described above are *zero offset*. Non zero-offset procedures can be employed to obtain oblique views of the structures.

The most complete techniques for a given array is to fire one transducer at a time and to receive with all the transducers in such a way to generate a set of time sections. This method produces a very large amount of data for imaging, and is employed in geophysical prospecting.

In the following, we will refer to the first two methods, which are of interest in echographic ultrasonography.

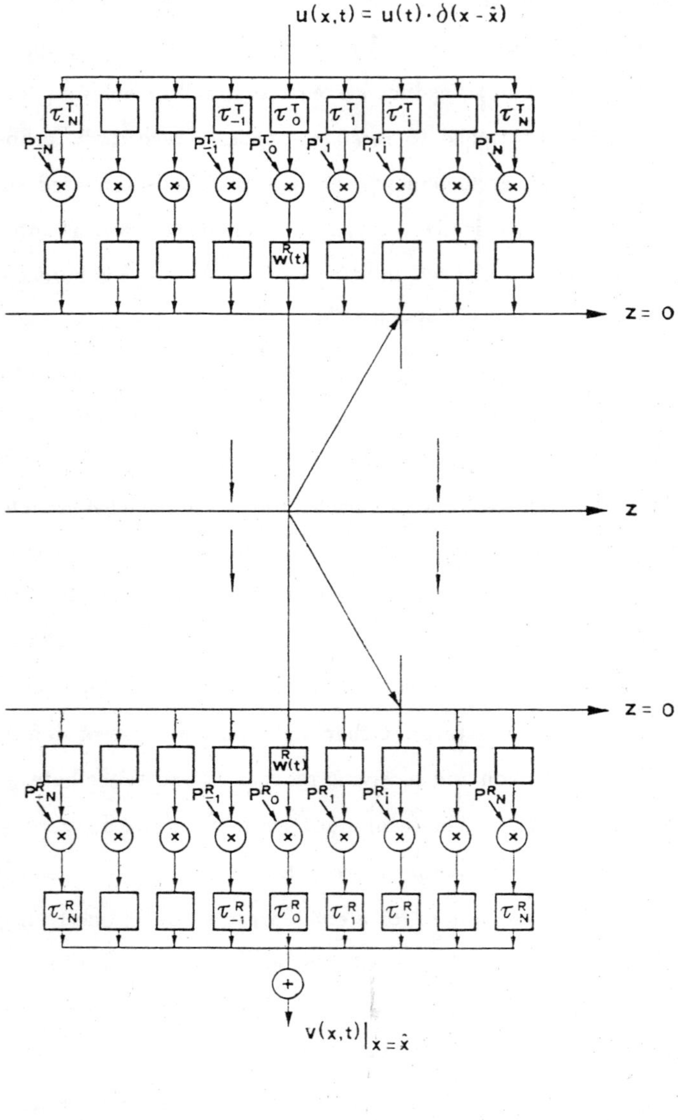

Figure 1: *The echographic B-mode scan system.*

Signal Processing in Echographic Systems

Let us refer again to the transducer arrangement depicted in Fig.1.

We assume that its geometry does not vary with time and displacement along the x axis. Further, we will admit that the system be linear, so that the superposition rule applies.

As a consequence, the whole system can be modeled as a cascade of some time-space linear shift invariant operators.

In Fig.2 the block diagram of the echographic system is depicted. A probe

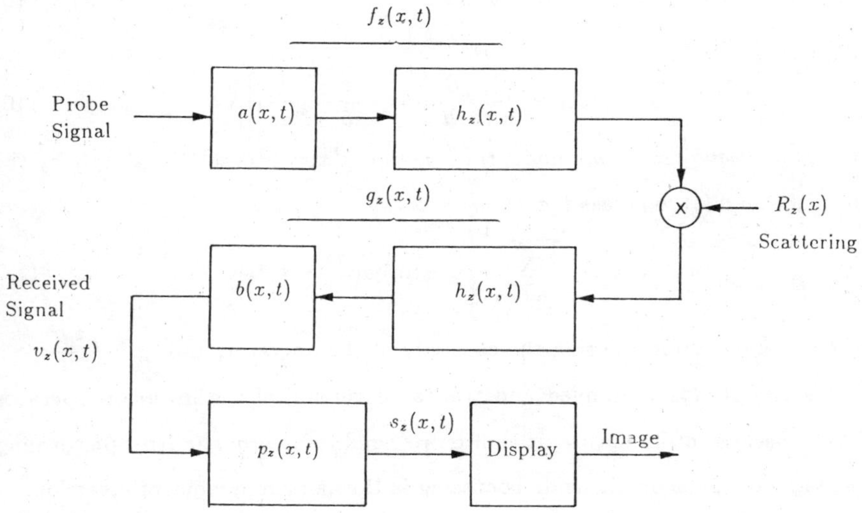

Figure 2: *Block diagram of the echographic system using linear operators.*

signal excites the transducer system in transmission mode, represented by $a(x,t)$. The exciting wave reaches the structures throgh the medium response $h_z(x,t)$ which depends on the depth z. The reflected waves are subject to the same transformation $h_z(x,t)$ returning back to the transducers set in receiving mode.

At the output of the transducers, the received signal $v_z(x,t)$ is then processed to give an image of the structures.

The response of the medium is directly obtained from the Rayleigh integral:

$$h_z(x,t) = \frac{\Delta}{2\pi c}\frac{1}{\rho}\frac{\partial}{\partial t}\left[\delta(t-\frac{\rho}{c})\right]\alpha(\frac{z}{\rho}) \tag{8}$$

where Δ is the linear physical size of the transducer in the y dimension, $\rho = \sqrt{x^2+z^2}$ and $\alpha(\frac{z}{\rho})$ is the directivity factor (tipically $\alpha(\frac{z}{\rho}) = \frac{z}{\rho}$).

Letting for the sake of simplicity $\alpha(\frac{z}{\rho}) = 1$, and applying the Fresnel approximation:

$$\rho \simeq z + \frac{x^2}{2z} \tag{9}$$

we have:

$$h_z(x,t) \simeq \frac{\Delta}{2\pi cz}\frac{\partial}{\partial t}\left[\delta(t-\frac{z}{c}-\frac{x^2}{2zc})\right] \tag{10}$$

In the *plane wave excitation* mode, the function $a(x,t)$ is described simply by the time function $w(t)$, whereas $b(x,t)$ is

$$b(x,t) = \sum_{n=-N}^{N} \delta(x-nd)w(t-\tau+\tau_n) \tag{11}$$

where τ is a delay for assuring the causality of the operator, and $\tau_n = \frac{n^2d^2}{2zc}$.

This operator is a sampled and truncated version of the inversion operator $h_z(x,t)$. Because of its nature, it is often referred to as *acoustic lens*, performing a focussing of the acoustic field. Focussing is thus a synonymous of inversion.

Notice that the inversion is directly performed by the lenses, so that the post-processor $p_z(x,t)$ reduces to $\delta(x,t)$.

The evaluation of the overall system response gives, in the time frequency domain:

$$S_z(\omega,x) = \left(\frac{\Delta}{2\pi cz}\right)^2 [j\omega W(j\omega)]^2\, e^{-j\omega(\tau+\frac{2z}{c})}(2N+1)\left\{\frac{sin\left[\frac{A}{2z}x\frac{2\pi f}{c}\right]}{\frac{A}{2z}x\frac{2\pi f}{c}}\right\} \tag{12}$$

where the aperture $A \doteq (2N+1)\cdot d$. The response in the time space frequencies is shown in Fig.3.

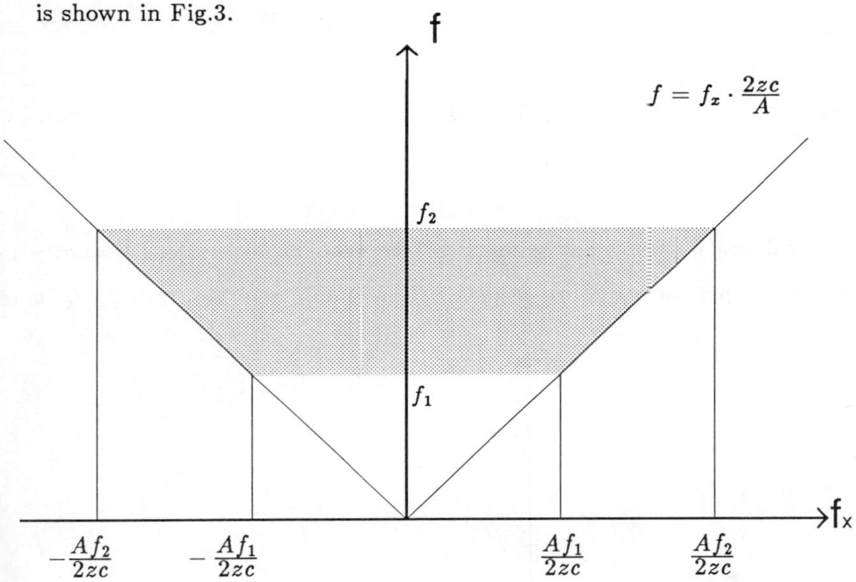

Figure 3: *The DF Echographic system response in the space-time frequencies domain.*

Thus, we can say that an echographic system sees the observed structures throgh a *directional filter*. This filter drops out the frequency components outside its bidimensional bandwidth. This fact prevents the complete field inversion and a perfect reconstruction of the structures, the information of which is spread onto the whole (f, f_z) plane. Moreover, the sampled nature of the operator may cause aliasing effects.

It is interesting to see the same response in the frequency domain of the output image. To compute this response, let us refer to the *dispersion equation*, which relates the spatial frequencies f_x^2, f_z^2 to the time frequency f of a wavefield:

$$f_x^2 + f_z^2 = \frac{f^2}{c^2} \tag{13}$$

Thus, in the plane f_x, f_z the spectrum is bounded by the curves:

$$f = f_1 \qquad \Rightarrow \quad f_x^2 + f_z^2 = \frac{f_1^2}{c^2}$$

$$f = f_2 \qquad \Rightarrow \quad f_x^2 + f_z^2 = \frac{f_2^2}{c^2}$$

$$f = \pm\frac{2zc}{A} f_x \quad \Rightarrow \quad f_z = \pm f_x \sqrt{(\frac{2z}{A})^2 - 1}$$

The typical spatial frequency window of the echographic system is shown in Fig.4. Again, perfect structure reconstruction should require a full f_x, f_z band

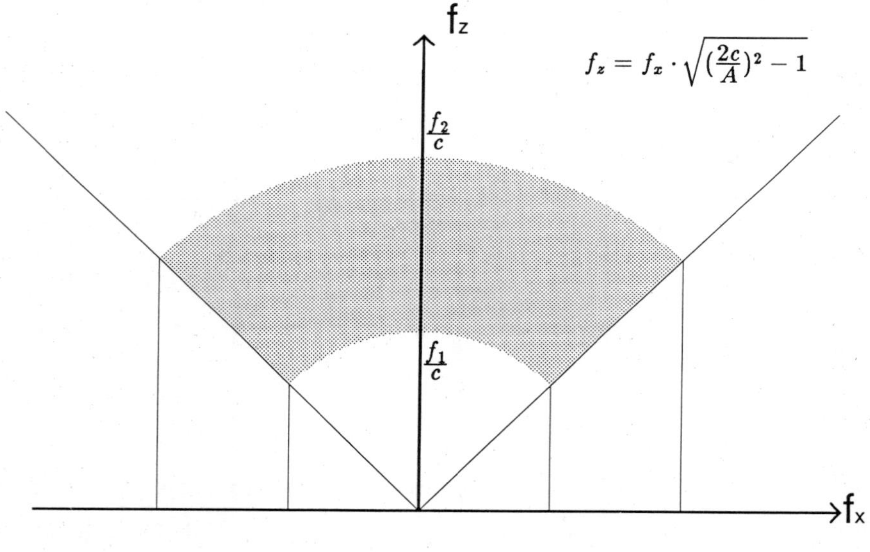

Figure 4: *The DF Echographic system response in the spatial frequencies domain.*

observation.

In actual echographic apparatus, the inversion operator is adjusted in receiving mode during the running time t, posing the parameter $z_0 = \frac{ct}{2}$ (dynamic focussing). In practice, the adjustement is not continuous, but $h_{z_0}(x,t)$ is kept constant for small time intervals. Thus, the inversion is perfectly performed only in a discrete number of points in depth.

Moreover, to improve the overall response, the illuminating energy is focussed in transmission mode too. This means that the inverse operator of $h_{z_0}(x,t)$ is applied during the excitation. Unlike in reception, this operator cannot be adjusted, because of the pulsed nature of the transmission, so that a perfect transmission focusing can be obtained at one depth only. In some cases, different experiments are made for different depth ranges in order to obtain a more accurate transmission focussing. It is not difficult to compute the echographic response for this Double Focusing (DF) mode.

Letting $a(x,t) = b(x,t)$, the response modifies as follows:

$$S_z(\omega, x) = \left(\frac{\Delta}{2\pi c z}\right)^2 [j\omega W(j\omega)]^2 e^{-j2\omega(\tau+\frac{z}{c})}(2N+1)^2 \left\{ \frac{\sin\left[\frac{A}{2z} x \frac{2\pi f}{c}\right]}{\frac{A}{2z} x \frac{2\pi f}{c}} \right\}^2 \tag{14}$$

Due to the square in the x domain, the bandwidth in f_x is doubled, so that we obtain a better lateral resolution.

In Fig.5 the profile of the (ω, f_x) spectrum for a fixed frequency f_0 is displayed for both the plane wave and for the double focusing echographic response, normalized to their maxima.

An even better result is obtained with the syntethic aperture (SA) technique. In this case, we pose $a(x,t) = b(x,t) = w(t) \cdot \delta(x)$, *i.e.* we activate one point at a time in the transducers array. This means that the inversion operator must be applied in cascade with the transmission/reception operation. In other words we must record the time section and then apply an off-line inversion procedure. This inversion operator is:

$$p_z(x,t) = \sum_{n=-N}^{N} \delta(x - nd)\delta(t - \tau_n) \tag{15}$$

where now, Because of the stretching in the x axis, $\tau_n = \frac{n^2 d^2}{zc}$.

Thus, the response is the same as in the plane wave case, provided that x is

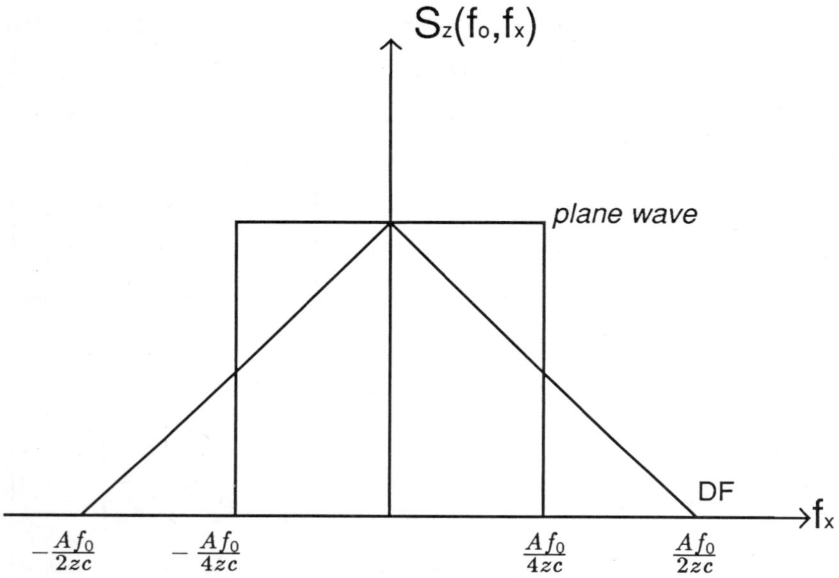

Figure 5: *Normalized profiles of the "plane wave" and the DF echographic system responses in the spatial frequency domain.*

replaced with $2x$:

$$S_z(\omega, x) = \left(\frac{\Delta}{2\pi cz}\right)^2 [j\omega W(j\omega)]^2 \, e^{-j2\omega(\tau+\frac{2z}{c})}(2N+1)\left\{\frac{sin\left[\frac{A}{z}x\frac{2\pi f}{c}\right]}{\frac{A}{z}x\frac{2\pi f}{c}}\right\} \tag{16}$$

As a consequence, in the plane f_x, f_z the occupation of the SA echographic system is the same as in the DF one. In particular, for a given frequency f_0, the normalized profiles alog f_x are compared in Fig.6.

It is evident that the high frequencies, corresponding to fine details in the x direction, are enhanced. This means that the lateral resolution is improved.

To better understand the difference of the two cases, let us observe the responses to a point reflector directly in the time-space domain in Fig.7.

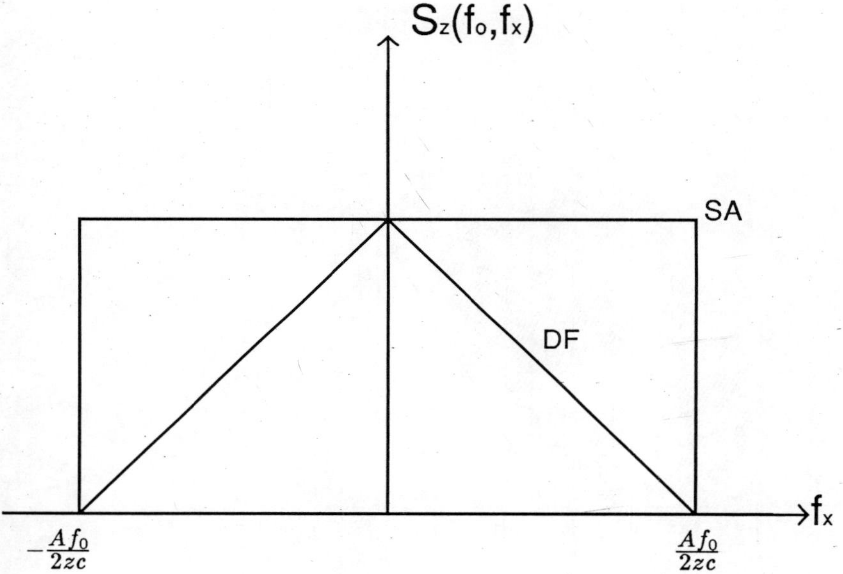

Figure 6: *Normalized profiles of the SA and the DF echographic system responses in the space frequency domain.*

We see that in the lateral direction the SA response is more sharp, even if some sidelobes are visible. The time resolution is the same, depending only on the time response of the transducers. Thus the depth resolution (which is determined by the time resolution) is the same for both systems.

Despite the superior resolution of the SA technique, the DF technique is employed in practice in medical echography. This fact is due mainly for the following reasons:

- the DF operates in line with acoustical lens, acting as a parallel high speed processor. This is a very efficient structure.

- the DF scans the space by focussing along one line at a time. Thus, a

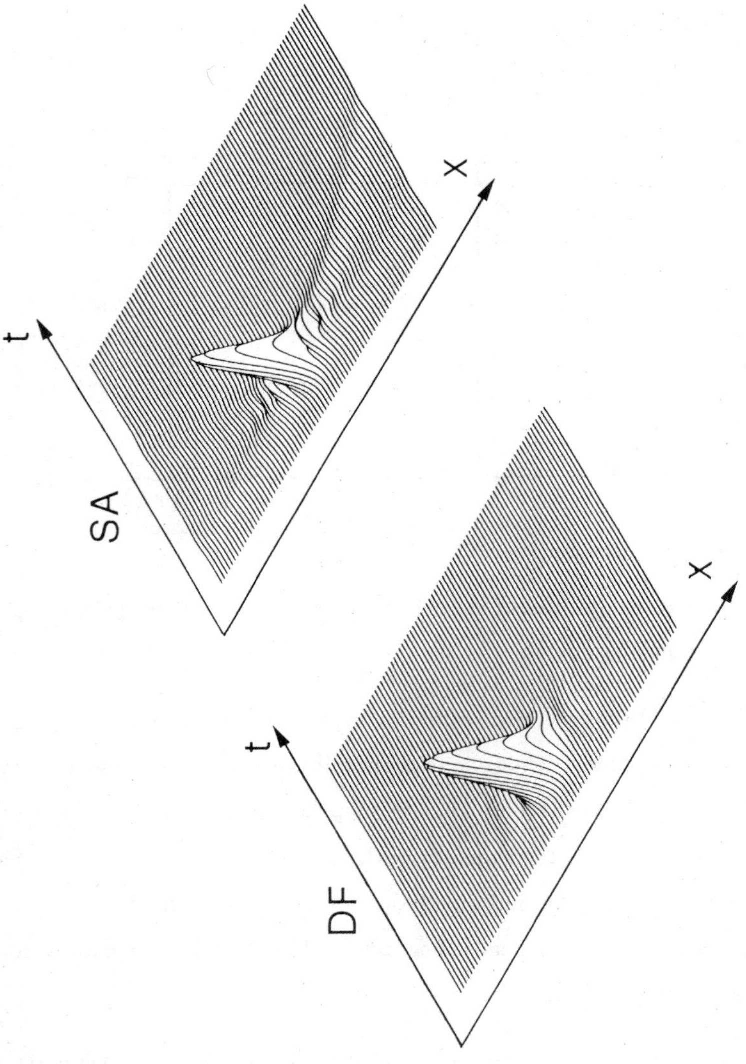

Figure 7: *The SA and the DF echographic system responses to a point reflector in the time-space domain.*

moving object is continously strobed during the lateral scan, unlike the SA case where each object is continoulsy observed, so that movement artifacts may occur. For instance this is relevant in hearth diagnostics.

- the DF technique illuminates the medium with the whole set of transducers into the aperture. this implies higher transmission power and higher signal to noise ratio of the output image.

However, it must be outlined that these reasons do not hold for echographic systems employed in non-destructive evaluation.

References

Jacovitti, G., Neri, A., Scarano, G., "Pulsed response of focused array in echographic B-mode systems", IEEE vol.SU-32, No.6 Nov. (1985)

Jacovitti, G., Neri, A., Scarano, G., "Ultrasonic B-mode scan imaging: a comparison with SAR technques", Digital Signal Processing-87, V.Cappellini and A.G.Costantinides eds, Elsevier Science Publisher B.V. (1987)

BROADBAND TOMOGRAPHY SYSTEM

L.Capineri*, G.Castellini**, L.Masotti*, S.Rocchi*
* Dipartimento di Ingegneria Elettronica
Universita' degli Studi di Firenze
V. di S.Marta,3 -50139 FIRENZE
** Istituto di Ricerca sulle Onde Elettromagnetiche C.N.R.
V. Panciatichi, 64 -50127 FIRENZE

Introduction.

The quantitative estimation of many pathological elements in tissue characterization gains an important clinical interest.

To validate the results coming from the traditional echographic equipment, a laboratory broadband system was built to compare different ultrasonic investigation techniques. The aim of the work is to estimate the ultrasonic limits.

It is possible to speak about limits, given two fundamental conditions which allow good results:
1) in "vitro" measurements,
2) front-end hardware with special features.

The first condition makes the measurement repeatable, the second one assures a good quality of the signals on which image formation is based.

The proposed methodology consists in comparing the results of three different techniques obtained with the same equipment.
The chosen techniques, reported below, are characterized by different physical assumptions:
1) Tomography based on time of flight and attenuation projections with straight-line propagation assumption,
2) Echotomography,
3) Diffraction tomography.

A good performance of the system is obtained not only taking care of the front-end hardware but supplying it with preprocessing algorithms implemented on DSP devices.

The developed algorithms are the envelope detection using Hilbert transform and the speckle reduction with Kalman filtering.

Finally, without speaking about theoretical considerations and numeric processing of the first two techniques [1],[2],[3] an original reconstruction algorithm on the diffraction tomography is presented. It is an image reconstruction directly obtained in time-space domain and based on the diffraction tomography. Numeric simulations point out the resolution and the experimental data give system performance.

Data acquisition system.

The images related to the three different techniques can be obtained with the same scanning and acquisition system. So the compared results have the same hardware characteristics.

The mechanical movements.

The method of data acquisition depends on the employed techniques, so a general set of possible movements is performed by the mechanical system. The positions of the transmitter and receiver are independently controlled by step motors and the resolution along the linear scans is 0.0083 mm, while the angular position of the object has the resolution of 1.8 °.

The front-end hardware.

The transmitter is a non focused probe with a central frequency f_c = 5.4 Mhz and Bw = 1.66 Mhz at 3 dB. The probe is excited with a double unipolar pulse generated by an excitation circuit employing an HEXFET device.

The receiving element is an hydrophone with Bw = 10 Mhz at 3 dB and a diameter = 1 mm. The received signal is preamplified with a low noise amplifier and sampled by a digital oscilloscope (GOULD MOD. 4072). The 8 bit samples are transferred by the IEEE 488 bus to the PC/IBM AT connected through an Ethernet link with the Department computing system.[4]

To verify the performance of the measurement system a linear scan of a test object has been made. The test object is a lattice membrane filled with 45% isopropylic alcohol solution embedded with a 2 mm diameter graphite stick. The graphite stick is resolved by the time of flight and attenuation profiles.

The attainable resolution is about 2 mm, that confirms the validity of measurement system.

Signal preprocessing block.

The numeric conversion at the radio frequency stage makes the system flexible and powerful. Many DSP routines can be implemented in quasi-real time. The envelope detection passing through Hilbert numeric transform is the main application [5],[6].

In fact signals obtained by using good transducers can not be considered narrow band signals. Therefore their envelope can't be obtained using rectifiers and simple smoothing filters. The only feasible solution is the direct calculation of the envelope from its mathematical definition, that is the magnitude of the analytic signal composed of the original radio frequency and its Hilbert transform.

It is also useful, before applying heavy post-processing techniques, to remove artifacts. They are introduced by interference phenomena of the coherent waves received from a single resolution cell, which contains too small structures to be resolved.

This leads to a random modulation, called " speckle ", of the interested region which should otherwise be displayed as an uniform gray level. For speckle reduction an adaptive one dimensional Kalman filter was implemented .

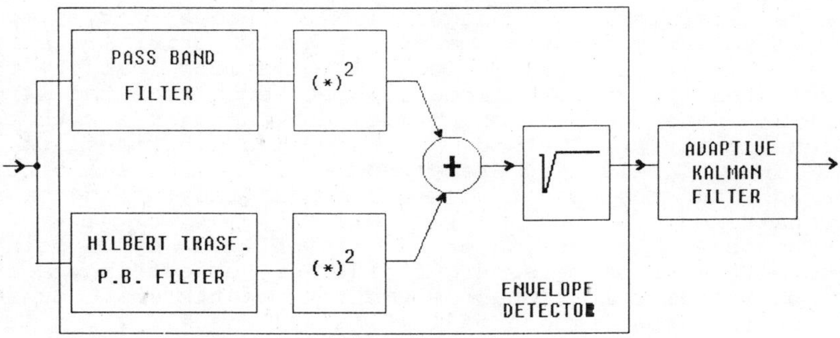

Figure - 1 - Signal preprocessing block.

The adaptivity of the Kalman filter results necessary to avoid loss of image resolution. The filter is applied to the envelope signal.

A suitable model for ultrasonic tracing based on the knowledge of the speckle first order statistic is developed [7].

The statistical features are extracted from zones which contain no discernable structures: zones of only speckle. The filter adaptivity for each point of the processed line has been obtained using a function of the local variation of the envelope signal. It results a monodimensional model which leads to fast execution of the filtering algorithm and easy implementation on the DSP devices for quasi-real time applications. The filtering is not affected by a-priori assignments of the necessary parameters, because the statistical features are extracted from the signals themselves [8].

In conclusion, an effective speckle's smoothing is obtained preserving the visual information associated with the investigated structures.

Signal processing for tomographic image reconstruction with broadband systems.

A tomographic technique employing wide band signals is presented.

A method to solve the inverse scattering problem is developed.

It is based on time of flight measurements that are correlated with unknown positions of the scatterers.
The method is direct, consequently it needs very simple software to carry out the images, without Fourier Transform and interpolations [9]. The informations about the investigated object are directly retrieved from the large amount of data collection, that are acquired with the complete data set method [10] applied in transmission mode. The signals are received over a large aperture compared with the object's size, so diffraction phenomena are considered.

The object in the water is insonified by a point-source with a pulsed pressure field. For "object" one means a 2D spatial distribution of elementary scatterers, bounded in a central area (square of l X l side, see fig 2).

The transmitter and the receiver move along two parallel lines, which distance is d (fig. 2).
The spatial sample is d_x (trasmitter) = d_y (receiver) = d/n, where n is the number of different transducer positions.

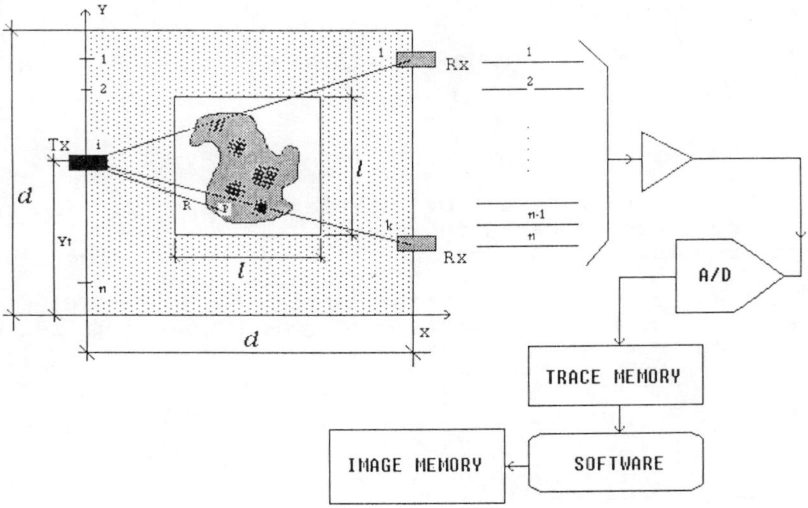

Figure - 2 -Data collection scheme

For each transmitter position, the receiver scans over the receiver line and for each couple of positions of the receiver and the transmitter, the signals from all the object's scatterers are sampled and stored (the traces are n^2). The same data acquisition process is repeated for the orthogonal direction. The proposed method takes into account the physical problem complexity of the ultrasound propagation in media with local inhomogeneities, like biological tissues.

However the validity of some approximations are assumed:

1) the scatterers are isotropic,
2) distance R is assumed such that the attenuation due to the propagation is the same for all the elements (the path differences are negligible),
3) Born approximation [11] is valid, so multiple scattering is neglected and the total field is given by the superposition principle.

After this first step, the inversion process is performed [12],[13].

The inversion process.

Let be the scattering function of the object $F(x,y)$ given by:

(1) $$F(x,y) = \sum_{1}^{N} f_k \, \delta(x-x_k, y-y_k)$$

where N is the number of scatterers and f_k is the scattering factor located in (x_k, y_k).
The transmitter in position $(0, y_t)$ is excited at time $t=0$, and $g(t)$ is the envelope of the pulsed incident field.
The incident field in (x,y,t) is defined by the solution of the wave equation for a homogeneous medium characterized by propagation velocity v :

(2) $$\nabla^2 P_i(x,y,t) - \frac{1}{v^2} \frac{d^2 P_i(x,y,t)}{dt^2} = g(t) \, \delta(x, y-y_t)$$

where the unperturbed solution is :

$$P_i(x,y,t) = g(t-R/v) / R \quad \{ g(t-R/v) = 0 \text{ for } t-R/v<0 \}$$

with $R = (x^2 + (y-y_t)^2)^{\frac{1}{2}}$

A scattered wave is emitted by each elementary scatterer with proportional amplitude to the incident field in (x_k, y_k) and f_k. The delay of this wave is given by $\tilde{t} = [x_k^2 + (y_k-y_t)^2]^{\frac{1}{2}} / v$.
The total field P received in the position (d, y_r) is given by:

(3) $$P(x=d, y=y_r, t) = P_i(x=d, y=y_r, t) + \sum_{1}^{N}{}_k P_i(x=d, y=y_r, t=\tilde{t}+t_k)$$

where $t_k = [(d-x_k)^2 + (y_k-y_r)^2]^{\frac{1}{2}} / v$

For each position the direct incident field is known, so the total scattered field P_s is given by the difference :

$$P_s = P - P_i$$

A family of ellipses with focusses in the position $(0, y_t)$ and (d, y_r) is defined by the function $e(x,y)$:

(4) $$e(x,y) = [x^2 + (y-y_t)^2]^{\frac{1}{2}} + [(x-d)^2 + (y-y_r)^2]^{\frac{1}{2}}$$

From equation (3),(4) it is possible to define the ellipses' family related to the delay $\tilde{t}+t_k$:

(5)

$$v\cdot(\tilde{t}+t_k)=e(x,y)*\delta(x-x_k,y-y_k)=\iint_S e(x,y)\cdot\delta(x-x_k,y-y_k)\,dx\,dy$$

$*$ ≡ spatial convolution

The quantity $\tilde{t}+t_k$ represents the time of flight of the scattered wave from the scatterer located in (x_k,y_k). The equation (5) defines the ellipse, on which the scatterer lies, for each couple of transmitter and receiver positions. The ellipse is the points' geometrical locus given by the possible positions of the scattering element. When data acquisition is completed, all the ellipses can be written in the image matrix with the pixel's weight equal to f_k. Only in the position (x_k,y_k) the weights are summed coherently.

The simulation of the method.

The performance of this method is obtained by computing the Point Spread Function (PSF) which gives the methods' attainable resolution.
In the simulation same assumptions are taken:
1) the trasmitter and the receiver are isotropic
2) the samples of the envelope g(t) have been calculated with an Hilbert filtering of the radiofrequency ultrasonic signal transmitted in water (Pulse width = 600 ns, Tsample = 100ns).

The simulation results are shown in figure 3.
In the right box the PSF of a single centered scatterer is presented. The other box aims at pointing out the method's resolution. In fact it shows the PSFs of two scatterers situated 2 mm apart, with about 25 dB dynamic range.

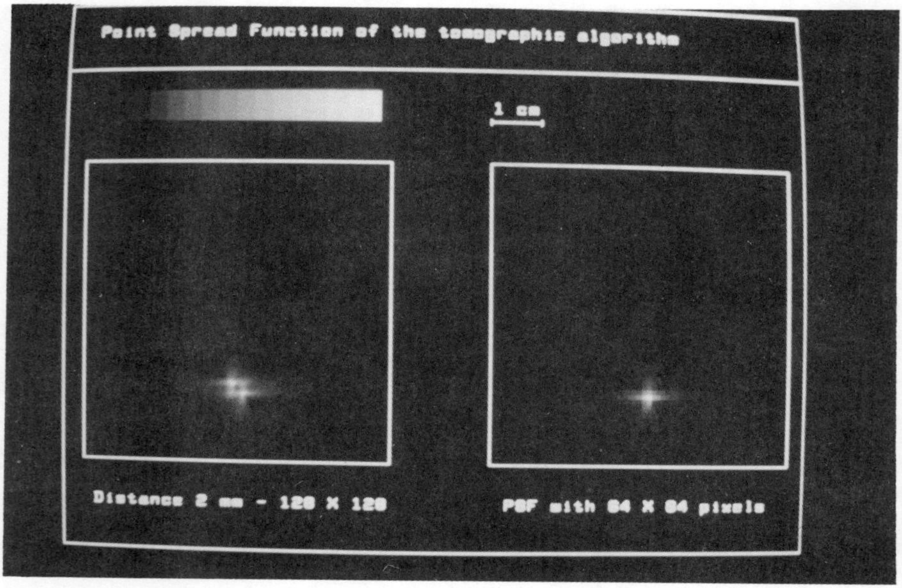

Figure - 3 -

Conclusions.

The work must be considered as a starting phase for the quantitative characterization of pathological elements in biological tissues.
The next steps will be:

1) Simulation based on more realistic models (non isotropic transducers and further conditions on the propagation medium)
2) Superresolution techniques using the suitable inverse bidimensional filters and noise rejection thresholds

At last, the experimentation on particular test objects will be performed to evaluate the performances and limits of the ultrasonic investigations.

REFERENCES

(1) Greenleaf J.F.,"Computerized tomography with ultrasound" Proc. of the IEEE vol.7,n.3,March 1983

(2) Lewitt R.M., "Reconstruction algorithm: transform methods" Proc. of the IEEE ,vol.71,n.3,March 1983

(3) Hiller D., Ermert H.,"System analysis of ultrasound reflection mode computerized tomography",IEEE Trans. on Sonics and Ultrasonics,vol.SU-31,n.4,July 1984

(4) Schafer M.E. ,Lewin P.A., "The Influence of front-end hardware on digital ultrasonic imaging", IEEE Trans. on Sonics and Ultrasonics , vol.SU-31 n.4, July 1984

(5) Gammel P.M., "Improved ultrasonic detection using analytic signal magnitude" Ultrasonics 19, 1981, pp. 73-76

(6) Bastogi A., Casini A., Castellini G., Constantinides A.G., Lotti F., Masotti L., Rocchi S., "Digital R.F. and video processing of echographic signals", Proc. of the third EC Workshop, Stuttgart, Nov. 1983

(7) Wagner R.F.,Smith S.W., Sandrik J.M., Lopez H., "Statistics of speckle in ultrasound B-scans", IEEE Trans. on Sonics and Ultrasonics, vol. SU-30,n.3, May 1983, pp.156-163

(8) Castellini G., Labate D., Masotti L., Rocchi S.,"Speckle reduction in ultrasound images by using an adaptive Kalman filter", IEEE Ninth Annual Conference of the Engineering in Medicine and Biology Society, 1987

(9) Nahamoo D., Pan S.X., Kak A.C., "Synthetic aperture diffraction tomography and its interpolation-free computer implementation", IEEE Trans. on Sonics and Ultrasonics, vol.SU-31,n.4, July 1984

(10) Berkhout A.J., Ridder Y., V.D.Wall L.F., "Acoustic imaging by wave field extrapolation" Acoustical Imaging vol.10, Plenum Press 1982, pp.513,565

(11) Soumekh M., Kaveh M., "Theoretical study of model approximation errors in diffraction tomography", IEEE Trans. on Ultr. and Ferr. and Freq.Contr. vol.UFFC-33,n.1, Jan.1983

(12) Azimi M., Kak A.C. "Distortion in diffraction tomography caused by multiple scattering", IEEE Trans. on Med. Imag., Vol. MI-2, n.4, Dec 1984

(13) Slaney M., Kak A.C., Larsen L.E., "Limitations of imaging with first order diffraction tomography", IEEE Trans. on Microwave Th. and Tech., vol. MTT-32, n.8, August 1984.

SHORT TERM REAL-TIME SPECTRAL ANALYSIS OF DOPPLER ULTRASOUND SIGNALS WITH A DIGITAL SIGNAL PROCESSOR AND MICROCOMPUTER

Schlindwein, F.S.[1,2] and Evans, D.H.[2]

1 - Federal University of Rio de Janeiro, Brazil, and

2 - Leicester Royal Infirmary, Dept. of Medical Physics & Clinical Engineering, Leicester LE1 5WW, ENGLAND

ABSTRACT

Doppler ultrasound velocimeters are widely used in the study of blood flow in arteries. Since the technique is noninvasive it permits normal as well as sequential studies. They work by detecting the change in frequency of a beam of ultrasound that is reflected or scattered from structures that are moving with respect to the ultrasound transducer. The Doppler shift frequency is proportional to the relative velocity between the source of reflection and the transducer. In the blood the main source of scattering is the red blood cells and because there are a large number of them within the sample volume and they move with a range of velocities, the Doppler shift signal is a wideband signal that is constantly varying with time. There are a number of alternative ways of processing the Doppler shift signal to obtain information about the evolution of velocities in the blood vessel, but the best is to perform real-time spectrum analysis and the most accepted form of displaying the velocity information is the spectrogram (sonogram).

This paper describes a system built for implementing real-time spectrum analysis of Doppler shift signals from arteries and the implementation of two different methods to calculate the sonogram of Doppler shift signals - the modified periodogram (based on the Fast Fourier Transform) and the autoregressive technique. The system is based on the Texas Instruments TMS320C25 Digital Signal Processor (DSP) chip and a Research Machines Nimbus, MS-DOS 80186 based, microcomputer. The audio signal is sampled at rates of up to 40.96kHz and the DSP is capable of outputting a 128 point power spectrum every 4.0ms. The sonogram is displayed in real-time on the computer screen. Details of the implementations of both the techniques will be discussed, as well as timing specifications of the spectrum analyzers, methods used to realize a system operating in real-time and comparative discussion of the two techniques for the application of blood velocity measurement.

1. INTRODUCTION

Doppler ultrasound, both pulsed and continuous wave, alone and in conjunction with ultrasonic imaging, is widely used as a noninvasive method for the assessment of blood flow. Doppler velocimeters work by detecting the change in frequency of a beam of ultrasound that is reflected or scattered from structures that are moving with respect to the ultrasound transducer. The Doppler shift frequency is proportional to the velocity of the blood within the sample volume, and since arterial blood flow is pulsatile, the Doppler shift signal has a spectrum that is constantly varying with time. Under ideal conditions the Doppler power spectrum has a similar shape to a histogram of the blood velocities within the sample volume, thus real-time spectral analysis of the Doppler signal produces information concerning the velocity distribution in the artery[1)].

The estimation of the power spectral density of a Doppler signal is normally performed by applying a Fast Fourier Transform (FFT) directly to the sampled signal and squaring the magnitudes of the output values. The signal is processed in individual frames of N samples (usually 128 or 256), and is usually weighted, using one of the many window functions known to minimize spectral leakage, before the FFT is calculated. This technique of transforming the data directly is normally referred to as the periodogram approach. The most common way of displaying the evolution of the spectral information is the Doppler sonogram, in which the Doppler shift frequency is charted along the vertical axis, time along the horizontal axis, and the signal power (corresponding to a particular velocity and time) as a change of intensity on a screen or on paper (fig.1).

The periodogram is computationally a very efficient method and produces fairly good results for the typical analysis regime used (128 frequency components every 5 or 10 ms), but it does have its limitations: The frequency resolution is the reciprocal of the time interval over which each frame is sampled, and the Discrete Fourier Transform implicitly assumes that the data is periodic and that an integer number of periods are used. Since this is generally not the case, the implicit periodicity causes a discontinuity that does not belong to the original signal, and the energy associated with this pseudo discontinuity 'leaks' to all the frequency coefficients, corrupting the FFT spectrum estimate. If, in the attempt to reduce the leakage, a weighting window is used, the price paid is a decrease in the frequency resolution of the power spectral density (PSD) estimate.

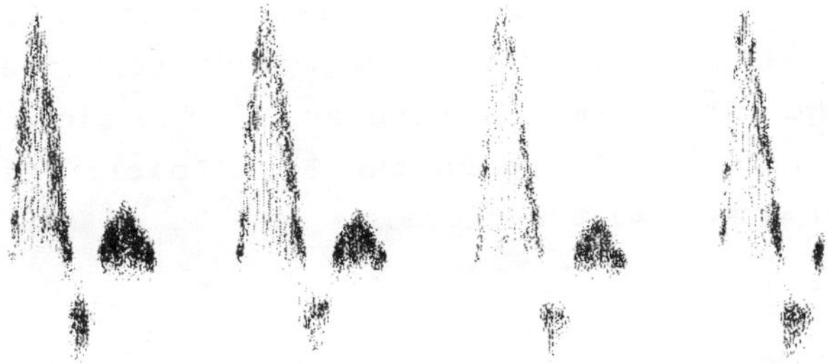

Figure 1 - Sonogram of a Doppler shift signal from the common femoral artery of a healthy subject. This sonogram was obtained using the present system running the AR model with order p = 20. The results were stored on a diskette file and then printed using a eight dot matrix printer with a pseudo grey scale of 32 levels.

An alternative estimate, the maximum entropy autoregressive PSD technique uses the autocorrelation function of the data to determine a number of autoregressive (AR) parameters, and these are used to extrapolate the autocorrelation function beyond its known values. The spectrum is then obtained from this extrapolated autocorrelation function. The extrapolation of the lags means that the signal is not assumed to be periodic outside the known interval, ie, the technique does not suffer from the undesirable effects of windowing. Also, it gives the method a better resolution than the traditional FFT approach. Kitney and Giddens[2)] have suggested that the ability of the autoregressive method to estimate the frequency contents from much shorter sections of data can make the technique attractive for Doppler signal analysis, especially for the description of the frequency content of unstable flow. The reader interested in a practical comparison between the FFT and the AR techniques applied to Doppler signals is referred to Kaluzynski[3)].

This paper describes a system built for implementing real-time spectrum analysis of Doppler shift signals from arteries and the implementation of two different methods to calculate the sonogram of Doppler shift signals - the modified periodogram (based on the Fast Fourier Transform), and the autoregressive technique.

3. WHY "SHORT-TERM"?

Blood flow in arteries is pulsatile. The fundamental frequency of this pulsatility is determined by the heart rate, ie, about 70 bpm, or approximately 1 Hz. The waveform corresponding to the mean velocity curve, however, is complex, and a certain number of harmonics are needed to describe it. Depending on the site where the Doppler signal is collected the number of harmonics needed can vary from about 20 to more than 100. This means that the frame length associated to each spectrum will be short.

Before implementing a real-time PSDE algorithm, a number of questions must be addressed. These include considerations about the frequency content of the signal, sampling frequency to be used, frame length, number of frequency components, and frame rate.

The frequency content of the signal will determine the sampling rate to be used, since the maximum frequency analyzed, f_{max}, is half of the sampling frequency, f_{sam}. Since f_{max} is not known a priori, it is useful to implement a number of ranges and leave the choice as to which one to use under operator control. The present system incorporates five choices of frequency range (table 1).

Table 1 - The five frequency ranges that can be selected by the operator, together with the corresponding sampling frequencies and frame lengths for both the implementations.

sampling frequency (kHz)	frequency range (kHz)	frame length (ms) AR	FFT
40.96	20.48	1.5625	6.25
20.48	10.24	3.125	12.5
10.24	5.12	6.25	25.0
5.12	2.56	12.5	50.0
2.56	1.28	25.0	100.0

The optimum length of each frame depends on the stationarity of the signal. Doppler signals are usually considered to be quasi-stationary for periods of up to 10 or 20 ms, but when turbulence occurs, and at some sites like the aortic arch where the spectrum of the signal is known to vary quite dramatically with time, the stationary period may be shorter. These considerations suggest the use of very short data segments. On the other hand, the use of too short a segment can lead to a statistically poor estimate. In the system described, 256 samples were used for the periodogram approach and 64 samples for the AR method. This corresponds to a frame length of 6.25 ms (FFT) or 1.5625 ms (AR), for the highest frequency range used (20.48 kHz), resulting in a theoretical maximum independent (non-overlapping) 'frame-rate' of 160 frames/s (FFT) or 640 frames/s (AR).

The number of frequency components of the FFT technique is determined by the number of samples (for N samples, N/2 frequency bins). For the AR-maximum entropy technique, it is a matter of choice, since the spectrum is continuous. We have chosen to compute 128 frequency bins, largely as a matter of compatibility with the FFT implementation.

Finally, the frame-rate is related to the length of the frame, but not dictated by it. As shown in fig. 2, there are

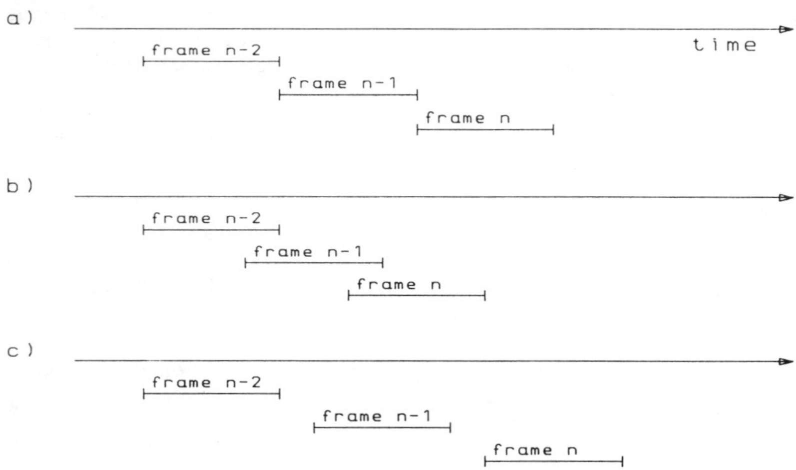

Figure 2 - According to the characteristics of the system, three different approaches can be employed for the partition of the continuous Doppler signal into frames for analysis: a) sequential frames, b) overlapping frames, or c) with gaps between consecutive frames.

three possible approaches to the implementation. The frames can be sequential, there may be an overlap of samples, or a gap between consecutive frames. The choice, in practice, depends on how fast the DSP can compute a single spectrum and how fast the computer can deal with the data. In our system, the choice was dictated by the speed of the computer and the fixed rate of 160 spectra per second was chosen.

3. THE FFT PERIODOGRAM POWER SPECTRAL DENSITY ESTIMATE

Spectral estimation techniques can be seen as attempts to fit the measured data to an assumed model. Conventional FFT power spectral density estimation (PSDE) is based on a Fourier series model of the data, ie, the process is assumed to behave as a sum of sinusoids (or exponential functions) and the coefficients of these sinusoids describe the process according to the model.

In the conventional FFT-periodogram approach, the signal x(t) is sampled at a fixed rate 1/T, and the DFT of finite "frames" of N samples is computed

$$X(k) = \sum_{n=0}^{N-1} x(nT)\exp(-2j\pi nk/N) \qquad k=0,1,2,...,N-1 \quad (1)$$

The power spectral density estimate is given by

$$S(k) = |X(k)|^2 \quad (2)$$

S(k) is periodic and discrete, but, since the DFT implicitly assumes that the **signal** is periodic and that an integer number of periods is used in the frame, the estimate is wrong whenever these assumptions are not true.

Normally each frame of N samples is multiplied by an anti-leakage function, w(n), before the FFT is performed (for a review of window functions see Harris[4]),

$$X_w(k) = \sum_{n=0}^{N-1} x(nT)w(n)\exp(-2j\pi nk/N), \quad k=0,1,...N-1 \qquad (3)$$

and the power spectral density estimate is given by

$$S_w(k) = |X_w(k)|^2 \qquad (4)$$

This use of anti-leakage windows in the time-domain signal creates two more errors in the estimate. Firstly, the estimate $S_w(k)$ of the true spectrum of x(t) corresponds to the circular convolution of the transform of the signal with the transform of the time-window. Since the transform of the window is not an impulse, the windowing process decreases the frequency resolution of the PSDE. The second problem is the fact that, by weighting differently samples that have the same statistical value, a bias is introduced in the estimation. Yuen[5] uses the expression "we are throwing data away" to describe this sort of weighting.

4. THE AUTOREGRESSIVE POWER SPECTRAL DENSITY ESTIMATE

Depending on the particular data to be analyzed, the Fourier model may not be the best to describe the process, ie, there might be an alternative model that is less prodigal in the use of parameters, or perhaps that does not suffer from the consequences of the assumptions made about the behavior of the data outside the measured interval.

The autoregressive model assumes that the current value of the process, x_n, can be described by a finite linear aggregate of the previous values of the process and the current value of a white noise driving process n_n. An autoregressive process of order p is defined as[6)]

$$x_n = n_n - a_1 x_{n-1} - a_2 x_{n-2} - a_3 x_{n-3} - \cdots - a_p x_{n-p} \quad (5)$$

The autoregressive model contains p+2 parameters which have to be estimated from the data: the coefficients, the mean of the samples, and the variance of the white noise. The estimation of these parameters for the AR model results in linear equations, which are computationally easy to implement.

4.1 Yule-Walker Equations and Recursive Algorithm

The classical way of evaluating the coefficients a(k) of an autoregressive process of zero mean and order p, is by using the Yule-Walker equations[7)]

$$R_{xx}(k) = - \sum_{n=1}^{p} a(n) R_{xx}(k-n) \qquad \text{for } k > 0, \text{ and}$$

$$R_{xx}(k) = - \sum_{n=1}^{p} a(n) R_{xx}(-n) \qquad \text{for } k = 0 \quad (6)$$

where $R_{xx}(k)$ is the estimate of the autocorrelation function of the process given by

$$R_{xx}(k) = \sum_{n=0}^{N-k+1} x(n+k)\, x(n) \;/\; N \quad (7)$$

and σ^2 is the variance of the driving white noise.

An alternative, and very efficient, recursive method for calculating estimates of the autoregressive parameters is provided by the Levinson-Durbin algorithm[7]. The method uses the important property that the coefficients of an AR(k) process can be computed from the parameters of the AR(k-1) model and k values of the autocorrelation function. The coefficients for the first-order process are first obtained, and from these, the algorithm proceeds recursively up to the desired order p.

In the following equations two indices are used for the coefficients a(k,i). The first is the order of the AR model and the second, the number of the coefficient.

The first order (Markov) AR process is described by

$$a(1,1) = - R_{xx}(1)/R_{xx}(0), \quad \text{and} \tag{8}$$

$$\sigma(1)^2 = [1 - a(1,1)^2]\, R_{xx}(0) \tag{9}$$

and then, the following recursion is used to compute consecutive superior orders from k=2 to p

$$a(k,k) = -[R_{xx}(k) + \sum_{i=1}^{k-1} a(k-1,i) R_{xx}(k-i)] \;/\; \sigma(k-1)^2 \tag{10}$$

$$a(k,i) = a(k-1,i) + a(k,k)a(k-1,k-i) \tag{11}$$

$$\sigma^2(k) = [1 - a(k,k)^2]\, \sigma(k-1)^2 \tag{12}$$

Once the desired order, 'p', is achieved, the power spectral density estimate of the data is given by

$$S(f) = \sigma(p)^2\, T \;/\; |\sum_{n=0}^{p} a(p,i)\, \exp(-j2\pi fnT)\;|^2 \tag{13}$$

where a(p,0) = 1

Notice that eqn (13) refers to a continuous function, S(f), computed from a finite discrete series of parameters,

$a(p,i)$ and $\sigma(p)^2$. In order to make the task feasible in a digital computer, the power spectrum is computed for a finite set of frequencies, k

$$S(k) = \sigma(p)^2\, T \,/\, \left| \sum_{n=0}^{p} a(p,i) \exp(-j2\pi nkT) \right|^2 \qquad (14)$$

where $a(p,0) = 1$.

4.2 Maximum Entropy Spectral Estimation

Maximum entropy spectral estimation is based upon an extrapolation of the known values of the autocorrelation function (or estimate) using the autoregressive model as the basis for the extrapolation. Supposing that p+1 values of the autocorrelation function, $R_{xx}(0)$ to $R_{xx}(p)$, are known, the maximum entropy extrapolation of the autocorrelation is

$$r_{xx}(n) = - \sum_{k=1}^{p} a(p,k)\, R_{xx}(n-k) \qquad \text{for } |n| > p \qquad (15)$$

This approach was suggested by Burg[7], and is named 'maximum entropy' because it is the one, amongst the infinite choices of extrapolations of the autocorrelation, that maximizes the randomness (entropy) of the unknown time series, producing the flattest (whitest) spectrum of all the alternatives for which the autocorrelation function in the interval $r_{xx}(0)$ to $r_{xx}(p)$ is equal to the known lags, $R_{xx}(0)$ to $R_{xx}(p)$. The maximum entropy extrapolation imposes the fewest constraints on the unknown time series[7].

When the maximum entropy extrapolation is used, it is convenient to choose an alternative representation for the power spectrum density, equivalent to eqn (13)

$$S(k) = T \sum_{n=-M}^{M-1} r_{xx}(n) \exp(-j2\pi knT) \qquad (16)$$

where M can be chosen as a power of two, in order to enable the computation be performed using an FFT. Of course, as M increases, so do the rounding errors, particularly when the process approaches nonstationarity[6)]. For this reason, it is not advisable to extend $r_{xx}(n)$ too much. The effect of the round-off errors can be minimized if the biased estimate of eqn (7) is used to compute the p initial values of R_{xx} because, due to the triangular window implicit in the biased estimation, the estimated values of the autocorrelation tend to zero as N grows. This is statistically very sensible, since it means that the more uncertainty there is in the estimation of a data point, the less weight it is given.

Since many of the problems of the periodogram PSD estimation are related to the assumptions implicitly made about the data outside the measured interval (that it is periodic), and the autoregressive method does not assume this, it does not suffer from 'window' distortions of the spectrum. Also, since eqn (15) can be used to extend the autocorrelation values beyond the ones estimated directly from the data, the resolution of the estimate of the power spectrum is not constrained to N/2 frequency values. This is particularly important in the estimation of frequency contents of non-stationary signals, since shorter sections of data are necessary for the AR PSDE compared to the FFT approach. The extra values obtained by the extrapolation are, by no means, equivalent to the sin(f)/f filtering interpolation obtained of the FFT when zero-padding is used.

5. HARDWARE OF THE SYSTEM

The hardware of the system[8)] comprises basically of three building blocks (fig.3): an analog interface; a Digital Signal Processor development board, based on the Texas Instruments TMS 320C25 DSP chip; and a Research Machines Nimbus microcomputer, based on the 80186 CPU.

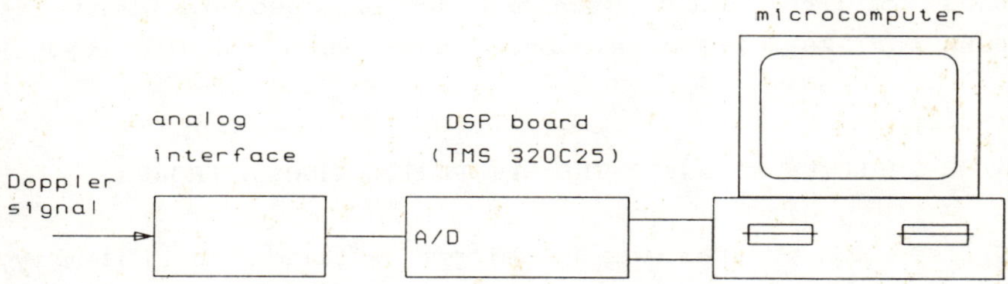

Figure 3 - Schematic configuration of the complete system. Apart from the analog interface which includes a programmable anti-aliasing filter, the hardware comprises of two commercial building blocks only, the LSI TMS 320C25 board and the RM-Nimbus PC2.

The analog interface consists of overload protection, a programmable antialiasing filter, and an operator controlled variable gain amplifier. The input signal to the system is a single channel Doppler waveform, containing both the forward and reverse outputs of a bidirectional Doppler velocimeter, mixed and offset about a user-set heterodyne frequency. The signal is AC coupled (f_o = 7.2 Hz).

The Loughborough Sound Images (LSI) DSP development board comprises of a TMS 320C25 chip clocked at 40 MHz and capable of executing 10 MIPS, 8 kbytes of zero wait states program memory, 8 kbytes of zero wait states data memory, a track & hold circuit, a 16 bit 17 μs A/D converter, and a programmable timer (used here to control the sampling rate).

The LSI 320C25 board can interrupt the microcomputer, and the interface arrangement is centered on dual-port access to the 320C25 memory, providing a very efficient means of transferring information to and from the microcomputer.

The Research Machines Nimbus was chosen in preference to a more standard IBM-PC compatible microcomputer largely because of its better standard screen specifications (16 colors for 320 x 250 pixels and a sub-BIOS to handle it).

6. THE IMPLEMENTATION OF THE FFT PERIODOGRAM APPROACH

The DSP program was developed entirely in Assembly, because no high level language would be fast enough to run the applications in real-time[8)]. The overall organization of the system software is detailed below. The main tasks of the DSP are:

a) to collect the Doppler signal samples (ISR),
b) to apply a window to the a frame of the last 256 samples,
c) to compute the FFT of the windowed data,
d) to square and add real and imaginary components of the transform to obtain the power spectrum,
e) to issue an interrupt request to the host computer, and,
f) to restart the process after T_{frame}.

The software is organized in a load-interrupt basis using a double buffer arrangement. The processor is allowed to run free until a new data sample is available. When this occurs the processor exits its main FFT calculation task and goes into a brief interrupt service routine (ISR), which loads the sample into a circular buffer of 256 words in the internal (fast) memory of the 320C25 chip. The processor then resumes its main task. At the end of each frame, the power spectrum and other pertinent results are copied to external data memory to be available to the computer, and an interrupt request is issued to the computer. A new frame may then be loaded from the circular input buffer into the linear calculation buffer, restarting the process. This interrupt-driven scheme to input samples reduces the overall speed of the FFT computation according to the expression

$$T_t = T_{proc} / (1 - T_{isr} f_{sam}) \tag{17}$$

where T_t is the total processing time (including ISR), T_{proc} is the processing time alone (without ISR), T_{isr} is the time corresponding to servicing the ISR, and f_{sam}, the sampling frequency (1/T). In the present implementation, T_{proc} is 1.574ms and T_{isr}, 2.7μs, resulting a total time of 1.769ms for the fastest sampling rate used (40 960Hz).

The FFT algorithm implemented performs the computation of 256 values by means of a 128 complex-point transform, using straight-line code and a mixture of radix-2 and radix-4 butterflies to achieve speed[8,9,10].

7. THE IMPLEMENTATION OF THE AR APPROACH

On the implementation of the AR technique[11], the main tasks of the DSP are:

a) to collect the Doppler signal samples (ISR),
b) to compute the autocorrelation values, R_{xx},
c) to obtain the autoregressive coefficients, a(p,i) and σ^2,
d) to expand the autocorrelation function from p+1 to 128,
e) to compute the power spectrum from the extrapolated r_{xx},
f) to issue an interrupt request to the host computer, and,

g) to restart the process after T_{frame}.

The sequence of partial results of the basic operations of the principal DSP program can be followed in fig.4. In the main program, the first task of the DSP is to compute the estimate of the autocorrelation function of the present frame (fig.4b). Firstly, the most recent 64 samples are copied from the circular input buffer into a calculation buffer, and then 64 values of R_{xx} are computed using the biased estimator according to eqn (7). The computation is directly implemented using the special purpose 'multiply-accumulate-data-move' instruction and takes 0.442 ms.

To use the full dynamic range allowed in the DSP, an auto-range scaling procedure was implemented in the estimation of the autocorrelation function. The correction of the scaling, back to the true values, is done by the DSP only after computing the power spectrum, ie, all the calculations are done using this 'pseudo-floating-point' approach. This procedure is very effective in avoiding deterioration of the values due to rounding-off errors.

The AR coefficients are computed according to eqns (8-12). Again, a pseudo floating point procedure is implemented to deal with the values. The program is written in such a way that the basic order of the AR model can be changed from p = 3 up to 32. At present, the system is programmed to use p = 20, but this choice was arbitrary, and further work must be done to select an optimum order. The computation of the 20 AR coefficients takes 0.672ms.

The extrapolation of the autocorrelation is performed according to eqn (15), again using the multiply-accumulate-data-move instruction of the DSP chip (fig.4c). The

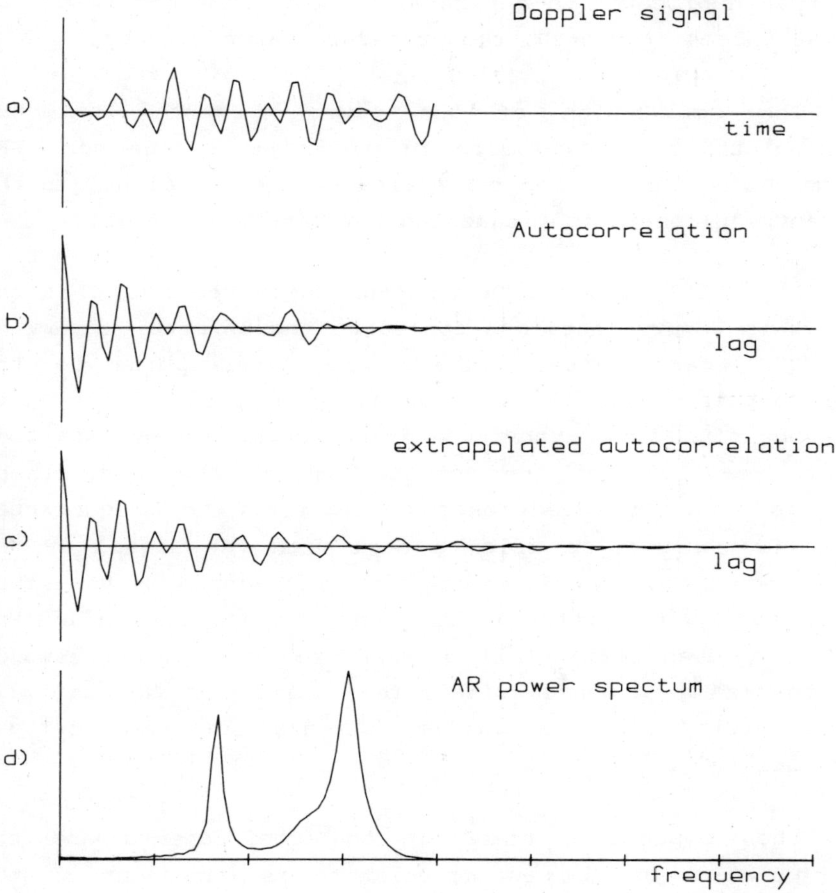

Figure 4 - The sequence of the principal DSP program. From the 64 samples of the frame (a), the autocorrelation function (b) is obtained, according to eqn 7. From the first 'p' values of Rxx, the autoregressive coefficients are computed, according to eqns 8-12. The autocorrelation function is then extrapolated from rxx(p+1) to rxx(128) according to eqn 15, resulting the waveform (c). Finally, rxx is mirror imaged, and the power spectrum (d) is calculated from the 256 values of the extrapolated autocorrelation according to eqn 16. In this diagram p = 20.

autocorrelation is extrapolated from $r_{xx}(p+1)$ to $r_{xx}(128)$, and then duplicated to generate an even function, resulting in 256 values. For p=20, the procedure takes 0.723ms.

The computation of the power spectrum from the extrapolated autocorrelation is performed by the same FFT algorithm of the periodogram approach. The FFT alone, in the present implementation, takes approximately 1.366 ms.

It is only after obtaining the power spectrum that the result is denormalized and copied to the external memory of the DSP board. After issuing the interruption to the microcomputer, the DSP is made to wait until 6.25ms have elapsed since the previous frame computations started, before re-initiating the computations of the next frame. This is done for three reasons. The first is to guarantee that the computer is able to deal with the signal at the rate that the DSP produces spectra. The second is to ensure that, even with different frequency ranges chosen by the operator, each screen-full of sonograms always corresponds to the same elapsed time. The third is to maintain compatibility with a system already in use in our hospital[8,12].

The computation times of the coefficients and the extrapolation of the autocorrelation is dependent on the chosen order, p of the AR model. For a model of order three, T_{proc} is 2.422ms, for p = 32 it is 4.055 ms.

8. MICROCOMPUTER SOFTWARE

The main task of the microcomputer is to display, in real-time, the sonogram corresponding to the power spectrum estimation computed by the DSP board. The spectra are read by an interrupt service routine, while the main program organizes the display and storage of the sonogram. The main tasks of the host processor are:

a) to read the power spectra from the DSP board (ISR),
b) to obtain the amplitude spectrum from the power spectrum,
c) to codify the amplitude spectrum in a colour-scale,
d) to display the sonogram on the screen, in real-time,
e) to store the sonogram on disk, for post-processing,
f) to interact with the operator (keyboard).

The RM-Nimbus microcomputer program was also developed in Assembly because of the speed requirements. Apart from the sub-BIOS calls that handle the graphic screen, the rest of the program uses standard MS-DOS system calls, ie, the system can be implemented with another MS-DOS machine, as long as the graphic handling is re-written.

When the microcomputer receives an interrupt from the DSP board, it reads one block of information by directly accessing the output buffer in the external memory of the DSP board. The spectrum is stored in one of eight buffers in the Nimbus memory, which are filled and processed in a circular fashion.

In the main program, the first task of the microcomputer is to take the square root of the power spectrum, to reduce its dynamic range. This is performed by using a look-up table. The sonogram displayed is the result of the average of two individual frames, ie, one column of the sonogram is written to the screen every 12.5 ms, over-writing circularly the previous sonogram. Together with the sonogram, the frequency envelope is found as described by

Gibbons[13]), and the minimum and maximum frequency bins exceeding a user-set threshold are indicated by bright dots on the screen. The sixteen colour look-up table can be chosen at will.

The operator can freeze the display (and processing) at any time to examine details of the sonogram. The frequency range (together with the antialiasing analog filter) is reprogrammed by the touch of a key to one of the five choices of table 1. The threshold used to find the envelope of the sonogram can also be changed on-line. As with the earlier systems, the operator has the option of storing the spectral information and the envelopes to disk.

9. RESULTS AND CONCLUSIONS

The system is capable of obtaining and displaying the FFT or the AR sonogram in real-time. Figure 5 is a video print-out of the screen showing the AR sonogram of a Doppler shift signal from the posterior tibial artery of a healthy subject, using the basic arbitrary order of p = 20.

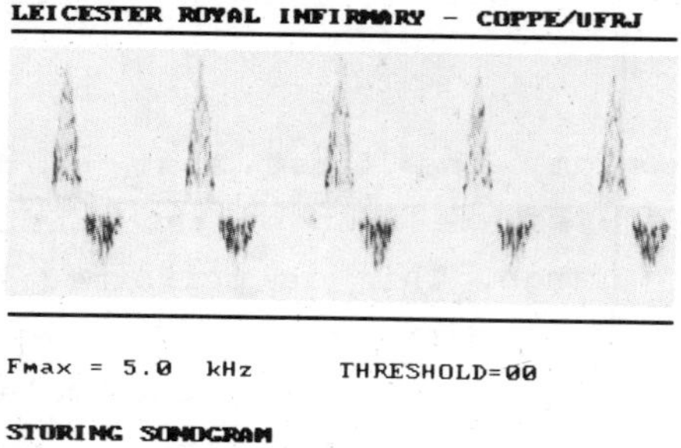

Figure 5 - Video print-out of the microcomputer screen with an AR sonogram in grey-scale. The frequency envelope is not shown for clarity (threshold = 0).

The processing times, T_{proc}, of the DSP procedures are: for the FFT technique, 1.574ms, and for the AR it depends on the order of the model. For order 20, it is 3.225ms.

We have shown that a modern DSP chip is quite adequate to perform real-time spectral analysis of Doppler ultrasound signals using either the FFT-periodogram approach or the maximum entropy AR technique. The greatest virtue of the system is its flexibility. Almost every aspect of the system, such as sampling rate, threshold of the frequency envelope, window function in the FFT approach, and order of the model in the AR technique are under software control.

It has been suggested by Kitney and cols.[2,14] that, under certain conditions, AR spectrum analysis could provide better spectral estimates of the Doppler signal than the usual FFT-periodogram approach. If this analysis is to be of real clinical value then the AR spectral estimation must be capable of being carried out in real-time. This study determines the feasibility of performing AR spectral analysis of Doppler signals in real-time using a modern DSP chip and a personal microcomputer.

The system produces spectra at a rate of 160 frames per second, but the TMS 320C25 is capable of running the present applications at a frame rate of about 560Hz (FFT) or 275Hz (AR). If a better temporal resolution is desired, then another configuration should be considered.

It is known that there is a theoretical optimum AR model order for each particular set of data, and some researchers have proposed techniques for finding this 'optimum-order'[14,15]. Since at present none of these techniques seems appropriate for implementation in real-time, the present system incorporates an arbitrary fixed order.

Acknowledgements - F.S.S. wishes to acknowledge the Conselho Nacional de Desenvolvimento Científico e Tecnológico (CNPq) and the Federal University of Rio de Janeiro (UFRJ) for financial support.

11. REFERENCES

1) Evans, D.H, McDicken, W.N., Skidmore, R. and Woodcock, J.P., "Doppler Ultrasound". John Wiley and Sons, Chichester (1988).

2) Kitney, R.I. and Giddens, D.P., "Linear estimation of blood flow waveforms measured by Doppler ultrasound". In MEDINFO 86, R. Salamon, B. Blum, M. Jorgensen (Eds.), Elsevier Science Publishers B.V. (North Holland) 672-677 (1986).

3) Kaluzynski, K. "Analysis of application possibilities of autoregressive modelling to Doppler blood flow signal spectral analysis". Med. & Biol Eng. & Comput., V.25, 373-376 (1987).

4) Harris, F.J. "On the use of windows for harmonic analysis with the discrete Fourier transform". Proceedings of the IEEE, vol.66, N.1 (1978).

5) Yuen, C.K., "Comments on Modern Methods for Spectrum Estimation". IEEE Transactions on Acoustics, Speech and Signal Processing, vol ASSP-27, N.3, pp 293-299 (1979).

6) Box, G.E.P. and Jenkins, G.M., "Time Series Analysis forecasting and control". Holden-Day, Oakland, California (1976) .

7) Kay, S.M. and Marple Jr., S.L., "Spectrum analysis - a modern perspective". Proc. IEEE, V.69, No.11, 1380-1419 (1981).

8) Schlindwein, F.S., Smith, M.J. and Evans, D.H., "Spectral analysis of Doppler signals and computation of the normalized first moment in real-time using a Digital Signal Processor". Med. & Biol. Eng. & Comp. 26, 228-232 (1988).

9) Brigham, E.O., "The Fast Fourier Transform". Prentice-Hall, Englewood Cliffs, New Jersey (1974).

10) Texas Instruments, "Digital signal processing applications with the TMS320 family. Theory, algorithms and implementations" (1986)

11) Schlindwein, F.S., and Evans, D.H., "A real-time autoregressive spectrum analyzer for Doppler ultrasound signals". Ultrasound in Medicine and Biology (in press).

12) Prytherch D.R. and Evans, D.H., "Versatile microcomputer-based system for the capture, storage and processing of spectrum-analyzed Doppler ultrasound blood flow signals". Med. & Biol. Eng. & Comp., 23, 445-452 (1985).

13) Gibbons, D.T., Evans, D.H., Barrie, W.W. and Cosgriff, P.S., "Real-time calculation of ultrasonic pulsatility index". Med. & Biol. Eng. & Comp., 19, 28-34 (1981).

14) Kitney, R.I., Talhami, H. and Giddens, D.P., "The analysis of blood velocity measurements by autoregressive modeling". J. Theor. Biol. 120, 419-442 (1986).

15) Akaike, H., "Fitting autoregressive models for prediction". Ann. Inst. Stat. Math. 21, 243-247 (1969).

PRACTICAL ASPECTS OF INCLUDING TRANSDUCER FIELD DISTRIBUTIONS IN SIGNAL PROCESSING FOR ULTRASONIC MEASUREMENT AND IMAGING

J R Blakey, R C Chivers and R A Bacon
Physics Department, University of Surrey, Guildford, Surrey, GU2 5XH, UK.

ABSTRACT.

Real transducers often behave in a non-ideal way and radiated field measurements must be used in order to obtain more realistic theoretical models. Two approaches are described here (1) the determination of effective geometrical parameters. (2) source reconstruction. Emphasis is placed on the limitations due to the nature of the measurement processes

1 INTRODUCTION

The key element in any ultrasonic measurement or imaging system is the transducer. In the simplest approximation plane wave irradiation is assumed and the problems are stated in essentially one dimensional terms. For most work this is inadequate and account needs to be taken of the diffractive nature of the ultrasonic beam caused by the finite aperture of the transducer.

The simplest way of doing this is to assume ideal behaviour of the transducers and to perform the 'diffraction corrections' that have appeared frequently in the literature. With the variety of boundary conditions that are involved in the manufacture of ultrasonic transducers, it is not surprising that it has often been shown that transducers do not behave in an ideal way. The fact that many transducers do appear to behave 'well' is not relevant when an investigator has a particular transducer at hand and is to use it for a series of measurements and/or imaging.

This paper discusses the ways in which direct measurements of the field radiated by a transducer can be used to obtain better approximations to a realistic model to be inserted into the analysis of the experiment which related the quantities actually measured to the parameters (such as velocity, attenuation or scattering cross-section) which it is hoped to measure. The first approach is based on the definition of effective geometrical parameters for the analysis, using a very few measurements of the field, while the second is based on extensive measurements of the radiated field distribution. In discussing these two approaches, emphasis is placed on the limitations that arise from practical measurement constraints.

2 EFFECTIVE GEOMETRICAL PARAMETERS

The diffraction corrections used for an axially aligned pair of transducers of the same nominal radius, a, have as their parameters the ratio of the radius to the wavelength, λ, and the ratio of the transducer spacing, d, to the wavelength (or to the radius). These are often combined into the dimensionless parameter $s = \frac{a^2}{\lambda d}$[1]. Given that, for velocity measurements, it is λ that is to be measured, it is clear that the use of any diffractive corrections must involve an iterative procedure until the magnitude of the corrections is considered to be insignificant. (This discussion had been extended elsewhere [2] to measurements of attenuation and of scattering). The measurement of d should present little problem, but the value of 'a' that it is appropriate to use will depend critically upon the closeness with which the actual behaviour of the transducer approaches the ideal (pistonlike) model assumed for it.

A number of approaches have been adopted for defining an effective radius. The most common involve either the lateral extent to which the beam drops to an arbitrary fraction of the (axial) peak value, or are based upon measurements inserted into a model. Our attention has focussed on the latter since it has the potential of being rather less arbitrary.

Over the last ten years advances in design and calibration of miniature hydrophone probes [e.g. 3-5] has opened up new possibilities in the quantification of ultrasonic fields. For a pistonlike disc source of radius a, the position of the last axial minimum is at a distance z_o from the disc surface, where z_o is given by $z_o = a^2/2\lambda$. Location of this minimum, which is spatially well-defined, in a fluid of known speed of sound, permits the calculation of an effective radiis for the transducer at a given frequency of excitation. Early experiments [6] showed their effective radius to be, in general, frequency dependent, and to differ by as much as 30% from the nominal geometrical radius of the transducer element.

With the development of techniques for making measurements of the phase distribution over planes in the field perpendicular to the axis [7, 8] it was possible to show, in a fairly comprehensive study [9] that the use of an effective radius gave greatly improved prediction of the actual fields radiated by four transducers. The effective radius was used in the theoretical pistonlike model to predict the phase and amplitude distributions in a given plane and these were compared with experimental measurements. The degree of agreement improved, the closer the effective radius was to its geometrical value.

The major advantages of this type of approach are several. It is relatively rapid and it requires no significant computational facilities. Most importantly it places minimal demands on the hydrophone. The hydrophone does not have tc be much smaller than a wavelength (although if it is comparable with 'a', problems will arise), and it does not have to have an accurate calibration either in terms of directivity or frequency response.

The extension of this approach to focussed bowl transducers has been successfully achieved [10, 11] although at least two measurements are needed to cover the two independent geometrical parameters that arise for such a transducer. The most appropriate appear to be the axial positions of the two minima closest to, but inside, the true focus of the transducer.

There are two disadvantages associated with present approaches based on effective geometrical parameters. The first is that they are essentially attempting to relate the real distribution to a particular theoretical model, which may not be the most appropriate model. The second is that there is no definite information on the degree to which the actual distributions fit the model, although it has been suggested [9] that the closeness of agreement of the effective radius and the geometrical radius may be a measure of this.

The choice of the most appropriate model has received little attention and although no work appears to have been performed it has been suggested [12] that one or two more amplitude measurements may indicate which of a small range of models is the most appropriate. The degree of agreement of the actual distributions with the selected model can only be discovered by detailed measurement (and this removes the simple value of the effective radius approach). This is discussed in succeeding sections.

3. MEASUREMENT OF FIELD DISTRIBUTIONS

With the availability of calibrated miniature hydrophones, the development of techniques for phase measurements, and advances in microprocessor technology, it is possible to measure the field distributions of a transmitter directly, for insertion into the measurement application. This appears, however, to be wasteful of effort in view of the fact that measurements over a single surface of regular geometry can be forward and back projected [13] to obtain the distribution over any other similar surface.

The remainder of this contribution is concerned with the practical aspects of this procedure, where the surfaces over which the measurements are made are parallel planes which are perpendicular to the axis of the transducer. Having made measurements over a plane there are a number of features which will limit the information available in any other phase (including the surface of the transducer). These are the finite aperture over which the measurements are of necessity taken, the finite size of the hydrophone probe (of growing importance as the frequencies used in ultrasonic applications increase), the spatial sampling of the measurements, the presence of 'noise' (of positional or electronic origin), and the relation of the phase of interest to the plane of measurement. Of these the finite area of the probe, and aspects of the noise will be discussed in some detail. The discussion assumes linearity and a harmonic time dependence.

3.1 The effects of finite probe receiving area

In acoustic near field reconstruction measurements the measuring probe can be regarded as an ideal point detector. At ultrasonic frequencies even miniature probes have dimensions which are significant compared to the wavelength. The output signal of such a probe represents some average value of the field over the probe's sensitive area and will not, in general, be the same as that which an ideal probe would give.

Limited calculation of the effect of a finite probe have been performed for the far field of a disk source [14] where it was shown that the effects of a probe of one wavelength in diameter are no more than the errors usually associated with the measurements. A more detailed discussion in needed for the near field.

Consider a plane radiator located in the plane z=0. The field distribution over a parallel plane at a distance z from the source in the near field can be expressed as:

$$f(x,y) = \iint_{-\infty\,-\infty}^{\infty\,\infty} F\, k_x,k_y)\, \exp\,[ik_x + ik_y]\, dk_x dk_y \tag{1}$$

which is what would be measured by an ideal point probe. A finite probe would yield an apparent measuring plane distribution given by:

$$f_m(x,y) = \iint_{-\infty\,-\infty}^{\infty\,\infty} F(k_x,k_y) B(k_x,k_y)\, \exp\,[ik_x x + ik_y y] dk_x dk_y \tag{2}$$

where $B(k_x,k_y)$ is the spatial frequency response of the measuring which can be obtained from phase and amplitude directivity measurements. probe, When the usual transform is applied an apparent source distribution $f_{sm}(x,y)$ will be reconstructed, given by:

$$f_{sm}(x,y)=\int_{-\infty}^{\infty}\int_{-\infty}^{\infty}F(k_x,k_y)B(k_x,k_y)\exp[ik_xx+iky]\exp[-ik_sz]dk_xdk_y \quad (3)$$

ie the same source distribution will be 'windowed' by the measuring probe spatial response function $B(k_x,k_y)$. It is evident that if the measuring probe response is known then it should be possible to recover the true source distribution.
For if we take the Fourier transform of (2) we have:

$$F(k_x,k_y)B(k_x,k_y) = \int_{-\infty}^{\infty}\int_{-\infty}^{\infty}f_m(x,y)\exp[-ik_xx-ik_yy]dxdy \quad (4)$$

In fact the determination of the true far field radiation pattern of microwave antennas from probe-corrected near-field measurements has been a standard technique for many years [15, 16]. It is a well-known consequence of the sampling theorem that complete information about the source is achieved by determining the field at a lattice of points with a spacing $\leq\lambda/2$ over the measurement plane. In the case of ultrasonic measurements the significant probe dimension may well exceed $\lambda/2$ [the probes discussed earlier have $D/\lambda \sim 0.75$ at 1 MHz]. In which case the successive areas covered by the probe will overlap. If the significant dimension of the probe is much greater than $\lambda/2$ the sampling will more closely resemble a continuous scan and the resolution will be effectively that of the probe significant dimension. The question arises: what amount of overlap [i.e. what is the maximum probe dimension?] given $\lambda/2$ sample spacing, will ensure an effective recovery of the true source distribution? The answer will be dictated ultimately by the system noise. In general terms, if the probe spatial frequency response $B(k_x,k_y)$ falls to very low values over a certain spatial frequency range then a correspondingly high gain will be required in the source recovery processing, with the consequent enhancement of noise errors in that range. In particular, if the response $B(k_x,k_y)$ should fall to zero, full compensation will be impossible. We can estimate when this is likely to occur. Consider the well known response of the form $\frac{J_1(\rho)}{\rho}$ The first null occurs for $\rho \simeq 3.8$. The null will appear in the non-evanescent spatial frequency response for probe diameters D greater than $\simeq 1.28\lambda$, suggesting that reconstruction using compensation should be possible for probe diameters of at least 1λ. This does not imply that the sample spacing can be correspondingly reduced since errors due to aliassing are then likely to be introduced.

3.2 Noise.

As already indicated system noise will give rise to errors in determining amplitudes and phases in the measured field, which in turn will yield an erroneous source reconstruction as well as determining the effective resolvable distance. [17].

Of immediate interest is the effect of probe positional errors in the sampling process, whereby the measured values of amplitude and phase at positions $ns+\delta_n$ will differ from those at the correct positions ns [where s is the sampling interval, n is an integer, δ_n the positional error at the nth sample]. This is equivalent to introducing a source of noise into the measurement and hence the reconstructions also. Without loss of generality we can consider 1-dimensional sampling of the field f(x) = with an ideal probe at a sampling interval $s = \lambda/_2$ $\{f_n(ns)\}$ The spectrum of the strictly uniformly sampled field is:

$$F_s(k_x) = \sum_{n=-\infty}^{\infty} f_n(ns) \exp[-insk_x]$$

When there are positioning errors the resulting spectrum is

$$\tilde{F}_s(k_x, x) = \sum_{n=-\infty}^{\infty} f_n(ns+\delta_n) \exp[-insk_x]$$

where δ_n is a random variable. Thus the spectrum itself is a random variable and we must apply statistical estimation procedures. Let $\hat{F}_s(k_x,x)$ be the estimated spectrum, ie

$$\hat{F}_s(k_x,x) = \Sigma\tilde{F}_s(k_x,x) = \Sigma\{\Sigma_n f_n(n s + \delta_n) \exp[-insk_x]\}$$

where $\Sigma(\tilde{F}_s(k_x,x))$ is the expectation value of $\tilde{F}_s(k_x)$.

By expressing $f_n(ns+\delta n)$ as the inverse Fourier transform aof the true spectrum F(w) we can write

$$\tilde{F}_s(k_x,x) = \left\{\sum_n \left[\int_{\infty}^{\infty} dk'_x \; F(k'_x)\exp[i(ns+\delta_n)k'_x]\right]\exp[-insk_x]\right\}$$

$$= \sum_n \exp[ins(k'_x - k_x)] \left\{\int_{-\infty}^{\infty} \exp[i\delta_n k_x] F(k'_x) dk_x .\right\}$$

$$= \left\{\exp[i\delta_n k_x]\right\} F_s(k_x)$$

$$= F_s(k_x) \int_{-\infty}^{\infty} \exp[i\delta_n k_x]\, p(\delta) d\delta$$

where $p(\delta)$ is the probability distribution function (pdf): of the positional errors. We note that the integral is a function of k_x, i.e.

$$\int \exp[i\delta_n k_x]\, p(\delta) d\delta = \Phi(k_x)$$

and this will modify the time spectrum in a manner which depends on the form of the pdf. For example a uniform pdf of errors will give a $\Phi(k_x)$ of the form

$$\frac{\sin(ak_x)}{a\,k_x} \quad \text{where } a = \frac{\beta k_x \lambda}{2} \quad \text{and } 0 < \beta < 1.$$

It is evident that the positional errors introduce an attenuation at high frequencies. Since the source reconstruction algorithm is linear and in particular only modifies the phases of the individual spatial frequencies it will yield the same signal/noise ratio at the reconstructed source plane. The signal/noise ratio is given by

$$\frac{S}{N} = \frac{\int_{-\infty}^{\infty} |F_s(k_x)|^2 dk_x}{\int_{-\infty}^{\infty} (|\hat{F}(k_x)|^2 - |F_s(k_x)|^2)} \simeq \frac{1}{\int_{-2\pi/\lambda}^{2\pi/\lambda} \left[1 - \mathrm{sinc}^2(ak_x)\right] dk_x}$$

Evidently the average noise spectrum increases with increasing frequency so that high frequency components will predominate. Thus an attempt to restore the true source distribution by inverse filtering will be achieved at the expense of degrading further the S/N ratio of the reconstructed image.

4 Preliminary Results

A series of model calculations has been commenced in order to evaluate the method prior to its experimental application. The field of a circular piston source, diameter 10λ, as determined by circular receiving probes of various diameters in planes at different distances from the source, have been calculated using the method described in reference [9]. Fignure 1 shows clearly the effect of probe smoothing on the field distribution at a distance equal to the source diameter (10λ).

The apparent source distributions reconstruscted from the field determined for the 0.67λ probe [figures 1 (a) and (b)], are shown in figure 2. Figures 2(a) and (b) show the effects of different sample spacings and the different sizes of sampled receiving area. In figures 2 (a) and (b) the sample spaciong of 0.5mm (0.33λ) and the fielkd was sampled out to tweice the source radius in the receiving plane. Ir is evident that the reconstructed field does not closely correspond to the ideal unioform source distribution.

In Figure 1 it is seen that there is still a significant field beyoind 15 mm and cutting off at this point appears to have introduced windowing artefacts into the reconstruction. In Figures 2 (c) and (d) the sample spacing has been reduced by a factor 2 and the sampled

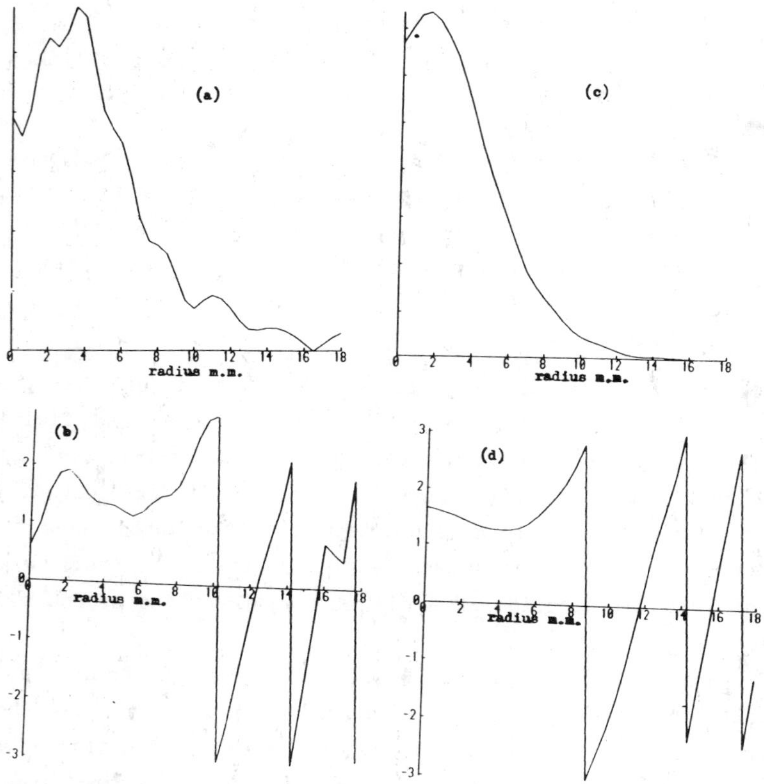

Fig 1. Smoothing effect of finite probe size on measured field distribution of a circular piston source.
(a) Amplitude (b) phase for 0.67λ dia. probe (c) amplitude (d) phase for 2λ dia. probe.

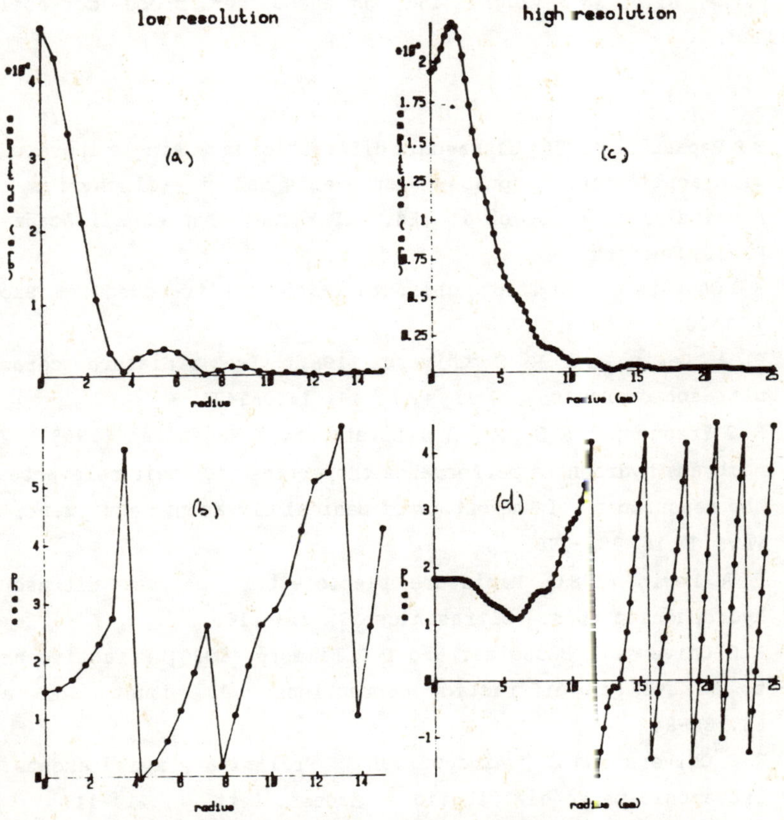

Figure 2. Source reconstruction without probe correction.
(a) amplitude (b) phase in source plane with sampling internal 0.5mm.
(c) amplitude (d) phase in source plane with sampling internal 0.25mm.

region in the receiving plane has been extended to a radius of 25mm. The results are much smoother and windowing affects are much less noticable.
The expected low pass filtering on the reconstruction due to the finite probe size is evident and the need for probe correction established.

References

[1] E P Papadakis (1975) Ultrasonic diffraction from single aperatures with application to pulse measurements and crystal physics. In Physical Acoustics vol 15 (Ed. W P Mason) Chapter XI, Academic Press, New York.

[2] R C Chivers (1981) Tissue characterization. Ultrasound Med. Biol. 7, 1-20.

[3] P A Lewin and R C Chivers (1981) Two miniature ceramic ultrasonic probes. J. Phys. E. 14, 1420-1424.

[4] R C Preston, D R Bacon, A J Livett and K Rajendran (1983) PVDF membrane hydrophone performance properties and their relevance to the measurement of the output of medical ultrasonic equipment. J. Phys. E. 16, 786-796.

[5] P A Lewin (1981) Miniature piezoelectric polymer ultrasonic hydrophone probes. Ultrasonics 19, 213-216.

[6] R C Chivers, L Bosselaar and P R Filmore (1980) Effective area to be used in diffraction corrections. J. Acoust. Soc. Am. 68, 80-84.

[7] R C Chivers and J D Aindow (1980) Preliminary measurements of ultrasonic phase distributions Acoust. Lett. 4, 114-117

[8] J D Aindow and R C Chivers (1982) Measurement of the phase variation in an ultrasonic field J. Phys. E. 15, 83-86.

[9] J D Aindow, A Markiewicz and R C Chivers (1986) Quantitative investigation of ultrasonic disc sources J. Acoust. Soc. Am. 78, 1519-1529.

[10] J Adach and R C Chivers (1986) Effective parameters for a spherical cap transducer. Acustica 62, 66-74.

[11] J Adach (1986) Focussed ultrasonic fields in inhomogeneous

[12] R C Chivers (1988) Effective radius for ultrasonic transducers Proc. 35th Open Seminar on Acoustics, Bialowieza, Poland (in press).

[13] R Reibold (1987) Sound source and construction using Fourier optics Acustica 63, 60-64.

[14] A Markiewicz and R C Chivers (1983) Typical errors in using finite miniature ultrasonic probes for field memasurements. Acoust. Lett. 6, 142-147.

[15] D M Kerns (1970) Correction of near-field antenna measurements made with an arbitrary but known measuring antenna. Electron Lett. 6, 348-349.

[16] D M Kerns (1975) Scattering matrix description and near-field measurements of electroacoustic transducers. J. Acoust. Soc. Am. 57, 497-507.

[17] J D Maynard, E G Williams and Y Lee (1985) Near-field acoustic holography:I. Theory of generalized holography and the development of NAH J. Acoust. Soc. Am. 75, 1385-1393.

APPLICATION OF THE LOCAL KRAMERS-KRONIG DISPESION RELATIONSHIPS TO CALCULATION OF THE ULTRASONIC WAVES VELOCITY DISPERSION IN RANDOM MEDIA

E.Soczkiewicz (1) and R.C.Chivers (2)
(1) Inst.of Physics,Silesian Techn.Univ.Gliwice,Poland
(2) Physics Depart.Univ.of Surrey, Guildford GU2 5XH,UK

ABSTRACT

Propagation of ultrasonic waves in random coarse-grained medium has been considered.Using expressions for the scattering coefficients obtained by means of the parabolic equation method,the dispersion in velocity of the mean acoustic field has been estimated for various forms of the correlation coefficient of the medium refractive index fluctuations,using the local Kramers-Kronig relationship.

1. INTRODUCTION

Wave propagation in random media is encountered in a large variety of branches of physics: acoustics,biophysics,optics,radiophysics,atmospheric physics[1,2].The problems associated with random media may be grouped into three main categories: discrete random scatterers,random continua and rough surfaces[3]. Scattering of waves by the inhomogeneities of the medium or surface causes stochastic changes in the amplitude and phase of the waves,and in the case of transverse waves - of the direction of polarization.A variety of methematical methods have been employed in studies of wave propagation in random media.[1-9]

The choice of the one which is actually used in a given situation,depends to a large extent on the scale of the inhomogeneities in relation to the wavelength of the radiation,and whether multiple scattering needs to be considered at some stage in the analysis[1,9]. In further considerations we limit ourselves to discussion the mean a-acoustic field in randomly inhomogeneous medium,studying the average loss of amplitude of waves due to scattering by the inhomogeneities and their velocity dispersion,leaving aside problems concerning the average fluctuations that may be observed in the amplitude or phase of the waves that are received[7,9].

The main parameter whose variation is considered is the acoustical refractive index n ,that we separate into mean and fluctuating parts as follows:

$$n = 1+\mu . \tag{1}$$

Acoustical properties of randomly inhomogeneous media are characterize by the mean square fluctuations of the refractive index $\langle\mu^2\rangle$ and by the correlation coefficient of the medium refractive index variations:

$$N(\vec{r}_1,\vec{r}_2) = \frac{\langle\mu(\vec{r}_1)\mu(\vec{r}_2)\rangle}{\langle\mu^2\rangle} , \tag{2}$$

where brackets $\langle\ \rangle$ indicate an ensemble average. In further considerations we shall study in detail the so called coarse-grained random media,i.e.the case in that the mean scale of the inhomogeneities is large in comparison with the wavelength of ultrasound,so that the scattering coefficient of the mean acoustic field can be estimated by means of the so called parabolic equation method[7].

Let:

$$K(\omega) = K_1(\omega) + i K_2(\omega) \tag{3}$$

denote the complex adiabatic compressibility coefficient of a medium,where ω is the cyclic frequency. The real and imaginary parts of $K(\omega)$ are related one to another by means of the Kramers-Kronig relationships[10]:

$$K_1(\omega) = \frac{2}{\pi} P \int_0^{\infty} \frac{\omega' K_2(\omega')}{\omega'^2 - \omega^2} d\omega' , \tag{4a}$$

$$K_2(\omega) = -\frac{2}{\pi} P \int_0^{\infty} \frac{\omega K_1(\omega')}{\omega'^2 - \omega^2} d\omega', \tag{4b}$$

where P denotes the principal value of the integral.

Introducing the complex wave number:

$$\tilde{k} = \frac{\omega}{c(\omega)} + i\,\alpha(\omega), \tag{5}$$

where $c(\omega)$ is the phase velocity and $\alpha(\omega)$ the attenuation coefficient,it is easy to obtain from (3) and (5) :

$$c(\omega) = \left(\varrho K_1(\omega)\right)^{-\frac{1}{2}} , \tag{6}$$

$$\alpha(\omega) = \frac{1}{2} \varrho\, c(\omega)\, \omega\, K_2(\omega), \tag{7}$$

where ϱ is the medium density and the inequality $\alpha \ll \frac{\omega}{c}$ has been assumed.Formulae (4) thus relate attenuation and velocity of waves,but they are of limited usefulness because they need a knowledge of $K_1(\omega)$ and $K_2(\omega)$ for all frequencies in the range $0,\infty$.

2. LOCAL KRAMERS-KRONIG RELATIONSHIPS

O'Donnell et al.[10] have applied the second formula (4) for $K_1(\omega) - K_1(\infty)$ obtaining:

$$K_2(\omega) = -\frac{2}{\pi} P \int_{-\infty}^{\infty} \frac{G(x) - G(\infty)}{e^x - e^{-x}} dx, \tag{8}$$

where the following notations have been used: $x = \ln\left(\frac{\omega'}{\omega}\right)$, $G(x) = K_1(\omega')$, $G(\infty) = K_1(\infty)$.After integration,the formula (8) goes into the form:

$$K_2(\omega) = -\frac{1}{\pi} \int_{-\infty}^{\infty} F(x) \ln\left(\mathrm{ctgh} \frac{|x|}{2}\right) dx \tag{9}$$

where $F(x) = \frac{dG(x)}{dx}$ .The function $\ln\left(\mathrm{ctgh} \frac{|x|}{2}\right)$ has a sharp maximum at the point x=0 and is equal zero outside narrow surroundings of x=0 ,so developing $\ln\left(\mathrm{ctgh} \frac{|x|}{2}\right)$

in powers of e^{-x} and $F(x)$ in Maclaurin series, the following formula has been obtained[10]:

$$K_2(\omega) = -\frac{4}{\pi}\sum_{n=0}^{\infty} F^{(2n)}(0)\left(\sum_{m=0}^{\infty}(2m+1)^{-2n-2}\right), \tag{10}$$

where $F^{(2n)}(0) = \left(\frac{d^{2n}F(x)}{dx^{2n}}\right)_{x=0}$. Using the formula[11]:

$$\sum_{m=0}^{\infty}(2m+1)^{-2n} = \frac{(2^n-1)\pi^{2n}}{2(2n)!}\,|B_{2n}|, \tag{11}$$

where B_{2n} are Bernoulli numbers, i.e. coefficients in developing the function:

$$\frac{t}{e^t-1} = \sum_{n=0}^{\infty} B_n \frac{t^n}{n!}, \tag{12}$$

one obtains from the formula (10):

$$K_2(\omega) = -\frac{4}{\pi}\left[\frac{\pi^2}{8}F(0) + \frac{\pi^4}{96}\left(\frac{d^2F(x)}{dx^2}\right)_{x=0} + \frac{\pi^6}{960}\left(\frac{d^4F(x)}{dx^4}\right)_{x=0} + \frac{51\pi^8}{483840}\frac{d^6F(x)}{dx^6} + \ldots\right]. \tag{13}$$

One can prove that the above series may be approximated by the expression:

$$K_2(\omega) = -\frac{\pi}{2}\,\omega\,\frac{dK_1(\omega)}{d\omega}, \tag{14}$$

if changes in the phase velocity and in the attenuation coefficient with frequency are not great over a limited frequency range. From (6),(7) and (14) it results:

$$\alpha(\omega) = \frac{\pi\omega^2}{2c^2(\omega)}\,\frac{dc(\omega)}{d\omega}. \tag{15}$$

or after integration:

$$\frac{1}{c(\omega_0)} - \frac{1}{c(\omega)} = \frac{2}{\pi}\int_{\omega_0}^{\omega}\frac{\alpha(\omega')}{\omega'^2}\,d\omega'. \tag{16}$$

If the value of ω is close to the reference frequency ω_0 the above formula can be further approximated to the form:

$$c(\omega) - c(\omega_0) = \frac{2c^2(\omega_0)}{\pi}\int_{\omega_0}^{\omega}\frac{\alpha(\omega')}{\omega'^2}\,d\omega'. \tag{17}$$

The local Kramers-Kronig formulae (15),(17) have been proved useful in studies relaxation processes in liquids[10], propagation of ultrasonic waves in solutions[12], in composite materials[13,14], in random media[9,15].

3. THE PARABOLIC EQUATION METHOD IN CALCULATION OF THE ATTENUATION COEFFICIENT OF ACOUSTIC WAVES IN RANDOM MEDIA

The parabolic equation method has been proved useful in calculation of the attenuation coefficient of the mean acoustic field due to scattering in randomly inhomogeneous media in the case of small wavelength in comparison with the mean scale of inhomogeneities a, i.e. if $ka \gg 1$[17] where k is the wave number. General wave equation in random media[9,16]:

$$\frac{(1+\mu)^2}{c_0^2}\frac{\partial^2 P}{\partial t^2} - \nabla^2 P + \frac{\nabla(\Delta\varrho)}{\varrho_0}\nabla P = 0, \tag{18}$$

where $\varrho = \varrho_0 + \Delta\varrho$ and $\Delta\varrho$ denotes fluctuation in the medium density, has been considered by Ross and Chivers[17] who showed that in the short wavelength approximation it reduces to:

$$\nabla^2 P - \frac{(1+\mu)^2}{c_0^2}\frac{\partial^2 P}{\partial t^2} = 0. \tag{19}$$

Putting $P(x,y,z,t) = p(x,y,z)\, e^{-i\omega t}$ for $x<0$ (the medium is assumed homogeneous in the half-space $x<0$) we obtain:

$$\nabla^2 p + k^2(1+\mu)^2 = 0 \tag{20}$$

for monochromatic wave, where $k = \frac{\omega}{c_0}$. Assuming $p = A_0 e^{ikx}$ for $x<0$ and $p = A(x,y,z)e^{ikx}$ for $x>0$ (the medium is inhomogeneous in the half-space $x>0$) we obtain:

$$\frac{\partial^2 A}{\partial x^2} + 2ik\frac{\partial A}{\partial x} + \frac{\partial^2 A}{\partial y^2} + \frac{\partial^2 A}{\partial z^2} + k^2(2\mu+\mu^2)A = 0 \tag{21}$$

Writing $A = A_0 + \Delta A$ where ΔA is the fluctuating part of the amplitude, and $\langle A\rangle = A_0 e^{-\alpha x}$, provided that $\alpha \ll k$, $\mu^2 \ll 1$, $ka \gg 1$, eq.(21) reduces to the parabolic type equation:

$$2ik\frac{\partial A}{\partial x} + \frac{\partial^2 A}{\partial y^2} + \frac{\partial^2 A}{\partial z^2} + 2k^2\mu A = 0, \tag{22}$$

since the term $\frac{\partial^2 A}{\partial x^2}$ can be neglected compared to others. Chernov[7] has shown that this is equivalent to taking into account only the forward waves.

Transforming coordinates into non-dimensional form: $kx \to x'$, $ky \to y'$, $kz \to z'$, and noting that for statistically homogeneous medium $\frac{\partial^2\langle A\rangle}{\partial y^2} = 0$ and $\frac{\partial^2\langle A\rangle}{\partial z^2} = 0$, we obtain from(22):

$$\frac{\partial\langle A\rangle}{\partial x'} = i\langle\mu A\rangle\,. \tag{23}$$

Chernov used a method of succesive approximation to solve the parabolic equation. With accuracy to the first approximation A_1 it is:

$$\langle\mu A\rangle \approx \langle\mu A_1\rangle = \frac{1}{2\pi}\langle\mu^2\rangle\langle A_0(x')\rangle\int_{x_0'}^{x'}\frac{d\xi}{x'-\xi}\iint_{-\infty}^{\infty}N(x'-\xi, y'-\eta, z'-\zeta)\,e^{i\frac{(y'-\eta)^2+(z'-\zeta)^2}{2(x'-\xi)}}\,d\eta\, d\zeta \tag{24}$$

Calculating the integrals by the method of steepest descent, one obtains :

$$\langle\mu A_1\rangle = i\langle\mu^2\rangle\langle A_0(x')\rangle\int_{x_0'}^{x'}N(x'-\xi, 0, 0)\,d\xi\,. \tag{25}$$

Introducing relative coordinates and extending the range of integration to $0, \infty$ we obtain:

$$\langle\mu A_1\rangle = i\alpha'\langle A_0(x')\rangle\ , \tag{26}$$

where:

$$\alpha' = \langle\mu^2\rangle\int_0^{\infty}N(\xi)\,d\xi\,. \tag{27}$$

For the intensity attenuation coefficient ($\gamma = 2\alpha$) we have in the original coordinates:

$$\gamma = 2\langle\mu^2\rangle k^2\int_0^{\infty}N(\xi)\,d\xi\,. \tag{28}$$

4. DEPENDENCE OF SCATTERING COEFFICIENT AND WAVE VELOCITY DISPERSION ON THE FORM OF CORRELATION COEFFICIENT

In theoretical considerations concerning acoustic wave propagation in random media there are often used the following forms of correlation coefficients of acoustic refractive index fluctuations[7,8]:

$$N(\xi) = e^{-\xi^2/a^2}, \tag{29}$$

$$N(\xi) = e^{-\xi/a}, \tag{30}$$

where a - the so called radius of correlation is the mean distance over which the refractive index fluctuations remain correlated. One can prove that in the case of correlation coefficient given by (29) the refractive index changes in continuous manner with changes in coordinates, while for the form (30) it changes in a discontinuous fashion[7,18]. From equations (28) and (29) it results:

$$\gamma = \sqrt{\pi}\langle\mu^2\rangle a k^2, \tag{31}$$

whereas for the expression (30) we obtain:

$$\gamma = 2\langle\mu^2\rangle a k^2. \tag{32}$$

Using the local Kramers-Kronig relationship (17) and formulae (31),(32), where $\gamma = 2\alpha$, we obtain the following expression for velocity dispersion:

$$c(\omega) - c(\omega_0) = B\langle\mu^2\rangle a \frac{c^2(\omega_0)}{c_0^2}(\omega - \omega_0), \tag{33}$$

where $B = \frac{1}{\sqrt{\pi}}$ in the case of Gaussian form (29) of correlation coefficient N and $B = \frac{2}{\pi}$ for the exponential form (30).

We see that for the Gaussian form of N both the scattering coefficient and velocity dispersion are smaller than in the case of N given by the formula (30), and that they depend on the ratio a/λ. The above study concerns the small wavelength limit $ka \gg 1$. In the long wavelength limit $ka \ll 1$ there are other dependences of γ and Δc on ω [9].

REFERENCES

1. Uscinski,B.J.,The Elements of Wave Propagation in Random Media.New York: McGraw- Hill,1977.
2. Rytov,S.M.,Kravtsov,Y.A.,Tatarsky,W.I.,Introduction to Statistical Radio-Physics.Moscow: Nauka,1978.
3. Ishimaru,A., Wave Propagation and Scattering in Random Media.New York: Academic Press,1978.
4. Sobczyk,K.,Stochastic Waves.Warsaw: P.W.N.,1978.
5. Chivers,R.C.,Phase and Amplitude Fluctuations in Propagation of Acoustic Waves in Lossless Inhomogeneous Continua with Velocity,Density and Bulk Modulus Variations.I.F.T.R.Report 41,78,Warsaw 1978.
6. Soczkiewicz, E.,Chivers,R.C.,Propagation of Acoustic Waves in Random Media - a Comparison of Methods and Results.Ultrasonic Int.87 Conf.Proc.,543,London 1987.
7. Chernov,L.A.,Wave Propagation in Random Media.Moscow: Nauka,1975.
8. Soczkiewicz, E.,Quantum Field Theory Methods in Studies of Ultrasonic Waves Propagation in Random Media. Ultrasonic Methods in Evaluation of Inhomogeneous Materials.Alippi,A.,Mayer,W.G. ed.Dordrecht: Martinus Nijhoff Publishers 1987,p.163.
9. Soczkiewicz, E.,Chivers,R.C.,Propagation of Acoustic Waves in Random Media Using the Parabolic Equation and Green's Function Methods. Acustica 65,149 (1988)
10. O'Donnell,M.,Jaynes,E.T.,Miller,J.G.,Kramers-Kronig Relationship between Ultrasonic Attenuation and Phase Velocity. J.Acoust.Soc.Am. 69,696 (1981).
11. Gradsztein,J.S.,Ryżik,I.M.,Tables of Integrals,Sums Series and Products.Moscow: Nauka,1971.

12. O'Donnell,M.,Jaynes,E.T.,Miller,J.G.,General Relation ships between Ultrasonic Attenuation and Dispersion. J.Acoust.Soc.Am. 63,1937 (1978).

13. Beltzer,A.I.,Kramers-Kronig Relationships and Wave Propagation in Composities.J.Acoust.Soc.Am. 73,355 (1983).

14. Beltzer,A.I.,Brauner,N.,Waves of Arbitrary Frequency in Random Fibrous Composites.J.Acoust.Soc.Am. 76,962 (1984).

15. Soczkiewicz, E.,Chivers,R.C.,Acoustic Wave Propagation in Random Media Using the Parabolic Equation and Green s Function.Proc.I.O.A. 9,part 3,193 (1987).

16. Chivers,R,C.,Acoustic Wave Fluctuations in Inhomogeneous Media. J.Phys.D 13,1997 (1980).

17. Ross,G.,Chivers,R,C.,A Note on the Helmholz Equation for Acoustic Waves in Inhomogeneous Media.J.Acoust. Soc.Am. 80,1536 (1986).

18. Soczkiewicz, E.,Chivers,R.C.,Scattering of Acoustic Waves by Turbulence.Proc.I.O.A. 8,part 5,20 (1986).

INVERSE PROBLEM FOR SCATTERING BY ELASTIC CYLINDERS

Alain Cand, and Gérard Quertin
Groupe de Physique des solides de l'E.N.S.- Tour 23 -
Université Paris 7 - 2,Place Jussieu-
75251 Paris Cedex 05 - France-

ABSTRACT

Actually many ways are investigated to attempt to solve many kinds of inverse problems.In this paper, two general solving methods applied to inverse problem for backscattering by elastic homogeneous cylinders at high frequencies are presented. Their respective efficiency are discussed.

INTRODUCTION

The solution of the inverse problem is interesting for many different reasons according to the considered direct problem (Electromagnetism, Geophysics, Acoustics, ...). The main domains which are interested in the inverse problem for backscattering by elastic cylinders are UnderWater Acoustics and Non Destructive Evaluation.

A direct problem is characterized by the knowledge of all parameters (characteristics, properties, ...) of the system and the study of its behaviour in known conditions.

An inverse problem is exactly the inverse study of the previous one. The responses of the system are known for a fixed sollicitation. The search induces the system's characteristics from these ones.

In our particular case, the system is an elastic homogeneous cylinder immersed in water which scatters short ultrasonic pulses at large ka (ka≥200) [k is the wavenumber in water of an ultrasonic plane wave, and a is the radius of the cylinder]. Since only high freqency phenomena are considered, they can be described by a geometrical theory of Acoustics.

In the direct problem the acoustical and geometrical characteristics of the surrounding medium (water) and the scatterer (cylinder) are known data whereas the characteristics of scattered echoes, mainly their arrival times after emission and their amplitudes,are to be determined. Conversely, the inverse search's aim is to calculated the acoustical and geometrical characteristics of the scatterering set-up from the scattered signal experimentally measured.

I) DIRECT PROBLEM : THE RESULTS OF THE GEOMETRICAL THEORY[1,2,3]

1) Geometrical Configuration of the System

The axis of the transmitter-receiver transducer and the scatterer's one are converging. They are perpendicular to each other as seen in Fig.1.

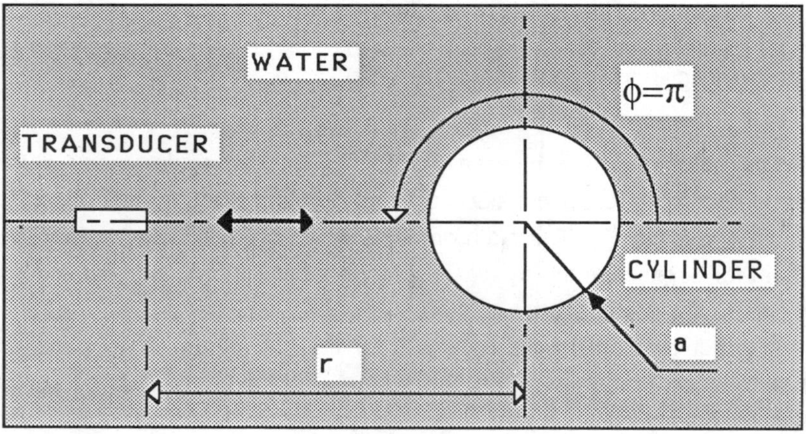

Fig.1 Set-up

The acoustic characteristics of water, the surrounding medium, are its density ($\rho_F = 10^3$ kg/m^3) and the sound's velocity inside it (c_F=1480 m/s). The cylinder's ones are its density (ρ_C), the velocity of shear waves (c_S) and the velocity of compressional waves (c_L). The geometrical characteristics of the experimental set-up are the radius of the cylinder (a) and the radial distance between the probe and the cylinder's axis (r).

After the scatterer was irradiated by ultrasonic pulse, the transducer receives a set of backscattered echoes characterized by their arrival times after emission ($T(n,m,\alpha_i)$) or after reception of specular echo ($t(n,m,\alpha_i)$) and their amplitude ($A(n,m,\alpha_i)$) (see Fig.2).

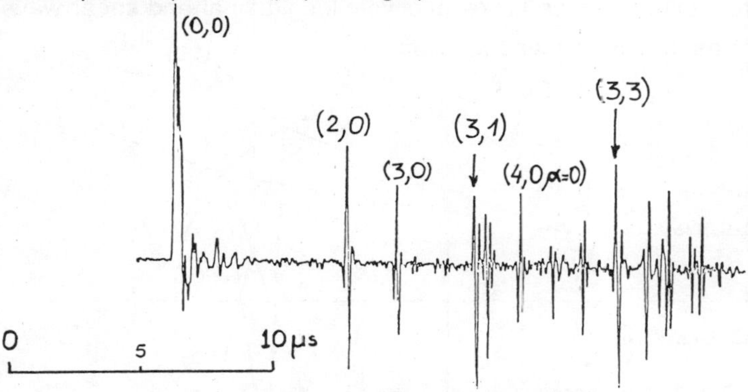

Fig.2 Set of backscattered echoes

2) The (n,m) "Modes" and the Characteristic Equation

Using a geometrical theory of acoustics (Rays Theory), we can study the path of an incident ray inside a cross-section of the cylinder (see Fig.3). A (n,m) "mode" generated at the angle of incidence of the ray upon the cylinder α, observed at the angle ϕ (here ϕ=180°), is a ray which follows a path inside the cross-section of the target as a zig-zag line with or without mode conversions at each reflection or refraction upon the cylinder-fluid interface. The total number of traverses inside the cylinder , (n), and the number of traverses as shear waves, (m), define (with α) a (n,m) mode. Then (n-m) is the number of traverses as longitudinal waves.

The Characteristic Equation or "modal equation" (1) provides us for a particular (n,m) mode a set of angles of incidence upon the boundary of the cylinder $\{\alpha_i\}$ which allow to receive the scattered echo (n,m) at the observation angle ϕ (here $\phi=\pi$ radians, backscattering case)[1,2].

$$f_{nm}(\alpha,\phi,l) = 0 \tag{1}$$

where l is the number of paths arround the cylinder's axis.($l=0,1,\ldots,l_{max}\leq n/2$) and where:

$$f_{nm} = -2\sin^{-1}(N) + 2\sin^{-1}\left(\frac{N}{R}\right) + 2(n-m)\sin^{-1}\left(\frac{N}{\xi_L}\right) + 2m\,\sin^{-1}\left(\frac{N}{\xi_S}\right) - (n-2l-1)\pi \pm \phi \tag{2}$$

with : $N=\sin\alpha$, $R=r/a$, $\xi_L=c_F/c_L$, $\xi_T=c_F/c_S$.

ξ_L and ξ_T are refraction indexes for respectively longitudinal and shear waves in cylinder, $\sin^{-1}$ as the inverse sine function.

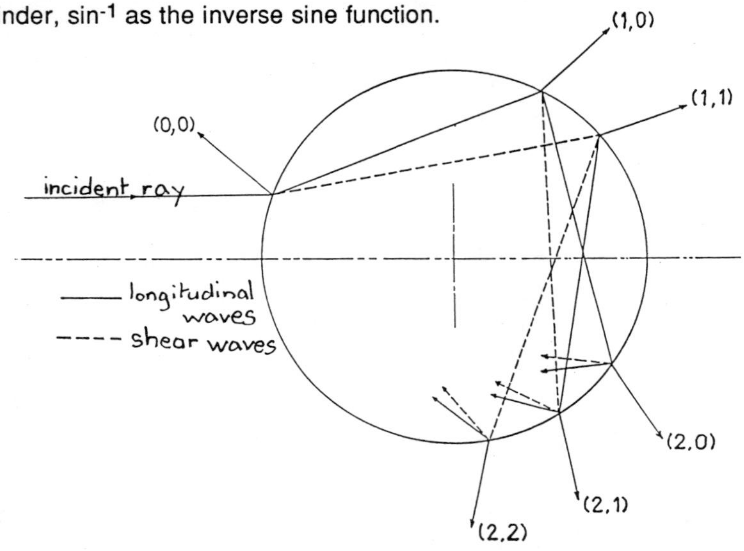

Fig.3 : Examples of (n,m) modes

3) The Arrival Times of Backscattered Echoes

The arrival time of backscattered echo after emission (3) is for a fixed (n,m,α) echo:

$$T_{(n,m)} = 2\frac{a}{c_F}\left\{ (R^2-N^2)^{\frac{1}{2}} + (n-m)\,(\xi_L^2-N^2)^{\frac{1}{2}} + m\,(\xi_S^2-N^2)^{\frac{1}{2}} - (1-N^2)^{\frac{1}{2}} \right\} \tag{3}$$

$$T_{(0,0)} = 2\frac{(r-a)}{c_F} = 2\frac{a}{c_F}\,(R-1) \tag{4}$$

$T_{(0,0)}$ is the arrival time of specular echo after emission (4). Then the arrival time of a (n,m) echo after reception of specular echo (5) is :

$$t_{(n,m)} = T_{(n,m)} - T_{(0,0)} \tag{5}$$

In order to choose the relevant echoes for Inverse Problem solving, we can notice that $t_{(n,m)}$ is a function depending on c_F, c_L, c_S, a and r. The three last previous parameters are the searched solutions of the Inverse Problem.

4) The amplitude of backscattered echoes.

The relative amplitude of backscattered (n,m,α) echo (6) is directly proportional to the scattered pressure on the probe.

$$A(n,m,\alpha) = \frac{P_S(n,m,\alpha)}{P_S(\varepsilon_0)} = 2\,(D\,(\varepsilon_0)\,)^2\, G(n,m,\alpha)\,[\,L(n,m,\alpha)\,B(n,m,\alpha)\,]^{-\frac{1}{2}} \tag{6}$$

where : $P_S(n,m,a)$ is the pressure due to (n,m,α) echo received upon the transducer.

$P_i(\varepsilon_0)$ is the pressure which would be incident upon the center of the cylinder in its absence.

$D(\varepsilon_0)$ is the directivity factor of the transducer.

B(n,m,α) is the geometrical spread in the plane perpendicular to the cylinder's axis.

L(n,m,α) is the geometrical spread in the direction parallel to the axis.

G(n,m,α) is the coefficient of the transmission and reflection losses.

As for our previous remark, A(n,m,α) is a function depending on the same parameters plus ρ_C, the density of the cylinder. Hence, we need to know A(n,m,α) to find this characteristic of the scatterer in Inverse Problem Resolution.

II) INVERSE PROBLEM : THE LIST METHOD

1) The Description of the Method.

The List Method was achieved by Dr. F. Luppé [4,5]. It uses a list of common materials, 3 hypothesis deduced from it, a comparison criterion and some relevant echoes.

MATERIALS	Density ρ_C(kg/m^3)	c_L (m/s)	c_S (m/s)	ξ_L	ξ_S	ξ_S/ξ_L
Cast Iron	6900	3500	2200	0.423	0.673	1.591
Porcelaine	2400	5600	3500	0.264	0.423	1.600
Alumina	3600	9000	5500	0.164	0.269	1.636
Tungsten Carbid	13800	6660	4000	0.222	0.370	1.665
Zinc	7100	4170	2410	0.355	0.614	1.703
Stainless Steel	7800	5820	3170	0.254	0.467	1.839
Magnesium	1700	5770	3050	0.256	0.485	1.892
Nickel	8800	5630	2960	0.263	0.500	1.902
Aluminium	2700	6320	3130	0.234	0.473	2.021
Copper	8900	4760	2330	0.311	0.635	2.043
Tungsten	19300	5410	2640	0.274	0.561	2.049
Bronze	8900	4550	2140	0.325	0.692	2.126
Brass	8600	4700	2110	0.315	0.701	2,227
Silver	10400	3650	1610	0.405	0.919	2.267
Platinum	21400	3960	1670	0.374	0.886	2.371

Table 1: Non-Exhaustive List of Velocities of Longitudinal and Shear Waves in some Common Materials and the Connected Refraction Indexes

First the List is built with the refraction index for longitudinal waves ξ_L or the ratio ξ_S/ξ_L of the cylinder's solid material growing up (see the non-exhaustive joint list). Some basic properties of the relative repartition of velocities for these materials allow us to state the 3 following hypothesis :

$$1 < \xi_S/\xi_L = c_L/c_S < 3.8 \qquad \text{(H1)}$$

$$0.16 \le \xi_L = c_F/c_L \le 0.94 \qquad \text{(H2)}$$

$$0.16 \le \xi_S = c_F/c_S \le 0.94 \qquad \text{(H3)}$$

and to extrapolate them as properties of all common solid materials. These hypothesis limit the size of space search to more likely physical solutions and then increase the rapidity and the efficiency of the method.

For each possible solution, the calculated arrival time of a particular (n,m) echo $t^C_{(n,m)}$ have to be compared to the measured one $t^M_{(n,m)}$. This operation needs a comparison criterion (7). A set of parameters (a, R, ξ_L, ξ_S) may be accepted as a possible solution if $t^C_{(n,m)}$ and $t^M_{(n,m)}$ verify :

$$| t^C_{(n,m)} - t^M_{(n,m)} | \leq \Delta T \left(1 + \frac{t^C_{(n,m)}}{t^M_{(2,0)}} \right) \tag{7}$$

where ΔT is the experimental error of measurement on arrival times.

The (n,m) echoes used are (0,0),(2,0),(3,0),(3,1), and (4,0) with α=0 for the last one. The (0,0) echo fixes the reference for measurements of $t_{(n,m)}$. The 3 following ones, (2,0),(3,0) and (3,1) permit us to evaluated the parameters a, R, ξ_L, ξ_S and then a, r, c_L, c_S where the two first echoes (m=0) give c_L and (3,1) gives c_S. The last echo, (4,0), provides us a means to find ρ_C and then to select the real material constituting the cylinder.

The procedure proceeds in 3 main stages corresponding to the use of the 3 echoes (2,0), (3,0), (3,1) successively. The intervals of variation of refraction indexes, (H2) and (H3), are sampled by step of 0.01. For each of the samples ξ_L inside interval (H2) a list of (a,R,ξ_L) is built with:

$$a = \frac{c_F}{4\xi_L} t^M_{(2,0)} \tag{8}$$

and

$$R = \frac{r}{a} = 1 + 2\,\xi_L \frac{T^M_{(0,0)}}{t^M_{(2,0)}} \tag{9}$$

from $T^M_{(0,0)}$ and $t^M_{(2,0)}$. The number of (a,R,ξ_L) is then reduced when we keep only the sets (a,R,ξ_L) which verify the comparison criterion for (3,0) echo. On the last stage, as in the first stage, a list of sampled values of (a,R,ξ_L,ξ_S) is built by association of each elements of ξ_S list to the elements (a,R,ξ_L). Then, the comparison criterion is applied to (3,1) echo for the (a,R,ξ_L,ξ_S) verifying the (H1) hypothesis. The remaining sets of parameters form the possible solutions' list.

A relevant comparison's parameter have to be chosen to compare the possible solutions to the elements of the common materials' list. It has been shown that the choice of the ratio ξ_S/ξ_L is quite more judicious than ξ_L or ξ_S one, for this method, as comparison parameter.

The possible solutions which are compatible with some elements of the list of materials (i.e. ξ_S/ξ_L is the same at less than 10^{-1}) are kept if they are compatible for ξ_L and ξ_S (i.e. ξ_S and ξ_L are the same ones at less than 10^{-2}) .If after this selection stage more than a single set (a,R,ξ_L,ξ_S) is a solution, then the list method uses the measured values of (4,0) and (2,0) echoes' amplitudes, respectively $A^M{}_{S\,(2,0)}$ and $A^M{}_{S\,(4,0)}$ in order to find ρ_C and then to identify the material. Since the reflection coefficient for compressional waves (longitudinal-longitudinal reflection coefficient) in solids R_{LL} is function only of $A^M{}_{S\,(4,0)}/A^M{}_{S\,(2,0)}$ ratio, R and ξ_L, then ρ_C (10) is obtained by :

$$\rho_C = Z\left\{\frac{1-R_{LL}}{1+R_{LL}}\right\}\frac{\xi_L}{c_F} \tag{10}$$

where Z is the acoustic impedance of surrounding fluid. Finally the solution is the set of parameters for which ξ_L, ξ_S, ρ_C are comparable with the one obtained for a given material inside the list.

2) Some Results of the List Method

The numerical data of the problem are :
$\Delta T = 30$ ns, $c_F{}^M = 1480$ m/s $^{\pm 7\ m/s}$, $\rho_C{}^M = 1000$ kg/m^3, r = 160.0 mm $^{\pm 0.5\ mm}$

First we consider the scattering from a duraluminum cylinder of 20.48 mm diameter. We apply the first steps of the list method to (0,0), (2,0), (3,0) and (3,1) echoes obtained experimentally. Two set of parameters may be solutions after the first three stages and comparison with the List , the first one corresponding with Aluminum and the second one with Copper :

$a^A = 10.30$ mm, $r^A = 159.5$ mm, $\xi_S/\xi_L{}^A = 2.043$, $\xi_L{}^A = 0.230$, $\xi_S{}^A = 0.470$

and

$a^C = 7.40$ mm, $r^C = 156.6$ mm, $\xi_S/\xi_L{}^C = 2.047$, $\xi_L{}^C = 0.320$ and $\xi_S{}^C = 0.655$

To find the solution, the calculation of the density of the cylinder is needed. It provides : $\rho^A = 2700$ kg/m^3 for the first set and $\rho^C = 3400$ kg/m^3 for the second one.

The real solution is then Aluminum because there is compatibility only for it for density. Finally we find :
$(r, a, c_L, c_S) = (159.5$ mm, 10.30 mm, 6435 m/s, 3149 m/s$)$. The errors on determination of parameters are almost equal to 1%.and respectively eqal to $(\delta r, \delta a, \delta c_L, \delta c_T) = (0.3\%, 0.6\%, 1\%, 0.5\%)$.

Now let's consider a Steel cylinder of 19.95 mm diameter. In this case it remains only one set of parameters after the first three stages and comparison with the List which corresponds with Stainless Steel :

a = 10.10 mm, r = 158 mm, ξ_S/ξ_L = 1.800, ξ_L = 0.250 and ξ_S = 0.450

then c_L = 5920 m/s and c_S = 3290 m/s and the errors on their determination are: $(\delta r, \delta a, \delta c_L, \delta c_T)$ = (1.3%,1.3%,1.2%,0.3%).

We can notice that the errors on determination of parameters are almost equal to 1%, then the accuracy of the method is quite good but the necessity of using of a list of materials limits the use's area of the List Method.It is also unsatisfactory to have no direct access to the velocities. The method which are presented afterward attempt to solve these defaults.

III) INVERSE PROBLEM : THE "ARTIFICIAL INTELLIGENCE" METHOD

1) The Best-First Procedure : General Description[6,7,8,9]

The Best-First Procedure is a solving method using an heuristic search in a tree procedure. This procedure belong to the class of Heuristic Procedures. This process is a means to group the advantages of Depth-First Procedure and Breadth-First Procedure and it avoids the most of their disadvantages.

This method needs needs to build a search-tree, and to use an heuristic evaluation function F, an opened nodes set $\mathbb{O}$ and a closed nodes set $\mathbb{C}$.

Generally, a search-tree is the search-space. It is compound on the one hand by nodes which are the Problem's configurations and on the other hand by branches which are the paths which connect nodes to their sucessors (nodes created by generation rules from a parent-node which give termination criterion and tell how to generate successors of any position in the tree) in the Problem's configuration space. This search procedure tests first the best node or more promising node among the nodes generated up to the generation of the current node and contained in the set $\mathbb{O}$.

The heuristic evaluation function F allow to "measure the proximity" of a goal-node (or solution-node) from the current node. It helps to discover the solution. At each step of the search process, its use allows to consider, at first, the more promising node or nearest node from a solution. For example, the heuristic evaluation function are :

- for Chess : F is the advantage of pieces against the adversary's ones
- for Puzzle : F is the number of pieces in right positions or misplaced
- for Commercial Traveller's Problem : F is the sum of distances he covered over here

The opened nodes set $\mathbb{O}$ contains the generated and evaluated nodes for which their successors have not been generated because when these nodes was the current ones they was not the more promising nodes.

The closed nodes set $\mathbb{C}$ contains the nodes from which the successors have been generated and are not the solutions.

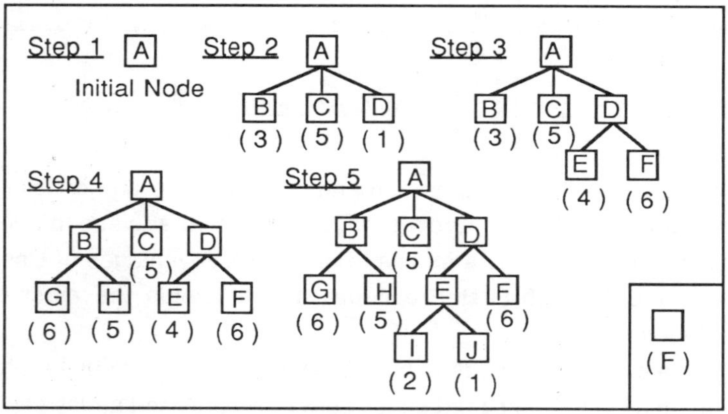

Fig.4 : Example of a Best-First Procedure

2) The A* Algorithm

2.1) Brief description

It's a particular Heuristic Procedure first describes by Hart and al. in 1968[6] and in 1972 [7] (see also ref. 8 and ref. 9), where the tree of search is a network of search and the heuristic evaluation function is shared into two parts, one which measures the cost from the initial node to the current one (G) and another one which estimates the supplementary cost to reach a goal-

The estimator of the heuristic evaluation function is F^*.(11)

$$F^*(n) = G(n) + H^*(n) \tag{11}$$

where n is the node to be evaluated, G(n) is the "cost" of path from initial node to the n node, $H^*(n)$ is the estimation of the additionnal "cost" to reach a goal-node from the actual n node.

2.2) Example of use of A* Algorithm : the 9-Pieces-Puzzle

This problem is an example taken from historically the main field of application of heuristic algorithms : the game's field. It is an excerpt of ref.9.

The aim of this problem is to transform by using only one piece plane translation the initial configuration N_0 into the goal configuration N_{Goal}. G(n) is the number of moves from N_0 to actual configuration n and $H^*(n)$ is the number of pieces misplaced.(see Fig.5).

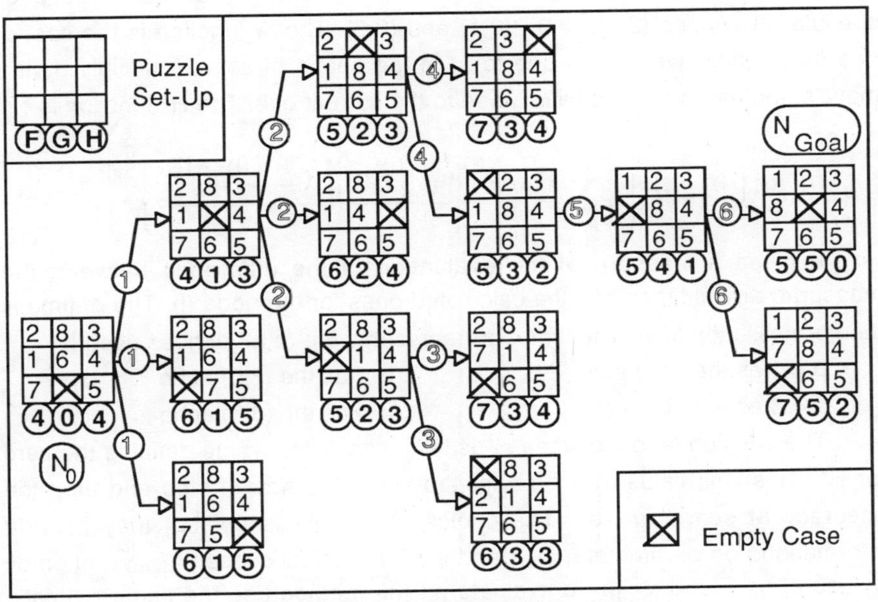

Fig.5 : A* Search for 9-Pieces-Puzzle

3) A* Algorithm and the Inverse Problem

The tree of search or space of search is a three dimensionnal space $[a,c_L,c_S]$, where a, c_L and c_S are the parameters of the search sampled from their minimum values (resp. a^{min}, $c_L{}^{min}$, $c_S{}^{min}$) to their maximum ones (resp. a^{max}, $c_L{}^{max}$, $c_S{}^{max}$) by step respectively Δa, Δc_L, Δc_S. Its nodes are enclosed between the planes which equations are : $c_L = 1.5\ c_T$ and $c_L = 3\ c_T$, where stay almost all physical solutions. The initial node is $(a^0,c_L{}^0,c_T{}^0)$. The efficiency and rapidity of the method depends strongly on its choice since more the initial node is far from the solution more the search may be unsuccessful. Each node of the inner search space is connected with six nearest neighbours and then the allowed moves from one nodes are only one step move to one of the five nodes which are not the current node's parent.

The heuristic evaluation function is built with the arrival times after specular of echoes (2,0), (3,0), (3,1) and (3,3). The G function is the cost of one move (here we fix this one to 1) in the space of search multiply by the move's number to join the initial node to the current one. The H* function is :

$$H^*(a^{(i)},c_L^{(i)},c_S^{(i)}) = \mathrm{INT}\left(\left\{\sum_{p=1}^{4}\frac{|t_{(p)}^{C}(a^{(i)},c_L^{(i)},c_S^{(i)}) - t_{(p)}^{M}(a^{(i)},c_L^{(i)},c_S^{(i)})|^2}{\Delta t}\right\}^{\frac{1}{2}}\right) \tag{12}$$

It gives a measure of the evaluation of the difference between the measured arrival times and the calculated ones for the node (i). The Δt time is the normalisation time. The index (p) represents the (n,m) echoes used :

p = 1 as the (2,0) echo   p = 2 as the (3,0) echo

p = 3 as the (3,1) echo   p = 4 as the (3,3) echo

The solution is found when H* is less than a limit value defining the zero for H*. This limit value may be adjusted when the size of step and then the accuracy of search varies. The choice of echoes is so that they provide informations on parameters, c_L, r and a by use of all chosen $t_{(n,m)}$ and on c_T by use of $t_{(3,1)}$ and $t_{(3,3)}$. To evaluate ρ_C the method use the same heuristic function of evaluation but with relative amplitudes $a_{(p)}$ rather than the arrival times $t_{(p)}$. The value of r (13) is not obtained directely as for a, c_L, c_T and ρ_C but by use of the measured arrival time after emission of (0,0) echo $T_{(0,0)}{}^M$ and the evaluated value of the radius a :

$$r = a + c_F \frac{T^M_{(0,0)}}{2} \tag{13}$$

In our simulations, we consider metallic cylinders about 10 mm diameter. We limited the space of search with the following borders :

$9\ mm \leq a \leq 11\ mm$ and $\Delta a = 0.1\ mm$

$2700\ m/s \leq c_L \leq 10500\ m/s$ and $\Delta c_L = 50\ m/s$

$1500\ m/s \leq c_S \leq 3600\ m/s$ and $\Delta c_S = 50\ m/s$

$2600\ kg/m^3 \leq \rho_C \leq 10600 kg/m^3$ and $\Delta \rho_C = 50\ kg/m^3$

and with : $1.5\ c_T \leq c_L \leq 3\ c_T$

We studied the same scatterers as for the List Method.

For the duraluminum cylinder of 20.48 mm diameter with the initial node $(a^0, c_L{}^0, c_T{}^0)$.= (11 mm, 6000 m/s, 3500 m/s) the A* search provided :
(r, a, c_L, c_T).= (160.1 mm, 10.3 mm, 6400 m/s, 3150 m/s)
and then the errors on the evaluation of these parameters are respectively $(\delta r, \delta a, \delta c_L, \delta c_T)$ = (0.1%,0.6%,0.5%,0.5%) which are less than 1% and they are at the most equal to the errors obtained by the List Method.

For the stainless steel cylinder of 19.95 mm diameter,from the initial node
$(a^0, c_L{}^0, c_T{}^0)$.= (10 mm, 6600 m/s, 2600 m/s) the A* search provided :
(r, a, c_L, c_T).= (160.0 mm, 10.0 mm, 5850 m/s, 3300 m/s)
and then the errors on the evaluation of these parameters are respectively $(\delta r, \delta a, \delta c_L, \delta c_T)$ = (0%,0.3%,0%,0%) which are less than 1% and smaller than the errors obtained by the List Method.

The Heuristic Search A* is much more rapid than the List Method to solve the Inverse Problem for backscattering by elastic cylinders at large ka and its accuracy is quite good and may be much better than the List Method one.

For example ,for our simulations, for a search with a fixed value of the radius of the cylinder its rapidity reach 10 times the List Method one.It is much more rapid for a search in the $[a, c_L, c_T]$ space. But the A* Searche's performances depend strongly of the chosen initial node from which the search starts and of the sampling of the continuous space $[a, c_L, c_T]$. Effectively, due to these reasons, the found solution may be a non-optimal solution but only an approach solution.

The step-to-be is the resolution of Inverse Problem for backscattering by elastic homogeneous cylinders at low ka (ka $\leq$ 30) of which the study needs to use the Resonant Scattering Theory and Frequency Analysis of back-scattered signals.

CONCLUSION AND ACKNOWLEDGMENTS

These two solving methods have been developped to resolve the particular Inverse Problem for backscattering by immersed elastic cylinders at large ka. They present good results to identify immersed cylindrical scatterers. The General Artificial Intelligence Solving Method presented here seems full of promise to be applied to another general classes of inverse problems.

We are grateful to the Direction des Etudes Recherches et Techniques for financial support.

REFERENCES

1. P.J. WELTON, M. de BILLY, A. HAYMAN, and G. QUENTIN, "Backscattering of short ultrasonic pulses by solid elastic cylinders at large ka", J.Acoust.Soc.Am. 67, pp. 470-476, (1980).
2. G. QUENTIN, M. de BILLY, and A. HAYMAN, "Comparison of backscattering of short ultrasonic pulses by solid spheres and cylinders at large ka", J.Acoust.Soc.Am. 70, pp. 870-878, (1981).
3. G. QUENTIN, and M. FEKIH, "Présentation et interprétation des expériences de diffusion par les cylindres élastiques aux valeurs élevées de ka", Revue du CETHEDEC - Ondes et Signal 72, pp. 91-101, (1982).
4. G. QUENTIN, and F. LUPPÉ, "Détermination des propriétées élastiques de cylindres par insonation à ka élevé", Revue du CETHEDEC - Ondes et Signal 78, pp. 65-72, (1984).
5. F. LUPPÉ, "Contribution à la résolution du Problème Inverse dans le cas de cylindres excités par des impulsions ultrasonore de haute fréquence. Etude du problème direct dans le cas de l'incidence oblique.", Thèse de Doctorat de 3ème Cycle, Université Paris 6, (1984).

6. HART, P, E, N.J. NILSON, and B. RAPHAEL, "A Formal Basis for Heuristic Determination of Minimum Cost Paths", IEEE Transactions on SSC, Vol.4, (1968).
7. HART, P, E, N.J. NILSON, and B. RAPHAEL, "Correction to 'A Formal Basis for Heuristic Determination of Minimum Cost Paths', ", SIGART Newsletter, Vol 37, (1972).
8. E. RICH, "Artificial Intelligence", Mac Graw-Hill, (1983).
9. J.-L. LAURIERE, " Résolution de problèmes par l'homme et la machine", Eyrolles, (1987).

ANALYSIS OF ULTRASONIC PULSES USING LIGHT DIFFRACTION

John W. Wolf

Ultrasonics Research Laboratory, Department of Physics,
Georgetown University, Washington, D.C. 20057, USA

Abstract

An experimental apparatus has been developed which analyzes light diffraction patterns produced by pulsed ultrasonic waves. The setup uses a long optical path which resolves diffraction orders produced by low MHz center frequency pulses with repetition frequencies as low as 40 kHz in water. The computer-controlled system digitizes diffraction patterns for in situ analysis of ultrasonic pulses.

Theory describes a diffraction pattern as a real time Fourier transform of the pulsed ultrasonic wave which produces it. For comparison, an empirical model of the transducer output in response to c.w. excitation with different frequencies was established using the optical apparatus. The response model is used to calculate the content of pulse Fourier spectra, which are input to the diffraction model to predict the shape of light diffraction patterns. The results demonstrate the feasibility to reconstruct the actual pulse shape from an analysis of intensity and frequency-related spacing of the orders in light diffraction patterns.

Introduction

Light diffraction by ultrasonic waves, long seen as a non-invasive means of analyzing ultrasonic fields, was first described by Raman and Nath[1] for the case of light diffraction by monochromatic, low MHz acoustic beams in transparent media. Their theory is limited in application since ultrasonic beams encountered in the laboratory are usually polychromatic. Multi-frequency acoustic beams were treated by Murty[2] but it was Hargrove[3] and Zitter[4] who formed the basis of light diffraction experiments involving simple ultrasonic pulses. The experiments of Häusler, Mayer and Schwartz[5] demonstrated a qualitative relationship between the harmonic content of an ultrasonic field and its light diffraction pattern.

For pulse diffraction, the complex theory requires a numerical solution. This was recently accomplished by Neighbors and Mayer[6] where a description of light diffraction by pulsed waves has been programmed into a computer model. A useful result of this theory, as suggested experimentally[5], is that, for appropriate pulse amplitudes, a light diffraction pattern represents the Fourier transform of the pulsed ultrasonic beam which produces it.

This paper describes an apparatus that optically probes pulsed ultrasonic beams. The apparatus first measures an ultrasonic transducer's response to c.w. excitation which is used to calculate ultrasonic pulse Fourier spectra and, with the model of Neighbors and Mayer, the structure of their light diffraction patterns. Light patterns then measured with the optical apparatus show a quantitative agreement with those predicted by theory.

Light Diffraction Theory

This section reviews the theory of Neighbors and Mayer[6] to describe light diffraction by pulsed ultrasonic waves which starts by assuming a plane wave optical beam of wavelength λ and width L, incident normal to the propagation direction of a periodic train of ultrasonic pulses. The time history of a pulse train with repetition and center modulation frequencies, f_p and f_o, is expressed by a Fourier sine expansion

$$\nu(x,t) = \nu \sum_{n=0}^{\infty} \alpha_n \sin[n(\omega_p t - k_p x) + \beta_n] \tag{1}$$

where: $\omega_p = 2\pi f_p$, $k_p = 2\pi/\lambda_p$, and λ_p = the pulse length in the medium. The values α_n and β_n are the amplitude and phase of the nth Fourier component of the pulse and ν is proportional to the pulse peak amplitude when the amplitude is small. The function $\nu(x,t)$ represents the light phase grating which reduces to one sine function for the case of a plane, monochromatic ultrasonic wave[1].

With due regard for theoretical limitations on the optical beam configuration, the approximate light intensity in the diffraction pattern is given[6] by

$$I_0 = \left| \prod_{n=1}^{\infty} J_0(\nu_n) \right|^2 \tag{2}$$

for the central order, and for the higher orders by

$$I_{\pm m} = \Bigg| J_{\pm 1}(\nu_m) \exp(\pm i\beta_m) \prod_{\substack{n=1 \\ n\neq m}}^{\infty} J_0(\nu_n)$$

$$+ J_2(\nu_{m/2}) \exp(\pm i2\beta_{m/2}) \prod_{\substack{k=1 \\ k\neq m/2}}^{\infty} J_0(\nu_k)$$

$$+ \sum_{\substack{n=1 \\ n<m/2}}^{\infty} \{J_{\pm 1}(\nu_n)\, J_{\pm 1}(\nu_{m-n}) \exp[\pm i(\beta_n + \beta_{m-n})] \prod_{\substack{k=1 \\ k\neq n,m-n}}^{\infty} J_0(\nu_k)\}$$

$$+ \sum_{n=1}^{\infty} \{J_{\mp 1}(\nu_n)\, J_{\pm 1}(\nu_{n+m}) \exp[\pm i(\beta_{n+m} - \beta_n)] \prod_{\substack{k=1 \\ k\neq n,n+m}}^{\infty} J_0(\nu_k)\} \Bigg|^2 \tag{3}$$

where: $\nu_n = \nu\, \alpha_n$, Π indicates the product over n, and all terms of noninteger summation indices vanish. The light patterns exhibit maxima at diffraction angles Θ that satisfy

$$\sin\Theta_m = \pm m\lambda/\lambda_p. \tag{4}$$

For very small pulse amplitudes $I_0 \simeq 1$ and

$$I_{\pm m} \simeq \tfrac{1}{4}\, (\nu_m)^2. \tag{5}$$

Equations 4 and 5 relate the dominant terms of the pulse amplitude spectrum directly to individual diffraction orders, producing an optical Fourier transform.

Optical Apparatus

The width and length of the optical beam depend on the desired acoustic frequency resolution. The diffraction order resolution is determined through Eq. 4 by the pulse repetition frequency and there must be a certain number of pulses within the illumination window at any instant in order to have sufficient light intensity in a diffraction pattern.

The apparatus uses He-Ne laser light with $\lambda = 6328$ Å and accommodates a light beam with an object to image plane distance of 7 m and width L = 9 cm, see Fig. 1. A cylindrical lens diverges the light beam in one dimension only, allowing all the laser light to pass through the acoustic beam, insuring proper normalization of a diffraction pattern. The acousto-optic interaction takes place within a channel of 4.5 cm width, bounded by glass windows allowing the light beam to intersect normal to the ultrasonic beam within the plastic interaction vessel. A second cylindrical lens collimates the light into separate beams just beyond the image plane. Light intensity measurements are performed with a photomultiplier tube detector positioned at the screen. A PC-controlled data acquisition system rotates the flat mirror, which passes a diffraction pattern across a slit that opens onto the detector tube element. Given a c.w. acoustic beam of 1 MHz, the optical system produces a diffraction pattern at the screen with an order separation of 12 cm.

Random system fluctuations influence the acquisition of data during a single pass of a light pattern. Therefore, the PC scans a single pattern three to five times, averages, normalizes to the peak light intensity, and stores the data in a disk file. The one PC fully manages experimental input, procedure and output. No operator has to be present in the laboratory which isolates the apparatus and minimizes fluctuations usually encountered with optical beams of long path length.

Acquiring a transducer response function typically involves applying a pulsed electric signal of known spectral composition to the transducer. Usually the shape of the pulse is sufficiently narrow so as to approximate a delta function whose Fourier spectrum is constant for all frequencies. Even though the transducer receives all frequencies with equal amplitude, in converting electrical energy to mechanical oscillation, it can use only those frequencies allowed by its response.

Measuring this response with another transducer, or probe, introduces an additional response to the response function being measured and distorts the acoustic field. Acousto-optic interaction of an ultrasonic signal with a beam of light produces no measurable distortion of the acoustic field; therefore, optical probing has the advantage of measuring the field noninvasively.

Optically probing the response of an ultrasonic source is also accomplished by sending many frequencies of equal amplitude to the transducer and measuring the resulting output. However, introducing a single delta function pulse to the source would not produce a measurable diffraction pattern. With the techniques used for the present experiments, one sends a specific range of frequencies, each equal in amplitude, to the ultrasonic source except that they are sent individually at different times.

Each frequency is input continuously, that is, a c.w. wave is produced which results in a Raman-Nath diffraction pattern. The data acquisition system digitizes each pattern and stores the data. The PC instructs the function generator supplying electronic signals to the transducer system to increase the input frequency a small increment while keeping the voltage amplitude constant. Incrementing the frequency continues until the nominal bandwidth of the transducer is measured.

The resulting series of light patterns are sampled to extract the peak intensity values of individual diffraction orders. These are converted by computer program into values of the Raman-Nath parameter, ν, which is assumed proportional to the output amplitude of the transducer. The data points are then smoothed using a

Blackman window filter[7]. The resulting normalized response curve for a nominal 1 MHz source is shown in Fig. 2. The curve represents the transfer function of the entire transducer system, including the r.f. power amplifier, connecting cables, and mounting of the transducer element.

Using the transducer response curve, pulse shapes are easily modeled from input parameters to the pulse function generator. The function generator outputs an electronic pulse signal of ideal shape which is then sent through the transducer system to produce the ultrasonic pulse in the water medium. A convolution of the measured transducer response and the generator output spectrum provides an accurate approximation to the actual ultrasonic pulse spectrum in the water.

Figure 3-a shows a typical output signal from the pulse function generator, showing both its time history and Fourier amplitude spectrum. This ideal shape is the same for all pulse trains output from the generator regardless of the values of f_p, f_o and burst number, (integral number of wavelengths defining the pulse width). The particular data used to generate Fig. 3-a is: f_p = 66.0 kHz, f_o = 0.988 MHz, and burst number 5. The amplitude spectrum is then used in a discrete convolution with the transducer response function of Fig. 2. The resulting pulse time history and spectrum are shown in Fig. 3-b. The new pulse spectrum can now be used by the light diffraction model to compare with measured light patterns.

Experimental Light Patterns

A measured pattern is acquired by positioning the ultrasonic source at the edge of the light beam and running the data acquisition system which digitizes the resulting diffraction pattern. Figure 4 shows a comparison of the measured light pattern to the model prediction, using the pulse spectrum of Fig. 3, and shows good agreement in most regions of the pattern. Slight deviations are attributed to measurement error in the response curve in regions far off the resonant frequency. The first set of satellite diffraction orders shows a direct correspondence to the pulse amplitude spectrum of Fig. 3-b as predicted by Eq. 5.

A second example is shown in Fig. 5 where the diffraction pattern is produced by the same 1 MHz ultrasonic source with pulse input parameters: f_o = 0.925 MHz, f_p = 132.2 kHz, and burst number 2. Both experiment and model show asymmetry in intensity about I_0 because this particular pulse contains harmonic terms in its spectrum due to nonlinear processes.

Figure 6 shows three experimental patterns produced by acoustic beams with f_o = 0.987 MHz. The first is by a continuous beam with ν = 2.00. This, evidently, is a Raman-Nath pattern for a 0.987 MHz wave in water with a repetition frequency of zero. The next two graphs show the first set of satellite diffraction orders in pulsed beam patterns for ν = 1.50 with repetition rates and pulse widths: b) 197.4 kHz, burst number 3, and c) 40.0 kHz, burst number 5. In the third pattern, the limiting spectral resolution of the optical apparatus is approached in that the diffraction order peaks are just resolved when the frequency is as low as 40 kHz.

Conclusions

An optical apparatus has been developed for analysis of pulsed ultrasonic beams with repetition frequencies as low as 40 kHz. With the apparatus, transducer frequency response is measured and provides a method to model ultrasonic pulse spectra in the liquid medium. The spectra were used for input to the light diffraction model whose output compares well with actual digitized diffraction patterns. The experiment shows a quantitative agreement with the diffraction theory and suggests a technique to noninvasively measure ultrasonic pulses whereby the light beam performs a real time Fourier transform of the ultrasonic signal.

Acknowledgement

This work was supported by the Office of Naval Research, U.S. Navy.

References

1. Raman, C. V. and Nath, N. S. N., Proc. Natl. Acad. Sci. A2, 406 (1935).
2. Murty, J. S., J. Acoust. Soc. Am. 26, 970 (1954).
3. Hargrove, L. E., J. Acoust. Soc. Am. 43, 847 (1968).
4. Zitter, R. N., J. Acoust. Soc. Am. 43, 864 (1968).
5. Häusler, E., Mayer, W.G. and Schwartz, M., Acoust. Lett. 4, 180 (1981).
6. Neighbors, T. H. III and Mayer, W. G., J. Acoust. Soc. Am. 74, 146 (1983).
7. Blackman, R, B. and Tukey, J. W., The Measurement of Power Spectra, Dover Publications, 1958.

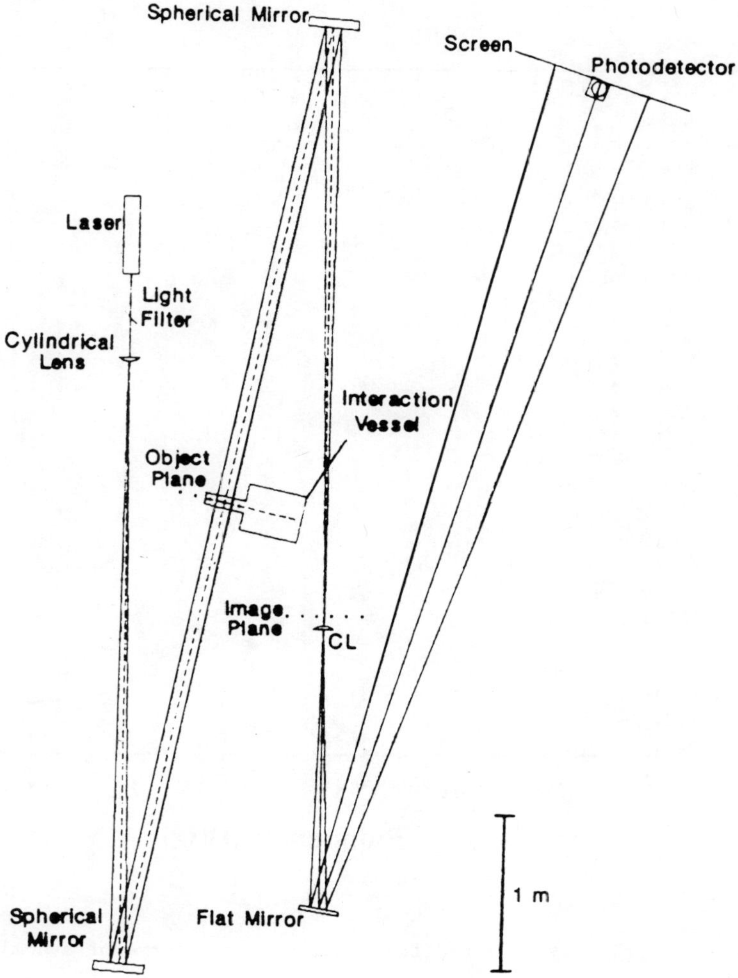

Fig. 1. Schematic of the optical apparatus, overhead view. The light path dimensions are approximately to scale.

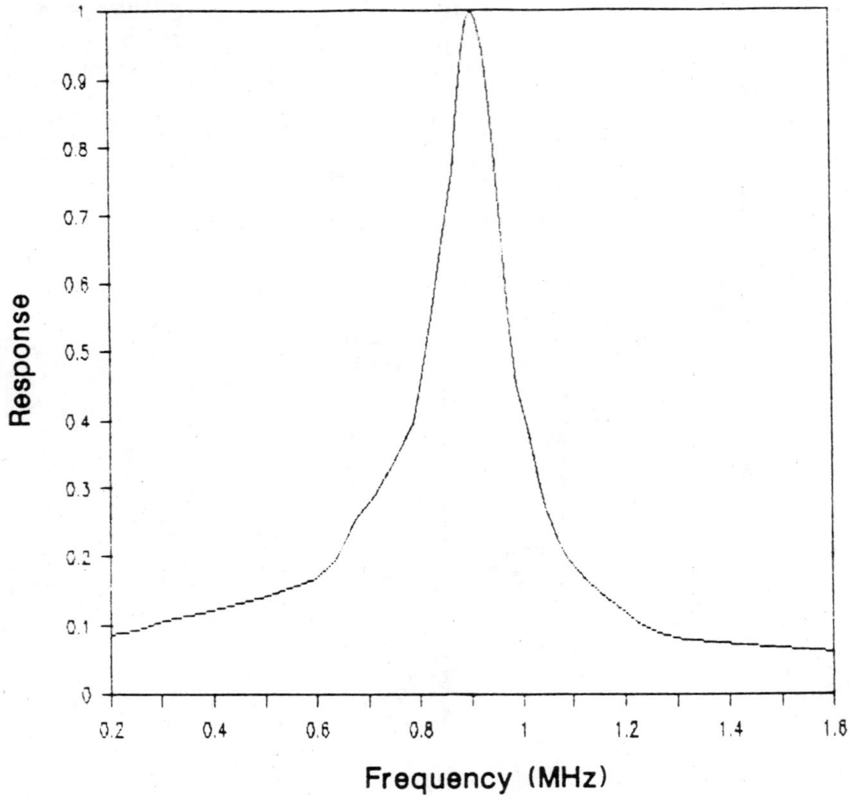

Fig. 2. Optically measured frequency response curve of a nominal 1 MHz, PZT disc of 2.5 cm diameter that is air-backed mounted in a plastic housing.

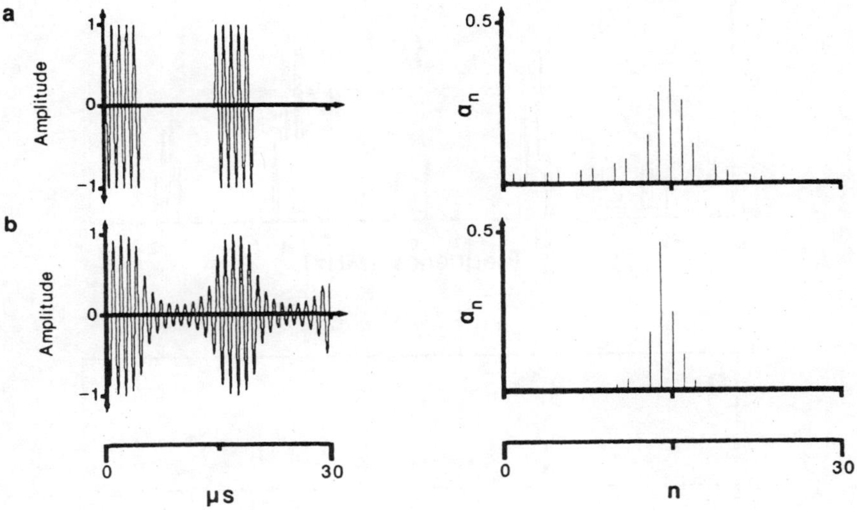

Fig 3. Pulse time history and Fourier amplitude spectrum with: f_p = 66.0 kHz, f_o = 0.988 MHz, burst number 5, a) ideal pulse function generator output, and b) the resulting convolution with the response curve of Fig. 2.

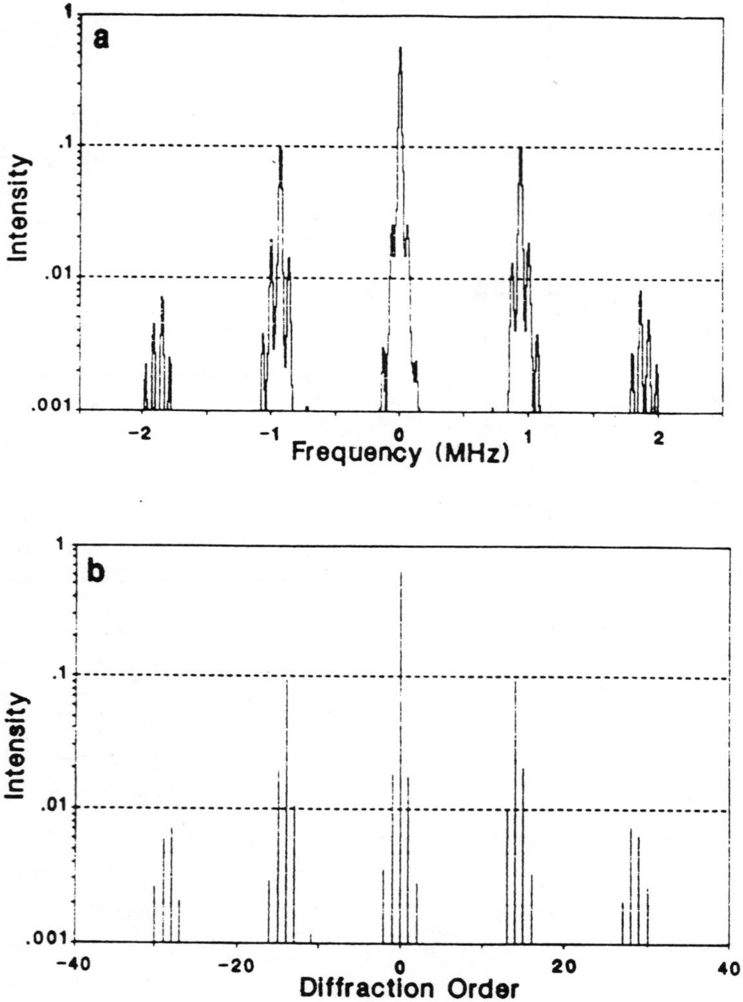

Fig. 4. a) Measured and b) modeled diffraction patterns produced by the pulsed beam of Fig 3-b with $\nu = 1.60$.

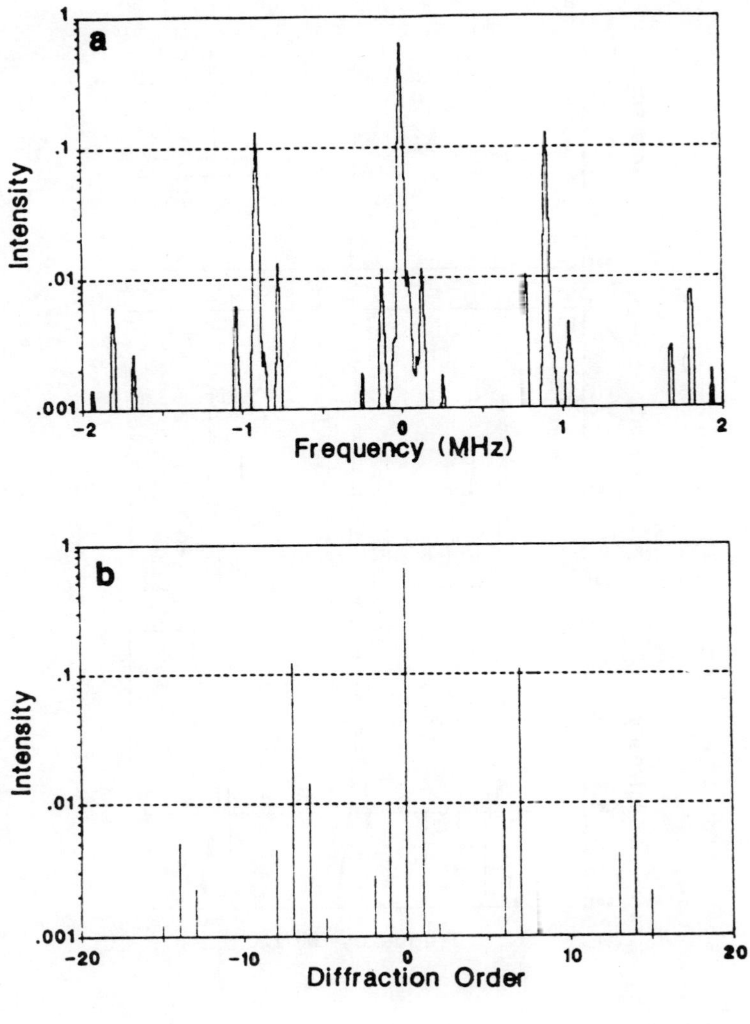

Fig. 5. a) Measured and b) modeled diffraction patterns produced by a pulsed beam with: f_o = 0.925 MHz, f_p = 132.2 kHz, burst number 2, and ν = 1.45.

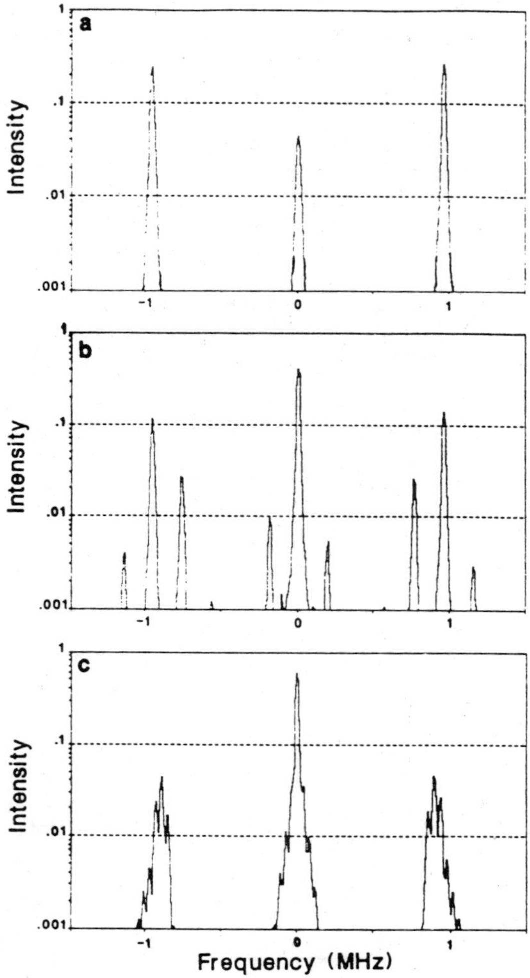

Fig. 6. Measured light diffraction patterns produced by acoustic beams with f_o = 0.987 MHz. a) is a c.w. beam with ν = 2.00 and the others are pulsed: ν = 1.50, b) 197.4 kHz, burst number 3, and c) 40.0 kHz, burst number 5.

ANALYSIS OF ULTRASONIC SPATIAL AND TEMPORAL MODULATION OF LIGHT BY THE PHASE VECTORS

Italo Gabrielli

Dipartimento di Fisica - Università degli Studi, Trieste, Italy
Istituto Nazionale di Fisica Nucleare - Sezione di Trieste, Italy

ABSTRACT
The analysis by the phase vectors of the spatial and temporal modulation of a collimated light beam crossing ultrasonic progressive waves is presented with reference to the interaction at 90 degrees of the light with a single ultrasonic beam and then applied to diffracting systems of two adjacent beams and some experimental verifications are quoted or proposed. The extension of the method to diffracting systems of one ultrasonic beam adjacent to an optical amplitude grating or to the study of the oblique ultrasound-light interaction is considered

1. INTRODUCTION

The diffraction of a collimated light beam by an ultrasonic (US) beam, after the experimental observations [1, 2] has been primarily studied by Raman-Nath (R-N) [3]. Later the analysis of more sophisticated light-sound interaction arrangements has been performed in the frame of an improved R-N theory, of the Fourier series analysis and of the wave equation [4].

All said methods need the use of computers to give predictions to be compared with the experimental results (diffraction pattern and light modulation waveforms). We propose to face directly the problem through the elementary phase-vector analysis (PVA), which presents the advantage of supplying a direct step by step reference to the physical mechanism of the studied interaction, also by computer-handled pictures, and avoids sophisticated mathematical treatment.

After a simplified introduction of the method, performed by applying it to the 90 degree interaction of the light with one US wave, we will discuss its application to the case of two adjacent US waves, to the case of one US wave adjacent to an amplitude grating and to the case of oblique interaction.

2. LIGHT-SOUND INTERACTION AT 90 DEGREES. THE PHASE- VECTOR ANALYSIS OF THE LIGHT INTENSITY OF THE ZERO-ORDER FRINGE

2. 1. General concepts.

We consider in the Fraunhofer arrangement (Fig.1) a collimated light beam crossing at 90 degrees several (N_W) US waves of wavelength (WL) Λ of one US beam which is described as an optical sinusoidal grating of amplitude V radians. We assume as V the R-N parameter $2\pi\ \Delta n\ L/\lambda$ where Δn is the maximum variation of the refraction index, L the actual width of the US beam and λ the WL of the entering monochromatic light.

Let us assume the usual coordinate system resulting from Fig. 1, which defines also the diffraction direction θ_n and the dimensions p and q of the rectangular cross-section of the incident light beam.

We suppose the US beam as frozen in time, its progressing with respect to the slit will be simulated later by increasing of $D\phi x$ the coordinate x, representing the sinusoidal grating (see eq. (2) and (2')).

We indicate by E_0 in arbitrary units, normalized to 1 (or 10 when considering light intensity percentages) the total amplitude of the light (entering through the slit of dimensions p x q). The total entering light intensity will be therefore $I_0 = E_0^2$, still in arbitrary units, normalized to 1 (or 100).

If we consider the US beam as splitted along the x-axis in a great number N of small elements dx, the light beam can be supposed made up by N elementary "light blades", each of section q x dx (see Fig. 1) and therefore carrying on the screen a vectorial contribution of amplitude:

$$\mathbf{dE_0} = (\mathbf{E_0}/N)\ dx, \tag{1}$$

with $\mathbf{E_0}$ assumed as a vector with zero phase reference.

Each light blade interacting with the phase grating at its abscissa x undergoes a dephasing:

$$D'\Phi\ (x,V) = V \sin kx, \qquad \text{where} \quad k = 2\pi/\Lambda. \tag{2}$$

We can treat this vectorial contribution on the screen as a phase vector (PV):

$$\mathbf{dE}\ (x,V) = \mathbf{dE_0} \exp\ (j\ V \sin kx). \tag{3}$$

The total amplitude $\mathbf{E}$ (V,0) of the light corresponding, on the Fraunhofer screen, to the zero order fringe ($\theta_0 = 0$, see Fig. 2a)) will be then:

$$E\,(V,0) \;=\; \int_0^p \mathbf{dE_0}\, \exp\,(j\,V \sin kx) \tag{4}$$

where the first argument indicates the R-N parameter characterizing the dephasing grating and the second one refers to the diffraction direction θ_0.

Of course the vectorial integral (4) is handled by the computers as a vectorial summation of phase-vectors $\Delta\mathbf{E_0}\,(x_i,V)$ and the complex value: exp [j(arg)] as the sum: cos (arg) + j sin (arg).

This calculation procedure is first considered as applied to one US WL and then extended to any number N_W of waves.

2. 2. Interaction with one ultrasonic wavelength.

We assume now first the slit dimension p reduced to one US WL Λ, i.e. $p = \Lambda$ (Fig. 2a). If we let the computer follow graphically the calculation of the light phase-vectors, according to formula (4), for the values V from 0 to 7 radians, it shows the curved lines of Fig. 3c), with the resulting vectors respectively indicated, as an oriented straight line joining the first to the last elementary phase vector on the vectorial integral.

Such entangled curves suggest to define them as "springs". Similar curves, related to greater dephasing effects (greater values of V and of θ_n), appear still more entangled with manifold turnings around and seem "snakes" or even "wool clews" (see e.g. Fig. 7 to 9).

For space saving reasons Fig. 3 and other figures have been drawn with the axes reversed (x-axis representing the cosine component of the vector is drawn vertical and directed downwards and y-axis, i.e. the sine component, horizontal). We note that all vectors present zero dephasing (corresponding also to n 2π, with n positive or negative integer) with respect to the phase of the vector for $n = 0$ and $V = 0$ (phase vector for the diffraction to the angular direction $\theta_0 = 0$, from a slit of opening Λ without ultrasound: imagine Fig. 2a), with the sinusoid removed). The length and the sign of each vector is indicated in the Figure. One can realize that the vector amplitudes correspond to the values of the corresponding Bessel functions J_0 (V) of order 0 and argument V. The results agree very well with the predictions of the R-N theory, according to which the light intensities I (0,V) of the zero-order fringes are given by:

$$I\,(0,V) \;=\; I_0\, J_0^2\,(V) \;=\; E_0^2\, J_0^2\,(V) \tag{5}$$

2\. 3. Interaction with several wavelengths of one ultrasonic beam.

If we suppose that the slit is open to several (say P) US WL , the relative intensity of the diffraction maximum at $\theta = 0$ remains unaltered, as each of the previously considered light amplitude vectors **E** (0,V), now labelled $\mathbf{E}_{1W}$ (0,V), as resulting from the interaction of a light blade of finite width Λ (i.e. with one US WL) arrives on the points of the screen corresponding to $\theta_0 = 0$ with the same phase and hence the overall light vector amplitude results:

$$\mathbf{E}_P\,(0,V) = \Sigma_P\ \mathbf{E}_{1W}\,(0,V) = P\,\mathbf{E}\,(0,V). \tag{6}$$

But such total light amplitude of the zero order fringe is now calculated with respect to an entering light amplitude increased P times with respect to the light interacting with one US WL and hence the relative light intensity is independent of P. What changes is the shape of the diffraction curve, which depends on the ratio $P = p/\Lambda$ (the greater the ratio, the narrower the bell), as it will be discussed in Ch. 5.

3. ANALYSIS OF THE NONZERO DIFFRACTION FRINGES

3. 1. Interaction with one ultrasonic wavelength.

We consider first the angular positions on the screen of the diffraction fringes of positive or negative orders $\pm$ n, due to a periodical grating of constant Λ, corresponding to angles:

$$\theta_n = \text{asin}\,(\pm\, n\, \lambda/\Lambda) \quad \text{(n integer)} \tag{7}$$

according to the classical diffraction theory, based on the Huygens-Fresnell principle. We propose a mathematical model which takes into account the geometric path difference between the different light blades (Fig. 2b) by introducing in the geometric path considered for the $\theta_0 = 0$ direction a superposed additional dephasing wedge, representing the corresponding optical path difference, as a function of the abscissa x of its interaction with the diffracting system (Fig. 2c). In this way the light amplitude of the nth fringe is calculated by considering elementary light blades:

$$\mathbf{dE}_0\,(x,V,\theta_n), \tag{8}$$

affected by a relative dephasing given by the above considered term $D'\Phi\,(x,V)$ (2), due to the interaction with the sinusoidal phase grating, and by the additional positive or negative dephasing:

$$D''\Phi\,(x,\theta_n) = \pm\, n\, 2\pi\, x/\Lambda \tag{9}$$

corresponding to the actual geometric path difference, with respect to the light blade located at x = 0. The considered light blade, when reaching the screen acts therefore as a vector:

$$\mathbf{dE}\ (x,V,\theta_n) \ = \ \mathbf{dE_0} \exp\left[\, j\, (V \sin kx \pm n\, 2\pi\, x\, /\Lambda)\right]. \tag{10}$$

The total amplitude $\mathbf{E}\ (V,\theta_n)$ is so calculated as:

$$\mathbf{E}\ (V,\theta_n) \ = \ \int_0^{\Lambda} \mathbf{dE_0} \exp\left[\, j\, (V \sin kx \pm n\, 2\pi\, x/\Lambda)\right] \tag{11}$$

with θ_n indicating the angular direction of the $\pm\, n^{th}$ diffraction fringe.

Fig.s 3a) to 3e) report the computer calculations of the phase vectors $\mathbf{E}\ (V,\theta_n)$ for the directions θ_n corresponding (2) to the diffraction orders from -2 to + 2, (including the discussed $\theta_0 = 0$, Fig. 3c)) and for values of V from 0 (slit of opening Λ, without ultrasound) to 7. Their amplitudes and signs are reported for each picture.

Also such amplitudes (if calculated with good accuracy by formula (11), reducing in these cases to its cosine term) agree, within an approximation better than 10^{-13}, with the numerical values of the corresponding Bessel functions J_n (V) (obtained by series expansion), so confirming to the PVA, in spite of its simplicity, its equivalence with the R-N method.

3.2. The extension to N_W wavelengths.

The consideration under Ch. 2-3 for $\theta_0 = 0$ can be repeated <u>if the diffraction directions θ_n (7) are considered</u>. In the case of several wavelengths each PV $\mathbf{E_{1W}}\ (V,\theta_n)$, representing the interaction with one US WL, results to be in phase with the preceding and with the next PV etc., .

In this case the integral (11), extended to a slit of opening $p = P\, \Lambda$ corresponds to a sum of partial PV each due to one WL, i.e :

$$\mathbf{E_P}\ (V,\theta_n) \ = \ \Sigma_P\, \mathbf{E_{1W}}\ (V,\theta_n) \ = \ P\, \mathbf{E}\ (V,\theta_n) \tag{12}$$

but it refers to an entering light amplitude P times greater and the relative amplitude and light intensities are therefore unchanged with respect to the case of one WL. Also in this case what changes is the shape of the diffraction bell.

One can find the integration "springs" in the case of $N_W = 8$ wavelengths later in Fig.s 7 to 9, in the general discussion of the shape of the diffraction bells (for the values of a variable θ, when coincident with the θ_n given by eq. (7)).

4. ANALYSIS OF MORE SOPHISTICATED DIFFRACTION SYSTEMS: TWO ADJACENT ULTRASONIC BEAMS IN SIMPLE WAVELENGTH RATIO

We consider now a system of two adjacent US waves of WL in simple frequency ratio f_1/f_2 being $f_1 = P f_0$ and $f_2 = Q f_0$, (where f_0 is a reference "fundamental" virtual frequency, P and Q are small integers) and of given R-N parameter V_1 and V_2.

Some recent experimental and theoretical investigations can be found [4, 5]. The problem has been faced earlier by our group by an improved R-N analysis, handled by the computer [6, 7]. The PVA allows a simpler treatment if the adjacent beams are placed close to each-other and can be considered as superposed. We present here the general procedure, applied e.g. to the frequency ratio 2/3 with $V_1 = V_2 = 2.41$, to refer us to a system experimentally tested [8, 9, 10].

The US phase grating ("fundamental grating") is the sum of the two sinusoidal waves, extended to $P = 2$ longer US waves ($\Lambda_1 = \Lambda_0 /P$) of frequency $f_1 = P f_0 = 2 f_0$ and to $Q = 3$ shorter waves ($\Lambda_2 = \Lambda_0/Q$) of frequency $f_2 = Q f_0 = 3 f_0$.

The diffraction directions θ_n, defined by eq. (7) are assumed as based on the "fundamental" virtual WL Λ_0 and the calculations are carried out on a single of said WL.

Table A reports the diffraction pattern (along the vertical lines) at increasing dephasing (in steps of $2\pi/16$) of two US parallel beams of said WL ratio for $V_1 = V_2 = 2.41$. The horizontal lines of the Table predict the temporal light modulation of a given diffraction fringe if two antiparallel US beams are used. Several interference fringes appear (orders ± 1, ± 5, etc.).

In Fig. 4 a), b) and c) three specimen sets of figures show the evolution of the PVA analysis with increasing of the mutual dephasing of the two sinusoids $\Delta\phi x$ (simulating two antiparallel US beams) i. e. the generation of the considered dephasing grating and in addition the evolution of the "snakes" generating the phase vectors, for the diffraction orders +1, 0 and -1. Each set is obtained with the dephasing $\Delta\phi x$, indicated as DEPHAS in the figures.

The change of the phase of the vectors with the dephasing between the US beams is an aspect of the research which should be further investigated (see Ch. 5).

5. THE SHAPE OF THE DIFFRACTION BELLS, THE EVOLUTION IN AMPLITUDE AND THE ROTATION OF THE LIGHT VECTORS WITH TIME

Let us now study the shape of the diffraction bells (which of course are narrower as the considered wave number interacting with the light is greater). They are asymmetric, even for the central (zero order) fringes. This is shown in Fig. 5 and 6, for one WL of one US beam. To evidence the shape of a diffraction bell, we let change the dephasing wedge amplitude (until now assumed corresponding to the angular directions θ_n (7) i. e. equal to n 2π), in steps of a fraction of 2π. Fig.s 5a) and b) report the "springs" so obtained without ultrasound, i.e. for the bare slit of width p, assumed = Λ, for computing reference. The angular exploration from $\theta = 0$ towards positive or negative values give, as expected, symmetric results and the closing of the "spring" (first diffraction zero) at the wedge amplitudes = $\pm 2\pi$, it has to be noted the progressive opposite vector dephasing of the light vectors. In Fig.s 5c) and d) the corresponding angular exploration is performed, but with one US wave of V = 1. From the pictures and from the reported vector lengths, a great asymmetry results: even the diffraction maximum does not occur at $\theta = 0$, but at an angle corresponding to a wedge greater than $\pi/2$. This effect is evidently due to a "wedge effect" of the US wave, assumed by equation (2). To clear the question let us assume now a sinusoidal phase grating with an additional phase term:

$$D'(x,V) = V \sin(kx + D\phi x). \tag{2'}$$

The result is shown in Fig. 6, still for V = 1. The first column refers to a positive diffraction wedge increasing from amplitude 0 to 4/32 of 2π for positive values, the second column for negative values. From the horizontal set of the first to the last line the phase $D\phi x$ is let increase from 0 to 4/8 of 2π. The first line analyzes the discussed asymmetry for $D\phi x = 0$, we note the evolution with $D\phi x$ (PH X in Fig. 6) of the asymmetry (reversing of the maximum from left to right-side pictures). The only vectors which remain numerically and angularly stable are the first ones in all sets.

The bell shape pulsates around its fixed central value, while the US wave propagates, whereas the integrated diffraction bell should remain constant.

To extend these considerations to several US WL, let me comment Fig.s 7 and 8 for the study of the diffraction bells at $D\phi x = 0$ and Fig. 9 at increasing $D\phi x$.

Fig. 7 reports the "springs", in the case of 8 WL and V = 1, when exploring the zero-order diffraction-fringe for positive (upper figures) or negative (lower figures) increments of θ in steps of $\pi/24$. One may compare the partial evolution of the "springs" related to a single US WL with the case of several WL, by looking again at Fig. 5 or 6. The zero-values (closed "spring") are reached of course at smaller wedge values (3, 6, 9, etc. x $2\pi/24$) than for one WL and so the secundary diffraction maxima appear. The bell-shape asymmetry is still evident, around the central value $J_0(1) = 0.765198$, which results independent of the wave number, so confirming the reliability of the PVA.

Similar analysis can be found in Fig. 8 for V = 2, the "springs" are more entangled for small values of θ, then the wedge bending effect prevails at greater values of θ, the zeroes occur as for V = 1, depending on the WL number, not on V.

In Fig. 9 the bell shapes of the diffraction orders n = 0, +1, +2 and +3 are analyzed, at increasing $D\phi x$, around the value $D\phi x = n\ 2\pi$.

The aim is the verification of the known variation, due to Doppler effect, of the frequency of the light, diffracted to the $\pm$ n order, by $\pm$ n ν, being ν the US frequency, quantistically interpreted as the addition (or subtraction) of a phonon to the interacting photon.. The Fig. 9 reports for 8 WL and for the order n = 0, 1, 2 and 3 the evolution of the diffraction bell, close to its maximum, with the US wave propagation. This is simulated by fixing as increasing steps of $D\phi x$ different fractions of 2π, selected to obtain an equal vector rotation of 45 degrees for all diffraction orders. From this PVA the mentioned frequency variation results, in its ondulatory aspect, from the rotation (with the progressing of the US waves) of the amplitude vector by an angular velocity proportional to n, while its amplitude remains unchanged (no temporal modulation by one progressive US beam).

6. APPLICATION OF THE PHASE-VECTOR METHOD TO THE OBLIQUE LIGHT-ULTRASOUND INTERACTION

Our program of phase-vector analysis, can also predict the diffraction pattern of a light beam crossing obliquely one US wave. Any analysis of such arrangements based on the R-N theory gives prediction of the diffraction pattern which differ from the experimental recordings in two fundamental aspects: 1) at diffraction angles for which the light beam interacts with an integer number of US WL the theory forecasts the complete vanishing of any diffraction effect [11)], whereas only marked diffraction minima are observed [12)]; 2) at any angle strong asymmetries are observed, whereas the theory predicts a symmetric diffraction pattern.

Both discrepancies are obviously due to neglecting the bending of the path of each light ray within the US field, effect which is difficult to face in the frame of the R-N analysis. It can be easily treated by the PVA, provided one knows the actual path of the light rays [13)], on which the mathematical model can be based (see Fig. 10 a)). Unfortunately only old and limited experimental data can be found and with uncomplete parameters of the experimental arrangement actually used. As a preliminary result in Fig. 10 b) a comparison of the PVA predictions with some experimental data is reported.

The agreement is good in a limited diffraction-angular range, as probably outside such limits some competitive effect occurs. New experiments by laser light in this neglected field should be welcome to clarify the interesting problem.

7. CONCLUSIONS

This survey of the PVA method and of the performed applications shows its simplicity when applied to the analysis of diffracting systems realized by the interaction at 90 degrees of a collimated light beam with two adjacent US waves or with one US wave, adjacent to an optical amplitude grating [14, 15], in addition to the study of the oblique incidence of the light on one US wave. The method can be extended of course to the study of the diffraction pattern produced by the interaction at 90 degrees of a collimated light beam with systems of optical phase gratings of any shape. The interest of this extension, in progress in our Department, is based on the possibility of analyzing non sinusoidal US waves (as distorted in their generation or propagation) by describing them as the joint dephasing effect of one sinusoidal wave adjacent to a phase grating of the suitable shape (triangular for the ramp distorsion, etc.). In this way, from the prediction of the diffraction pattern due to such simulation of a "distorted US wave" we will be in a position to make the diagnosis of the distortion of an actual US wave, by observing their diffraction pattern.

In addition to the extension to other cases of our verification for the frequency ratio 2/3 of the dependence on the mutual dephasing of the diffraction pattern [8], another interesting research can be suggested by our analysis of the temporal modulation of the light by two adjacent antiparallel US waves in other simple frequency ratios, i. e. the experimental check of the light-vector rotation, mentioned e.g. in Ch. 5. Such research could be done improving by laser light the optical experiments performed to verify the Doppler effect on the nonzero diffraction orders.

8. ACKNOWLEDGEMENT.

The Author is indebted to Mr. Livio Giuliani for his valuable help in the initial phase of the investigation on the "snakes" generating the phase vectors related to two US waves, to Mr. A. Felmer for his appreciated collaboration in the systematic research, to Dr. S. Zugna for the oblique incidence study, and to Mr. S. Rizzarelli for the text-editing.

REFERENCES

1) Debye P. and Sears F.W., Proc. Nat Acad. Sci. (USA), 18, 410, 1932.

2) Biquard P.and Lucas R., C. R. Acad. Sci. (Paris), 195, 121, 1932.

3) Raman C.V. and Nagendra Nath S., Proc. Ind. Acad. Sci.- I: A2, 406, 1935; II: A2, 413, 1935; III: A3, 75, 1936; IV: A3, 119, 1936; V: A3, 495, 1936.

4) Several experimental and theoretical papers are listed in the Proc. of the First (1980) and Second (1983) Spring School on Acoustooptics, Gdansk.
We refer particularly to the Ref.s quoted by:
a) Mertens R., Hereman W., Verhest F., pag. 33-51, 1980.
b) Ciuti P. and Gabrielli I., pag. 120-135, 1980.
c) Gabrielli I., Ciuti P., pag. 84-104, 1983.

5) Patorski K., Acustica, 52, 246-254, 1983.

6) Calligaris F., Ciuti P. and Gabrielli I., Acustica, 38, 37-43, 1977.

7) See pag. 87 and Table A, pag 98 of Ref. 4c).

8) Gabrielli I., Kwiek P., Markiewicz A., Sliwinski A., Acustica, 66, 281-285, 1988.

9) Ciuti P. and Gabrielli I., Acoustics Letters, 6, 116-119, 1983.

10) Kwiek P., Gabrielli I. and Sliwinski A., Acoustics Letters, 7, 1-5, 1983.

11) Ciuti P. and Gabrielli I., Nuovo Cimento, C86, 77-82, 1985.

12) O. Nomoto, Proc. Phys.-Math. Soc. Japan, 24, 613-639, 1942.

13) Gabrielli I. and Zugna S., Acustica, 62, 156-162, 1986.

14) Gabrielli I., Acustica, 21, 97-103, 1969.

15) Calligaris F., Ciuti P. and Gabrielli I., Acustica 35, 74-80, 1976 ,and: Jour. Acoust. Soc. Am., 61, 959-964, 1977.

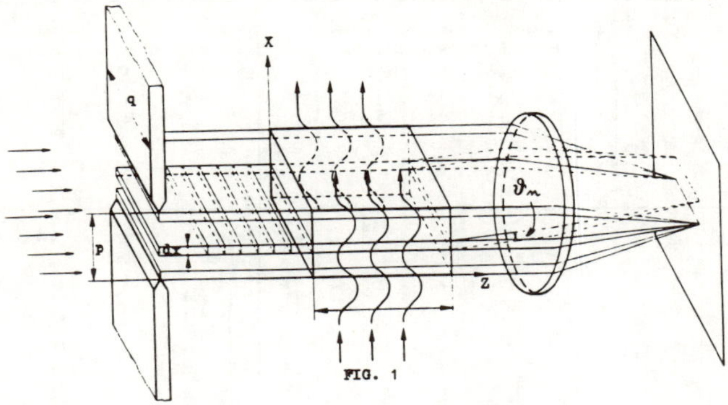

FIG. 1 - EXPERIMENTAL ARRANGEMENT OF THE DIFFRACTION OBSERVED IN THE FRAUNHOFER PLANE. THE BLADES OF LIGHT ARE SHOWN.

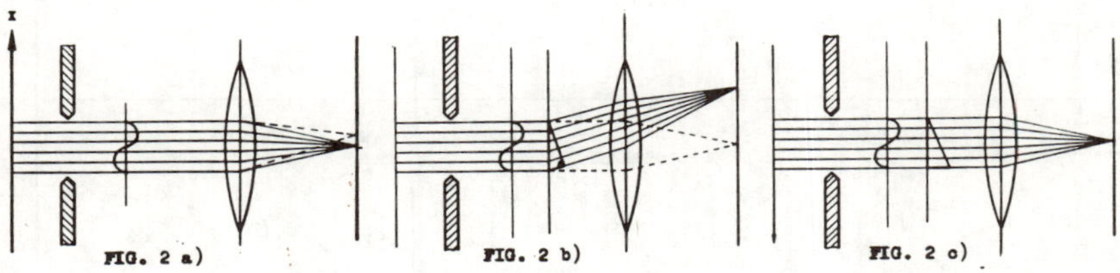

FIG. 2 a) - SCHEMATIC MODEL OF THE LIGHT DIFFRACTION ON ONE ULTRASONIC WAVE.
b) - THE ACTUAL PATH OF THE LIGHT CONVERGING TO A NONZERO - ORDER FRINGE.
c) - THE EQUIVALENT PATH OF THE LIGHT RAYS IN THE MODEL USED BY THE PHASE - VECTOR ANALYSIS: DEFINITION OF THE ADDITIONAL DEPHASING WEDGE.

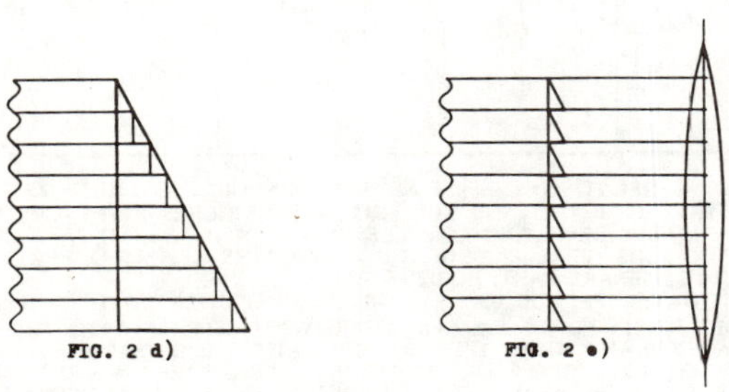

FIG. 2 d, e) - DIFFRACTION ON SEVERAL (8) ULTRASONIC WAVELENGTHS: THE DEPHASING WEDGE FOR THE DIFFRACTION ORDER +1.
d) - THE WEDGE MATHEMATICALLY CONSIDERED FOR THE ANGULAR DIRECTIONS OF EQ. (7): THE PROGRESSIVE DEPHASING DIFFERENCE IS 2π.
e) - THE EQUIVALENT SAWTOOTH WEDGES: ALL PARTIAL VECTORS, RELATED TO ONE WAVELENGTH, ARE EQUAL IN AMPLITUDE AND PHASE.

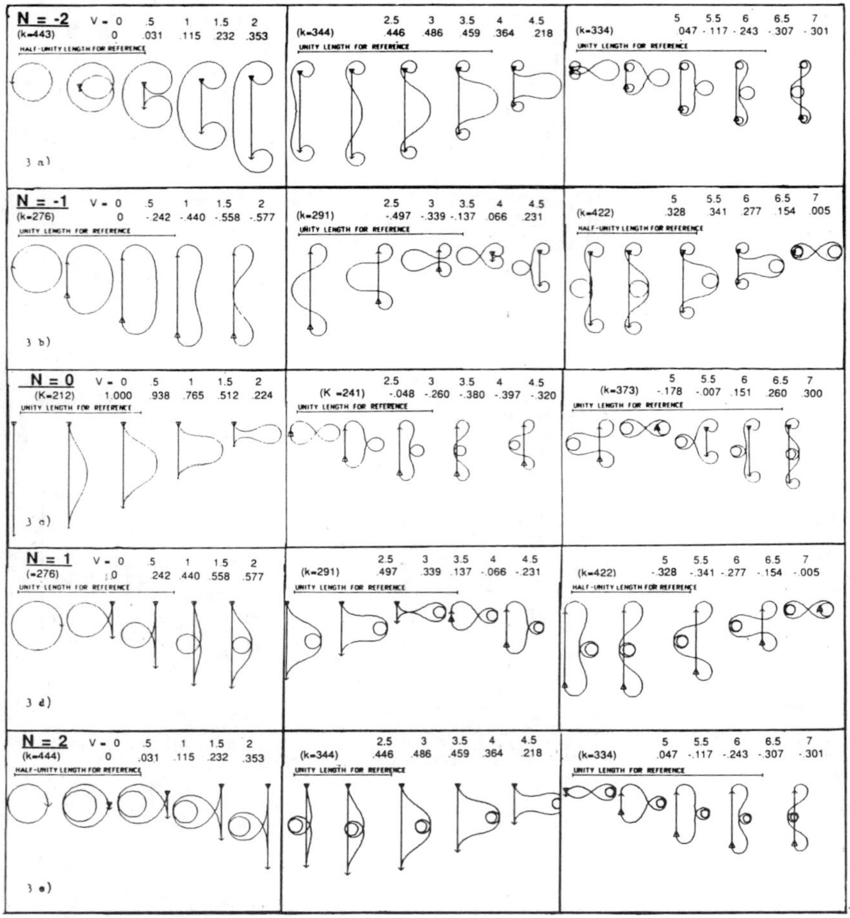

Fig. 3 - THE "SPRINGS" GENERATING THE PHASE VECTORS OF THE LIGHT AMPLITUDE OF THE DIFFRACTION ORDERS FROM -2 TO +2. THIS AND THE FOLLOWING FIGURES REPRESENT THE GENERATION OF THE PHASE VECTORS, USUALLY RESULTING COMPLEX IN A X - Y PLANE. TO SAVE SPACE IN FIGURES 3 AND FROM5 TO 9 THE X - AXIS (REAL PART OF THE VECTOR) IS REPORTED VERTICALLY DIRECTED DOWNWARDS, THE Y - AXIS (IMAGINARY PART) HORIZONTALLY.
THE LIGHT DIFFRACTED TO THE FRINGES OF N.th ORDERS, REPRESENTED BY THE PHASE VECTORS CONSIDERED HERE, RESULT DEPHASED BY $\pm n\pi$ (n INTEGER) WITH REFERENCE TO THE ENTERING LIGHT VECTOR ASSUMED AT PHASE = 0 AND HENCE THEY COINCIDE WITH THEIR REAL PART.
THE 15 PICTURES OF EACH LINE REFER TO THE R - N PARAMETERS INCREASING FROM 0 (NO ULTRASOUND: DIFFRACTION FROM THE BARE SLIT) TO 7 RADIANS OF MAXIMUM DEPHASING OF THE LIGHT VECTOR BY THE ULTRASOUND.
THE WRITTEN LINES REPORT THE RESULTING VECTOR LENGTHS, CORRESPONDING TO THE BESSEL FUNCTIONS $J_N(V)$. K INDICATES A GRAPHIC - SIZE NORMALIZATION FACTOR, REPORTED FOR EACH SET OF 5 "SPRINGS" GIVEN, AS THE GRAPHIC UNITY (OR HALF - UNITY) REFERENCE LINE, TO COMPARE THE FIGURES: EACH SET IS DRAWN AT ITS MAXIMUM POSSIBLE SIZE.

THE FIVE HORIZONTAL SETS REFER TO THE DIFFRACTION FRINGES OF ORDER N:

FIG 3	a)	b)	c)	d)	e)
N	-2	-1	0	+1	+2

T A B L E   A

TYPE OF GRATING: SIN. - SIN.   FREQUENCY RATIO: 2/3   R-N BEAM PARAM.: 2.41 - 2.41   CALCUL. ON 192 POINTS

ORDERS REFER. TO: Λ_0	Λ_1	Λ_2	MODULAT. %	MAX. INT.	MUTUAL BEAM DEPHASING (2.PI/16) 0	1	2	3	4	5	6	7	8	9	10	11	12	13	14	15	16
-10	-5	0	51.8	4.89	2.36	2.72	3.61	4.51	4.89	4.51	3.61	2.72	2.36	2.72	3.61	4.51	4.89	4.51	3.61	2.72	2.36
-9	0	-3	16.0	1.14	0.96	1.04	1.14	1.11	1.05	1.11	1.14	1.04	0.96	1.04	1.14	1.11	1.05	1.11	1.14	1.04	0.96
-8	-4	0	39.5	6.37	6.37	5.99	5.09	4.21	3.85	4.21	5.09	5.99	6.37	5.99	5.09	4.21	3.85	4.21	5.09	5.99	6.37
-7	0	0	88.8	11.38	1.28	2.53	5.88	9.67	11.38	9.67	5.88	2.53	1.28	2.53	5.88	9.67	11.38	9.67	5.88	2.53	1.28
-6	-3	-2	43.6	0.02	0.02	0.02	0.02	0.01	0.02	0.01	0.02	0.02	0.02	0.02	0.02	0.01	0.02	0.01	0.02	0.02	0.02
-5	0	0	54.9	10.28	10.28	10.06	8.67	6.07	4.63	6.07	8.67	10.06	10.28	10.06	8.67	6.07	4.63	6.07	8.67	10.06	10.28
-4	-2	0	14.5	5.35	4.57	4.93	5.35	5.13	4.84	5.13	5.35	4.93	4.57	4.93	5.35	5.13	4.84	5.13	5.35	4.93	4.57
-3	0	-1	80.9	2.12	2.12	1.27	0.40	1.20	2.00	1.20	0.40	1.27	2.12	1.27	0.40	1.20	2.00	1.20	0.40	1.27	2.12
-2	-1	0	46.7	4.74	4.53	3.51	2.53	3.65	4.74	3.65	2.53	3.51	4.53	3.51	2.53	3.65	4.74	3.65	2.53	3.51	4.53
-1	0	0	98.4	23.30	23.30	20.36	12.65	4.12	0.37	4.12	12.65	20.36	23.30	20.36	12.65	4.12	0.37	4.12	12.65	20.36	23.30
0	0	0	100.0	2.99	0.00	1.50	2.99	1.50	0.00	1.50	2.99	1.50	0.00	1.50	2.99	1.50	0.00	1.50	2.99	1.50	0.00
1	0	0	98.4	23.30	0.37	4.12	12.65	20.36	23.30	20.36	12.65	4.12	0.37	4.12	12.65	20.36	23.30	20.36	12.65	4.12	0.37
2	1	0	46.7	4.74	4.74	3.65	2.53	3.51	4.53	3.51	2.53	3.65	4.74	3.65	2.53	3.51	4.53	3.51	2.53	3.65	4.74
3	0	1	80.9	2.12	2.00	1.20	0.40	1.27	2.12	1.27	0.40	1.20	2.00	1.20	0.40	1.27	2.12	1.27	0.40	1.20	2.00
4	2	0	14.5	5.35	4.84	5.13	5.35	4.93	4.57	4.93	5.35	5.13	4.84	5.13	5.35	4.93	4.57	4.93	5.35	5.13	4.84
5	0	0	54.9	10.28	4.63	6.07	8.67	10.06	10.28	10.06	8.67	6.07	4.63	6.07	8.67	10.06	10.28	10.06	8.67	6.07	4.63
6	3	2	43.6	0.02	0.02	0.01	0.02	0.02	0.02	0.02	0.02	0.01	0.02	0.01	0.02	0.02	0.02	0.02	0.02	0.01	0.02
7	0	0	88.8	11.38	11.38	9.67	5.88	2.53	1.28	2.53	5.88	9.67	11.38	9.67	5.88	2.53	1.28	2.53	5.88	9.67	11.38
8	4	0	39.5	6.37	3.85	4.21	5.09	5.99	6.37	5.99	5.09	4.21	3.85	4.21	5.09	5.99	6.37	5.99	5.09	4.21	3.85
9	0	3	16.0	1.14	1.05	1.11	1.14	1.04	0.96	1.04	1.14	1.11	1.05	1.11	1.14	1.04	0.96	1.04	1.14	1.11	1.05
10	5	0	51.8	4.89	4.89	4.51	3.61	2.72	2.36	2.72	3.61	4.51	4.89	4.51	3.61	2.72	2.36	2.72	3.61	4.51	4.89

TOTAL LIGHT INTENSITY FOR A GIVEN BEAM DEPHASING (SUMMATION OVER 41 ORDERS FROM -20 TO 20)
/ 100.0 100.0 100.0 100.0 100.0 100.0 100.0 100.0 100.0 100.0 100.0 100.0 100.0 100.0 100.00100. 0100.0

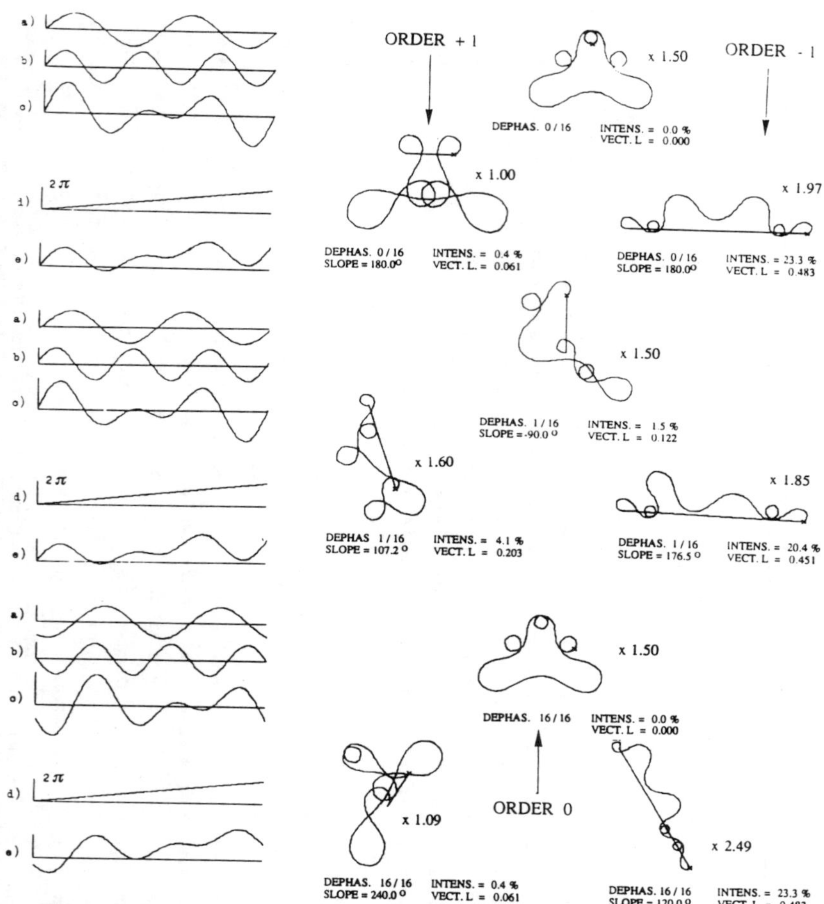

FIG. 4 - OVERALL $V_1 = V_2 = 2.41$ - FREQUENCY RATIO = 2/3.
ANALYSIS OF THE FRINGES +1 AND -1 IN THE LIGHT DIFFRACTION BY TWO ANTIPARALLEL PROGRESSIVE US BEAMS OF WAVELENGTH RATIO 3/2 AND V = 2.41.
- COLUMN 1: EVOLUTION, WITH THE MUTUAL BEAM DEPHASING, OF THE ULTRASONIC GRATINGS:
a) TWO WAVELENGTHS OF THE FIRST US BEAM
b) THREE WAVELENGTHS OF THE SECOND US BEAM
c) THE US DEPHASING GRATING, SUM OF a) AND b)
d) THE DEPHASING WEDGE, CORRESPONDING TO THE FIRST - ORDER DIFFRACTION FRINGE
e) THE COMPLETE DEPHASING GRATING, SUM OF c) AND d).
- COLUMN 2, 3, 4: RELATED EVOLUTION OF THE "SPRINGS" GENERATING THE LIGHT PHASE - VECTORS:

COLUMN	DIFFRACTION ORDER	
2	+1	(POSITIVE WEDGE, SEE d), COL 1)
3	0	(NO WEDGE)
4	-1	(NEGATIVE WEDGE)

FOR EACH FIGURE ARE REPORTED:
DEPHAS: THE MUTUAL US BEAM DEPHASING IN THE INDICATED FRACTION OF 2π
INTENS.: THE LIGHT INTENSITY OF THE FRINGE, IN PERCENT OF THE ENTERING LIGHT
SLOPE: THE SLOPE OF THE NONZERO VECTORS, IN DEGREES WITH RESPECT TO THE REAL AXIS, PHASE REFERENCE OF THE ENTERING LIGHT
VECT. L.: THE AMPLITUDE OF THE LIGHT PHASE VECTOR, WITH REFERENCE TO THE ENTERING LIGHT, ASSUMED AS UNITY.
THE PICTURES ARE DRAWN AT A SUITABLE SIZE. FOR A COMPARISON AMONG DIFFERENT "SPRINGS" THEY MUST BE ENLARGED BY THE INDICATED GRAPHIC COEFFICIENTS

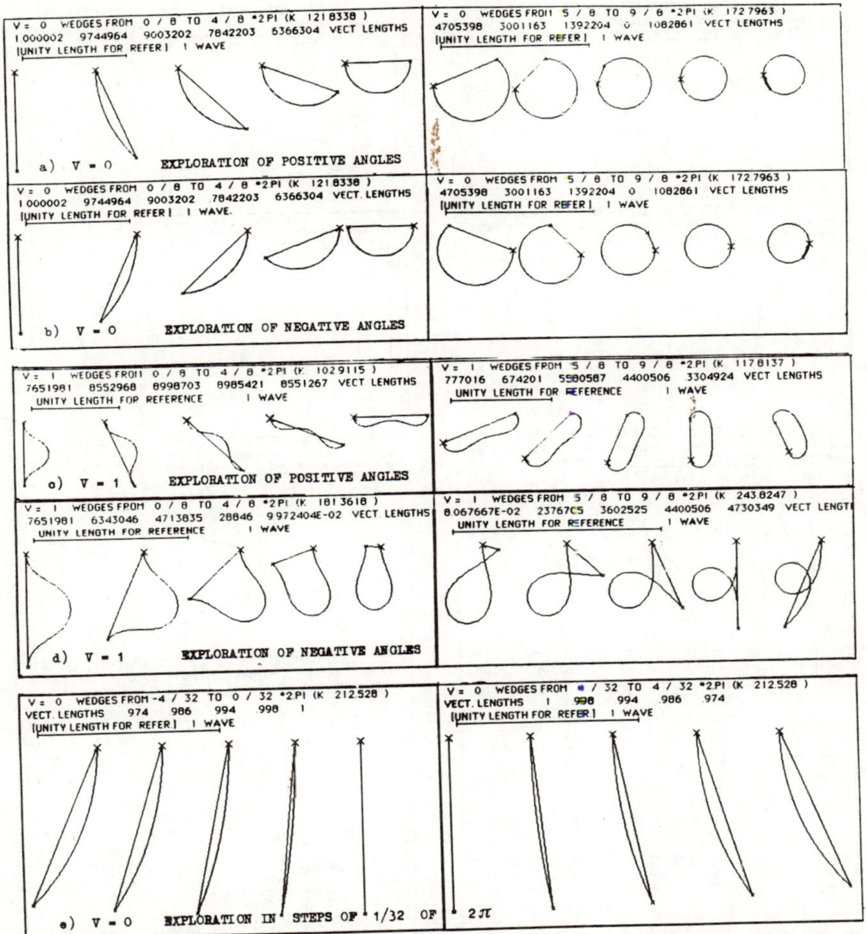

FIG. 5 - STUDY OF THE BELL-SHAPE OF THE FRINGE OF ORDER 0 IN THE LIGHT DIFFRACTION BY A BARE SLIT (V = 0: a) AND b)) AND BY ONE US WAVE OF R-N PARAMETER V = 1 : c) AND d). THE US WAVE IS IN SINE PHASE (EQ. 2). THE DEPHASING WEDGE IS LET TO INCREASE FOR POSITIVE (SETS a) AND c)) OR NEGATIVE STEPS (SETS b) AND d)), STARTING FROM 0 TO 9/8 OF THE ANGULAR DIRECTION OF THE FIRST-ORDER FRINGE: 2π.

THE BARE SLIT PRODUCES SYMMETRIC BELL SHAPES : SEE a) AND b). ONE US WAVE GIVES A STRONGLY ASYMMETRIC BELL, DISPLACING EVEN THE ANGULAR POSITION OF THE MAXIMUM VECTOR AMPLITUDE: SEE c) AND d).

FIG 5 e) REFERS STILL TO V = 0 (NO US). THE BELL SYMMETRY IS EVIDENCED IN SMALLER WEDGE STEPS (1/32 OF 2π).

HERE, AS IN FIG. 6 TO 9, THE REAL COMPONENT OF THE VECTOR IS DRAWN VERTICALLY DOWNWARDS, THE IMAGINARY HORIZONTALLY. THE LARGER CROSS IN THE FIGURES REPRESENTS THE TAIL OF THE VECTOR (ORIGIN OF THE CURVE).

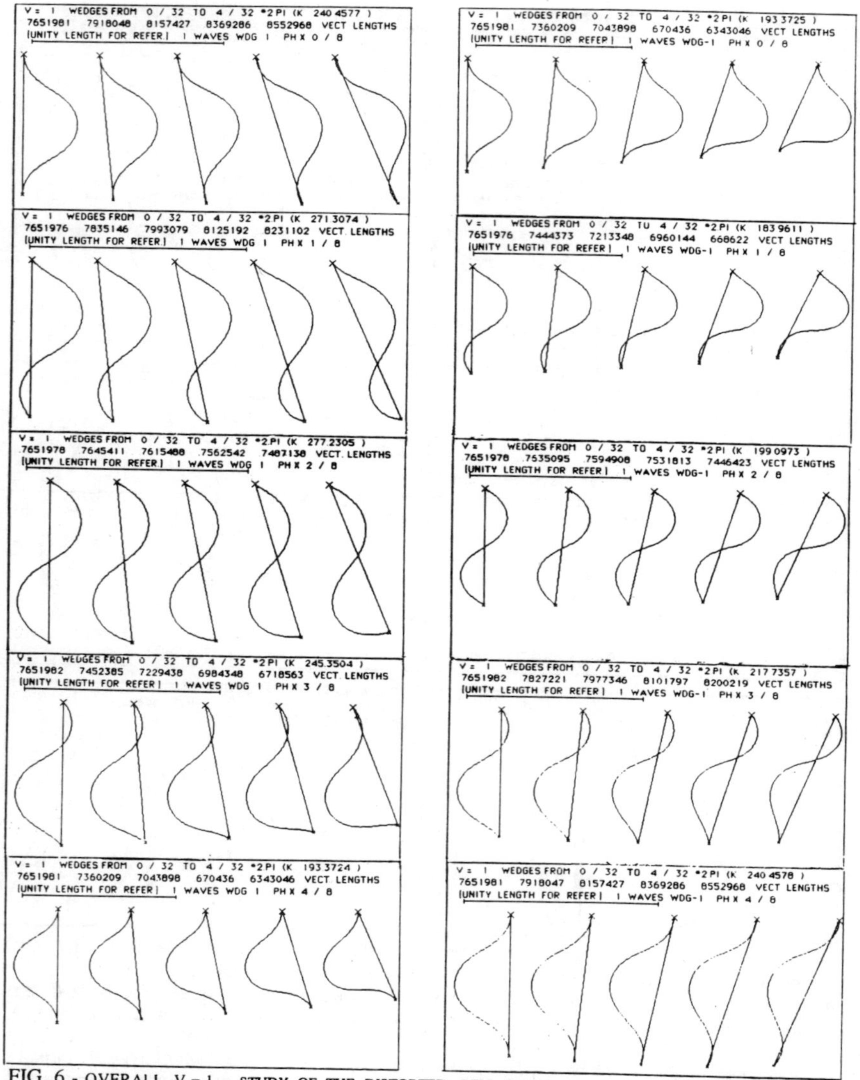

FIG. 6 - OVERALL V = 1 - STUDY OF THE DISTORTED - BELL SHAPE OF THE FIRST - ORDER DIFFRACTION FRINGE. THE LEFT (RIGHT) COLUMN REFERS TO A POSITIVE (NEGATIVE) DEPHASING WEDGE INCREASING IN EACH HORIZONTAL SET FROM 0 TO 4/32 OF 2π, WHILE ONE US WAVE IS LET ADVANCE (EQ. 2') FROM DϕX = 0 TO 4/8 OF 2π (PH X IN FIG.). ONLY THE FIRST VECTOR OF ALL SETS (WEDGE AMPLITUDE = θ = 0) REMAINS STABLE IN ANGLE (DEPHASING = 0) AND AMPLITUDE ($J_0(1) = 0.765198$, AS DISCUSSED IN CH.3-1), WHILE THE BELL PULSATES AROUND IT. THE MAXIMUM AMPLITUDE APPEARS FIRST IN THE POSITIVE AND THEN IN THE NEGATIVE SIDE OF θ = 0.

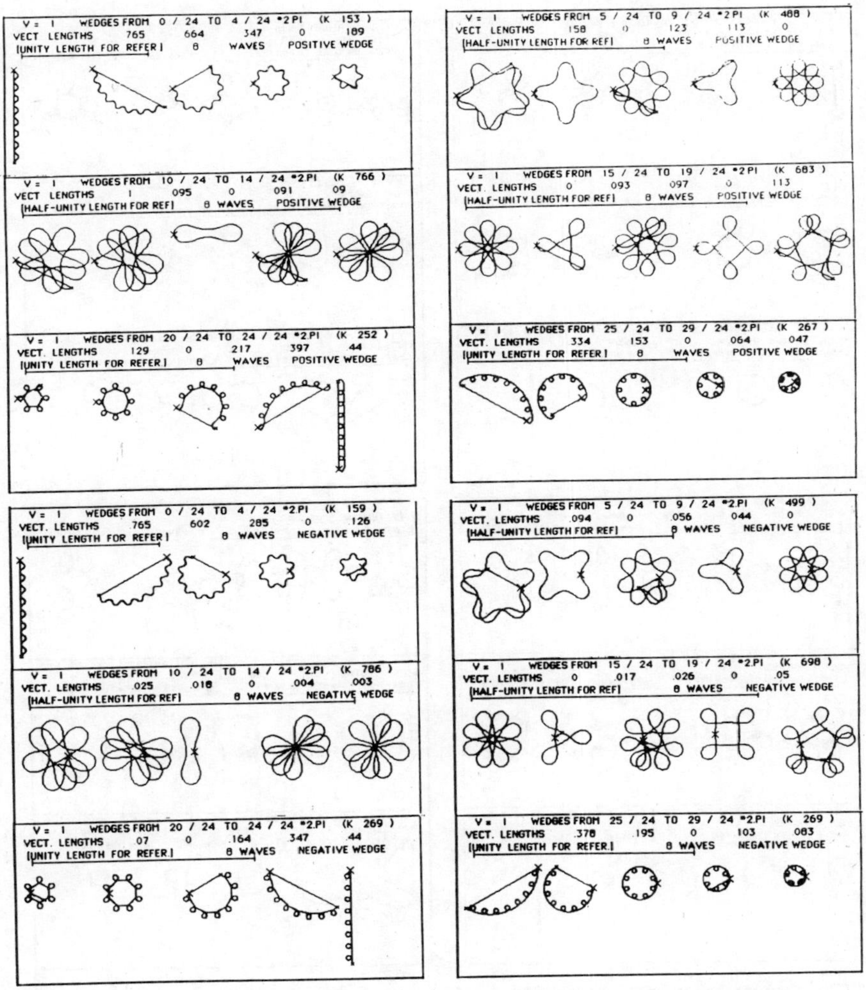

FIG. 7 - OVERALL V = 1 - THE EVOLUTION OF THE "SPRINGS" GENERATING THE LIGHT VECTORS OF THE ZERO - ORDER DIFFRACTION FRINGE FOR 8 WAVELENGTHS OF ONE US BEAM WITH V = 1. THE UPPER FIGURES REFER TO THE INCREASING OF THE DEPHASING WEDGE FOR POSITIVE VALUES, THE LOWER FIGURES FOR THE CORRESPONDING NEGATIVE VALUES, FROM 0 TO 29/24 OF 2 π, I.E. BEYOND THE POSITIVE OR NEGATIVE FIRST - ORDER FRINGE (24/24). THE "SPRINGS" CLOSE THEMSELVES AT MULTIPLES OF $\pi/4$, AND THEN THEY OPEN FOR THE NEXT DIFFRACTION MAXIMA. THE DEFORMED - BELL ASYMMETRY IS PRESENT, BUT THE CENTRAL VALUE IS NOW THE MAXIMUM AND ITS RELATIVE VECTOR - AMPLITUDE AGREES STILL WITH THE R - N THEORY PREDICTION I. E. TO $J_0(1) = 0.765198$. ALSO THE VECTOR AMPLITUDE ($J_1(1) = 0.440050$) FOR THE WEDGE OF 2 π CONFIRMS THE RELIABILITY OF THE PHASE - VECTOR ANALYSIS.

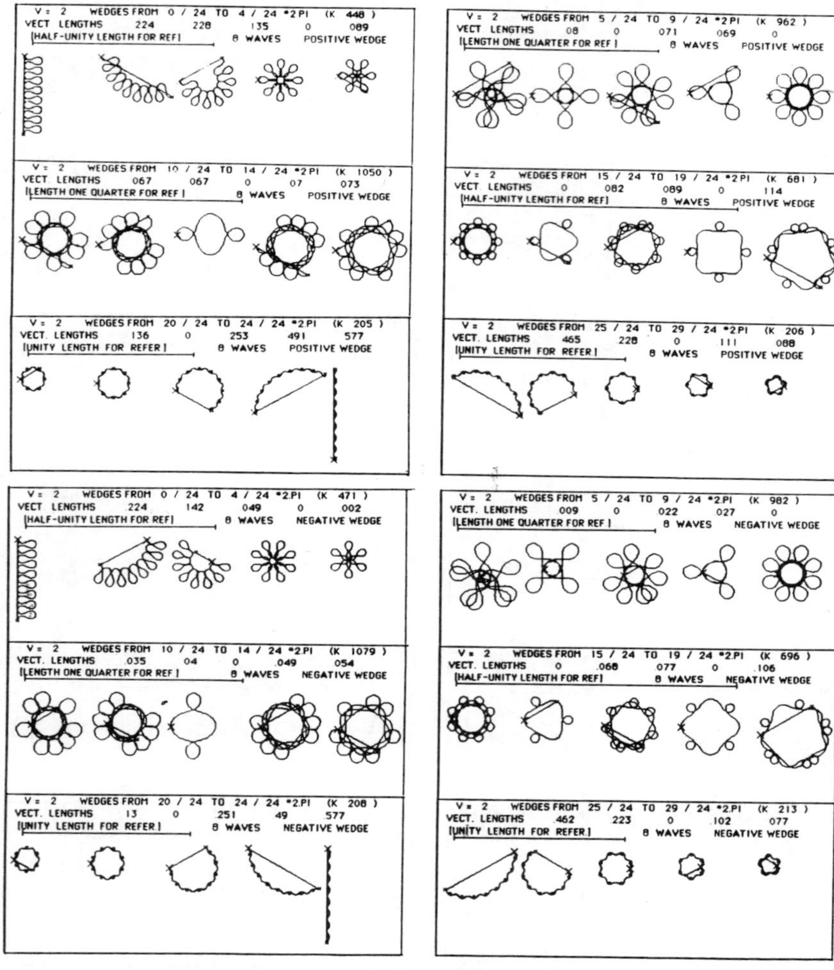

FIG. 8 - OVERALL V = 2 - THE EVOLUTION OF THE "SPRINGS" GENERATING THE LIGHT VECTORS OF THE ZERO - ORDER DIFFRACTION FRINGE FOR 8 WAVELENGTHS OF ONE US BEAM WITH V = 2. THE UPPER FIGURES REFER TO THE INCREASING OF THE DEPHASING WEDGE FOR POSITIVE VALUES, THE LOWER FIGURES FOR THE CORRESPONDING NEGATIVE VALUES, FROM 0 TO 29/24 OF 2 π, I.E. BEYOND THE POSITIVE OR NEGATIVE FIRST - ORDER FRINGE (24/24). THE "SPRINGS" CLOSE THEMSELVES AT MULTIPLES OF $\pi/4$, AND THEN THEY OPEN FOR THE NEXT DIFFRACTION MAXIMA. THE DEFORMED - BELL ASYMMETRY IS PRESENT, AS FOR ONE US WL (FIG. 6) ALSO HERE THE MAXIMUM - AMPLITUDE VECTOR IS NOT LOCATED AT $\theta = 0$, BUT IT OCCURS WITH A POSITIVE WEDGE CLOSE TO 3/128 OF 2π, WHERE ITS VALUE IS 0.238514. THE RELATIVE VECTOR - AMPLITUDE STILL AGREES WITH THE R - N THEORY PREDICTIONS BOTH FOR THE CENTRAL VALUE ($J_0(2) = 0.223891$) AND FOR THE FIRST - ORDER FRINGE ($J_1(2) = 0.576725$, WEDGE OF 2π), SO CONFIRMING THE RELIABILITY OF THE PHASE - VECTOR ANALYSIS.

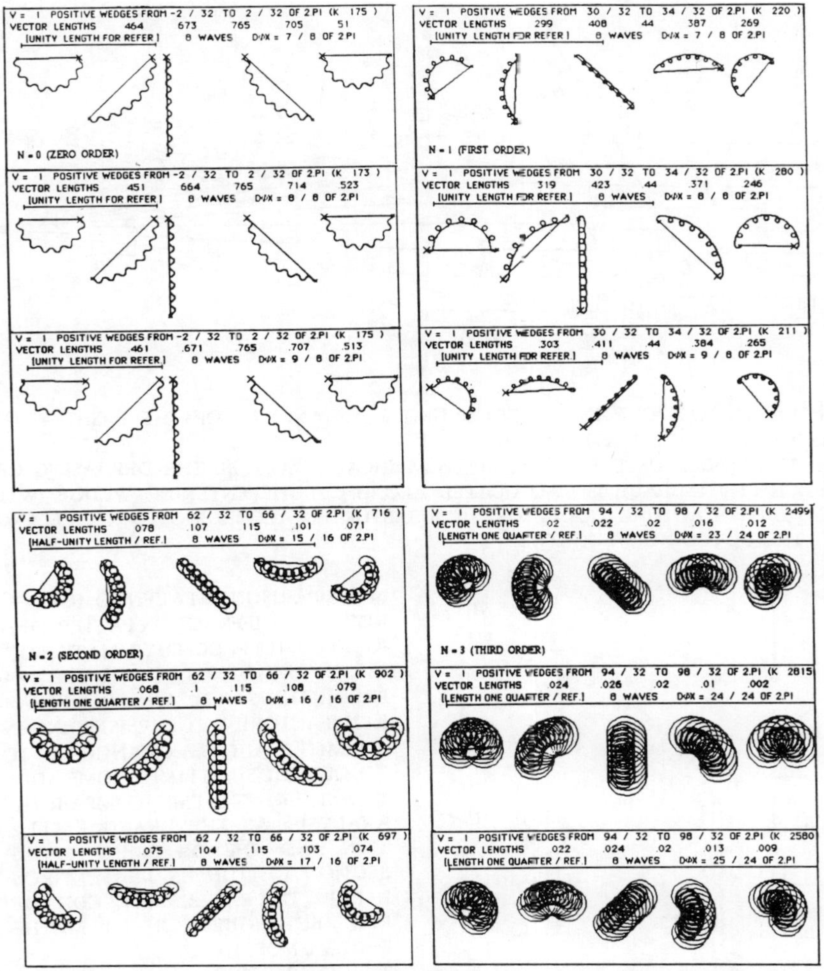

FIG. 9 - OVERALL V = 1 - ROTATION OF THE PHASE VECTOR OF THE DIFFRACTION ORDERS FROM 0 TO 3, WITH THE PROGRESSING OF 8 WL OF ONE US BEAM WITH RESPECT TO THE SLIT.
IN EACH HORIZONTAL SET THE 5 PICTURES REFER TO THE ANGULAR EXPLORATION OF THE FRINGES OF THE DIFFRACTION ORDERS N = 0, 1, 2 AND 3, RESPECTIVELY, BY DECREASING OR INCREASING THE DEPHASING WEDGE AROUND ITS CENTRAL VALUE IN STEPS OF 1/32 OF 2π.
THE THREE VERTICAL FIGURES IN EACH OF THE FRAMES 2), 3) AND 4) CORRESPOND TO THE PROGRESSING OF THE US WAVES OF AN AMOUNT DϕX, WHICH DETERMINES THE SAME ROTATION OF 45 DEGREES OF THE VECTORS. SUCH STEPS ARE 1/8 OF 2π FOR N = 1 (FIRST ORDER), 1/16 FOR N = 2, 1/24 FOR N = 3, SO CONFIRMING THE VARIATION OF THE LIGHT FREQUENCY BY $\pm N\,\Omega$, BEING Ω THE US FREQUENCY. THE PICTURES OF FRAME 1) VERIFY THAT THE LIGHT FREQUENCY REMAINS UNALTERED IN THE ZERO - ORDER FRINGE: NO ROTATION OCCURS.

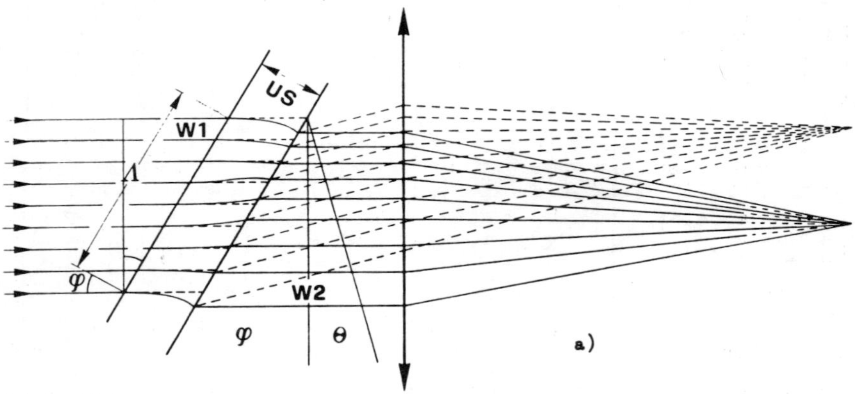

FIG. 10 - OBLIQUE INTERACTION UNDER AN ANGLE φ OF THE LIGHT WITH ONE ULTRASONIC WAVELENGTH.

a) THE MODEL USED FOR THE MATHEMATICAL ANALYSIS. THE DEPHASING OF EACH LIGHT BLADE IS CALCULATED ALONG: 1) THE ENTERING WEDGE (W1), 2) ITS OWN ACTUAL TRAJECTORY (US), DISTORTED ACCORDING TO REF.13, AND 3) THE OUTGOING WEDGE (W2).

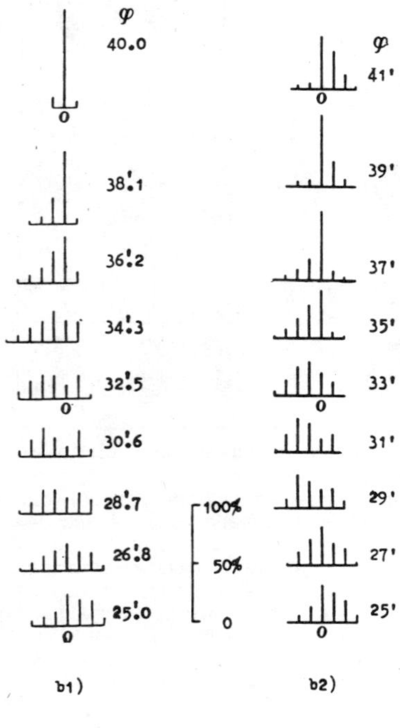

b) COMPARISON BETWEEN SOME EXPERIMENTAL DIFFRACTION PATTERN (b1) (O. NOMOTO: PROC. PHYS.-MATH. SOC. JAPAN, 24, 613-639, 1942) AND THE PVA PREDICTIONS (b2), WHICH APPEAR SATISFACTORILY CORRESPONDING IN A LIMITED ANGULAR RANGE (25' TO 37'). OUTSIDE SUCH LIMITS SOME ADDITIONAL EFFECT SEEMS TO IMPAIR THE ANALYSIS, AS THE BRAGG REFLECTION AND THE DIFFERENT CROSS-SECTION OF THE US BEAM, CIRCULAR IN THE EXPERIMENT, CONSIDERED RECTANGULAR IN THE MATHEMATICAL MODEL.

THE GEOMETRICAL AND PHYSICAL PARAMETERS ARE THE FOLLOWING: US VELOCITY = 1500 m/s, US FREQUENCY = 4.71 MHz, AND HENCE US WL = 0.318 mm, REFRACTION INDEX OF THE MEDIUM = 1.534, LIGHT WL = 5.46 x 10^{-7} m, US-BEAM DIAMETER (US-BEAM WIDTH L IN THE MODEL) = 6 cm, Δn (OSCILLATION AMPLITUDE OF THE REFRACTION INDEX) = 1.85 x 10^{-5}.

CONTINUOUS WAVE ULTRASONICS - an old method with new applications

A. KULIK, J.E. BIDAUX, G. GREMAUD, S. SATHISH

Ecole Polytechnique Fédérale de Lausanne
Institut de Génie Atomique
CH-1015 LAUSANNE, SWITZERLAND

ABSTRACT

In most of the actually performed experiments in the field of ultrasounds, pulse-echo method is used. This method has two inherent drawbacks :

1) wide bandwidth measurement systems are needed which have poor Signal to Noise ratio,

2) spread of the frequency content of the signal leading to problems when dispersive media are studied.

Modern Network Analyzers available on the market allow us to work with narrow band low noise receivers. Equivalent pulse echo response can be calculated using inverse Fourier Transform, giving excellent time resolution, and good Signal to Noise ratio. Such systems have been used for studying phase transition in cobalt, and also in an acoustic microscope. Continuous Wave Acoustic Microscopy allows us to measure Rayleigh wave velocity at single frequency, and also to perform phase-sensitive imaging.

INTRODUCTION

Measurements of the Ultrasonic Attenuation and Velocity are commonly used for characterizing the properties of materials and for nondestructive testing of different structures. In most cases pulse-echo technique is used. This method has inherent drawbacks :

- wide bandwidth measurement systems are needed, so that poor Signal to Noise Ratio (SNR) is obtained,

- spread of the frequency content of the signal, which could create problems while studying dispersive media

- minimal sample thickness is limited by the pulse width, and often by the temporary saturation of the receiver amplifiers (so-called "dead zone").

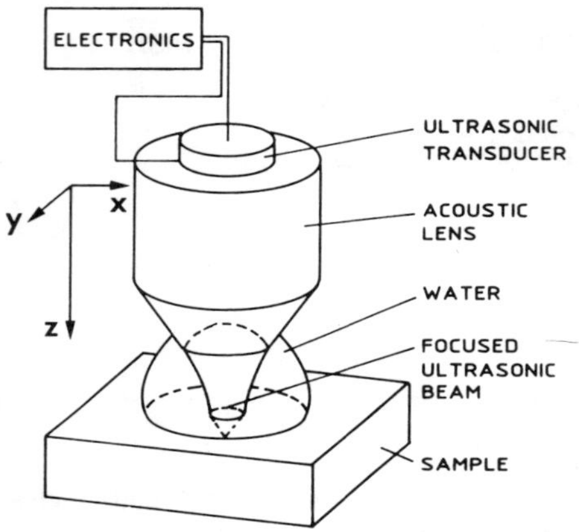

Pulse echo method is used in Scanning Acoustic Microscope [1], with high frequency acoustic lenses of radius ranging about 50 μm. Such small water path is chosen in order to minimize high attenuation in the acoustic coupling water. Using such small radius lenses implies that acoustic echos from the lens cavity are very close to the useful echos from the sample. For separating useful information, very short acoustic pulses must be used. Usually, only amplitude of the echo from the sample is recorded as a function of **x** and **y** positions, and phase information is lost.

Fig. 1. Principle of the scanning acoustic microscope.

Acoustic microscopes are often used, not only for imaging, but also for localized measurements of the velocity of the Surface Acoustic Waves (SAW) [2]. SAW, in the case of layered media, can be highly dispersive [3]. Therefore, use of very short acoustic pulses (wide bandwidth) spreads the received signals, which disturbes the measurements.

Continuous Wave (CW) Ultrasonic measuring systems were used in 1960's [4] and became not very popular afterwards, probably due to the lack of suitable equipments on the market. Actually, Vector Network Analyzers seem to be well adapted for CW Ultrasonic applications.

Using a Network Analyser, we designed a setup allowing measurements of the US attenuation and velocity as a function of temperature [5]. Another setup was constructed for CW Acoustic Microscope [6].

CONTINUOUS WAVE ULTRASONIC SYSTEM

For studying the phase transition in cobalt, high temperature (300 - 850K) CW ultrasonic system has been constructed (fig. 2) [5].

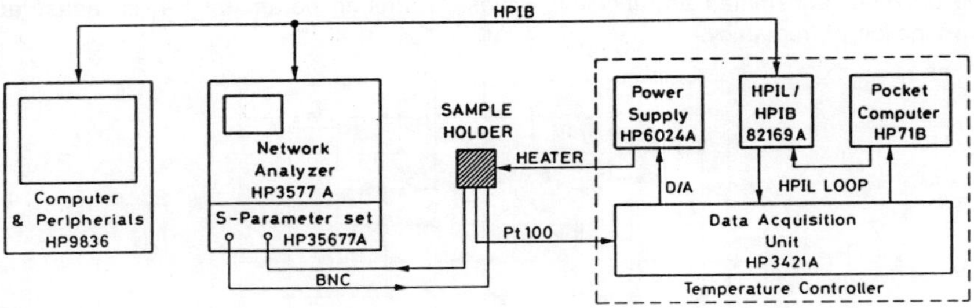

Fig. 2. Block diagram of the CW Ultrasonic System.

Mechanical system

The composite resonator (transducer-sample-transducer) is placed inside a sample holder (fig. 3).

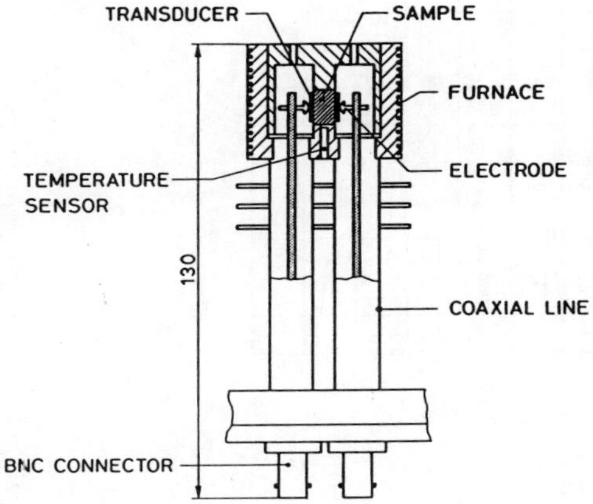

The mechanical construction has to insure good electrical screening between the two sides of the sample in order to prevent electrical cross-talk, which could mask acoustical signals. Sample holder was made from OFHC copper to guarantee good temperature uniformity.

Wire wound heater and platinum temperature sensor are provided for temperature control. During the measurements, whole the sample holder is kept under a secondary vacuum.

Fig. 3. Construction of the sample holder for CW Ultrasonics.

Electrical System

The theory of the CW transmission setup is well described in [4]. The composite resonator (sample with two piezoelectric transducers) is connected to a Network Analyzer. This instrument measures S_{21} scattering parameter, defined as the ratio between transmitted and incident waves. Scattering parameter S_{21} is measured as a function of frequency.

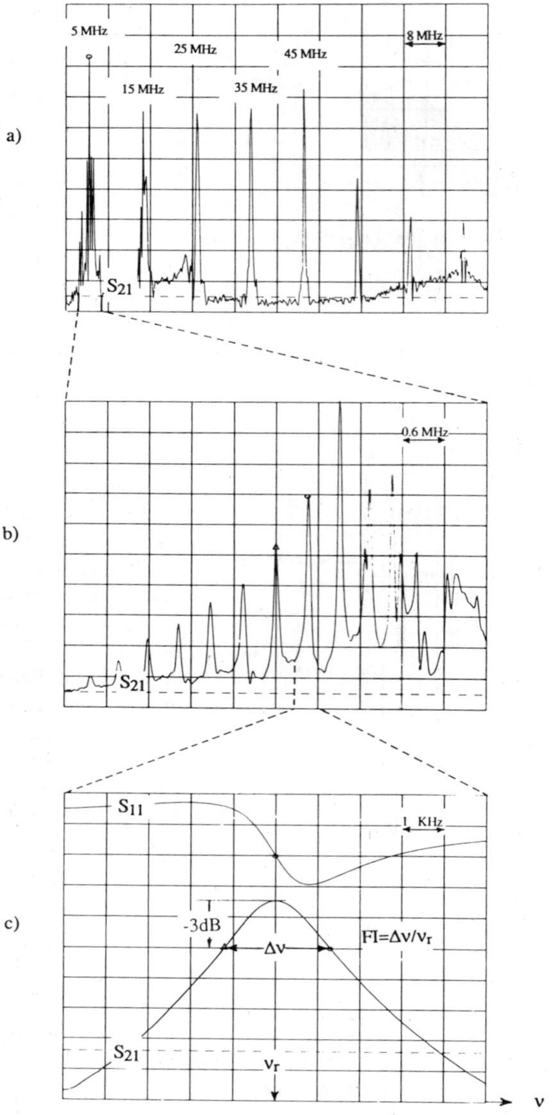

Fig. 4. Typical frequency response in transmission (S_{21}) for different frequency spans. a) Frequency span = 80 MHz. b) 6 MHz. c) 10 kHz.

Using a wide frequency span (fig. 4a), one can observe a succession of peaks as a function of frequency, corresponding to the fundamental frequency and the odd overtones of the transducers. More narrow span (fig. 4b) exhibits fine series of peaks closely spaced, corresponding to longitudinal resonances (standing waves) convolved with the envelope of the bandwidth of the transducers. Distance between adjacent peaks allows us to calculate the ultrasonic velocity [4].

Narrowing the frequency span further (fig. 4c), one can measure single resonance width, which furnishes the attenuation in the sample.

Distorsions of the transducer envelope (fig. 4b) can be explained as the effect of the static capacity of the transducers, which in principle can be compensated. Existing mismatch of the impedance (50 Ω - Network Analyzer, high impedance Lithium Niobate transducer) insures very small electrical attenuation background of the system.

Typical bandwidth of the used receivers was about 100 Hz, which is several orders of magnitude narrower than in a typical pulse echo ultrasonic system. Excitation voltage of the generator is about 1 V, which is sufficient, due to the mentioned narrow bandwidth, in order to obtain good SNR.

Temperature control system

Anelastic measurements, specially of the phase transition, can be very sensitive to the heating or cooling rate. Work in vacuum leads to long thermal time constants.

A temperature controller was constructed using a commercially available equipment (fig. 2). A low cost Data Acquisition Unit is used for measuring the resistance of the temperature sensor. The value is sent to the HP71B Pocket Computer through HPIL Interface and used to calculate the temperature. A linear setpoint ramp can be created if needed. Using PID algorithm, an appropriate value of the heater power is calculated and sent back to the Data Acquisition Unit. A digital to analog converter card is addressed and the output voltage is amplified using Power Supply, and finally used for heating.

The typical deviation between programmed and actual temperature, during 1K/min ramps, does not exceed 0.05 K, and, in the steady state, 0.01 K.

Results

System was used for studying phase transition in cobalt [7] (fig. 5,6). Results in the MHz range do not depend on US frequency. Strong variations of the ultrasonic attenuation (fig. 5) confirm that CW ultrasonics is well adapted for the study of highly attenuating materials. Rapid changes in US velocity or resonant frequency (fig. 6 a,b) in the vicinity of the phase transition shows adaptability of the method for studying transient phenomena.

A detailed description of the CW ultrasonic system has been published elsewhere [5].

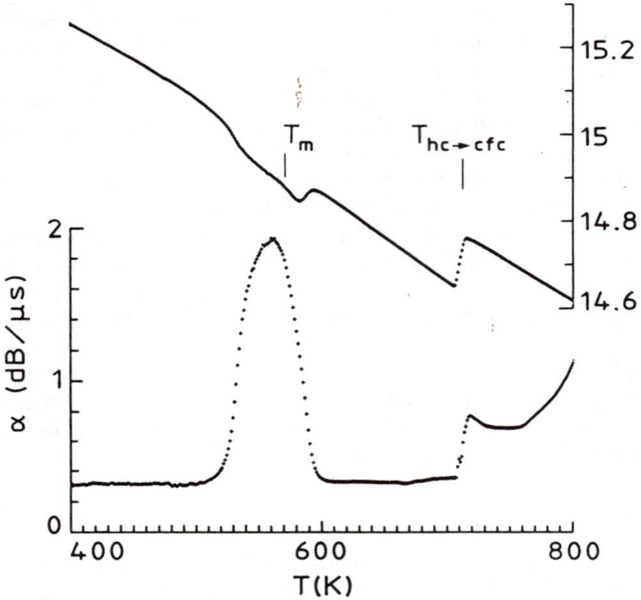

Fig. 5. Attenuation and resonant frequency of 99.98% cobalt as a function of the temperature (heating). T_m : magnetic transformation, $T_{hc \rightarrow cfc}$: hc to cfc phase transition temperature.

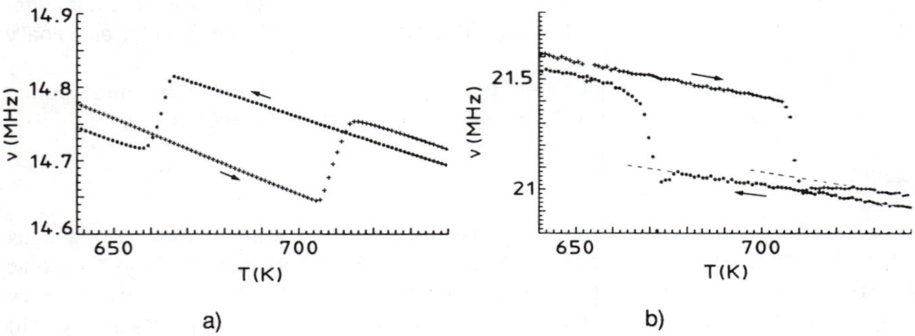

Fig. 6. Resonant frequency in vicinity of the phase transition during cooling and heating in monocrystalline cobalt. a) US propagation parallel to C-axis. b) US propagation 45° with respect to C-axis.

SAMCRUW - SCANNING ACOUSTIC MICROSCOPE USING CONTINUOUS REFLECTED ULTRASONIC WAVE

CW Acoustic microscope was designed using an acoustic lens, an **xyz** scanning movement and a Network Analyzer [6] controlled by a computer.

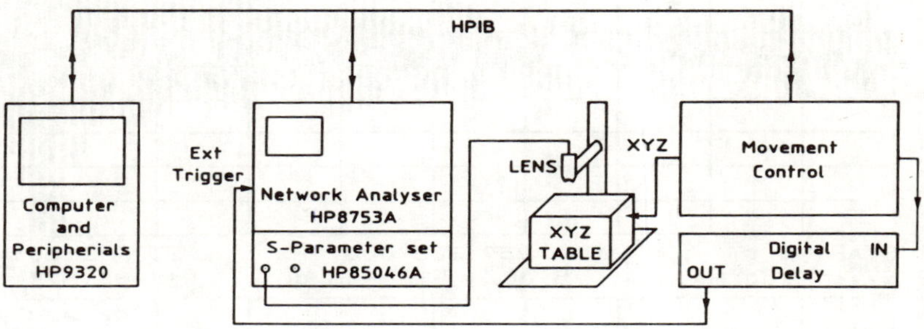

Fig. 7. Block diagram of the CW scanning acoustic microscope.

<u>Realisation</u>

A High Frequency Network Analyzer is used as CW Ultrasonic source and receiver (fig. 7). The system is working in reflection mode, so that S_{11} scattering parameter is measured. This parameter S_{11} is defined as the ratio of reflected to incident waves, and can be measured as a function of :

- the distance between the lens and the sample (CW V(z) measurements at constant position **x, y** and constant frequency),
- the **x, y** positions on the sample (CW imaging at constant frequency),
- the frequency f (all three spatial coordinates **x, y, z** are kept constant).

Using a specialized Fourier Transform processor, the Equivalent Pulse Echo Response can be calculated.

<u>CW V(z) measurements</u>

In pulse echo acoustic microscopy, when measuring the amplitude of the echo received from the sample as a function of the distance between the lens and the sample, an oscillating curve is obtained, which is called V(z).

This V(z) curve is the result of interferences between the specular reflected waves and those propagating partially as a leaky SAW in the sample. From the main period of the V(z) curve, SAW velocity can be calculated, as well as its attenuation from the rate of decay [2].

CW V(z) curve (fig. 8) differs from the pulse-echo V(z) by a highly oscillating component coming from the standing waves in coupling water. That signal is modulated in amplitude by the classical pulse echo V(z) curve.

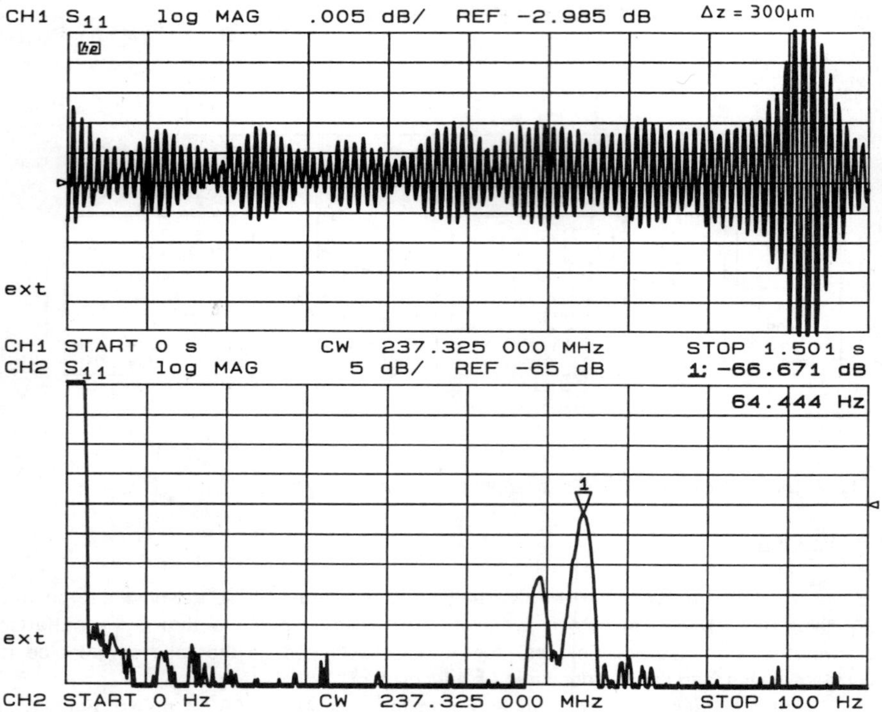

Fig. 8. Continuous Wave V(z) curve measured on Tungsten-Carbide sample and its Fourier Transform.

Despite of the lack in theoretical analysis of the CW V(z), basic understanding cited above was verified experimentally. Fourier Transform (fig. 8) exhibits the main peak related with the standing waves in the coupling water. That signal is (slightly imperfectly) modulated in amplitude by V(z). The interesting main V(z) periodicity can be found from the distance between the water peak and its sidelobe. The potential advantage of the CW V(z) measurements is related to the fact that SAW velocity measurements can be carried out in dispersive media, because the working frequency is perfectly known.

CW imaging

Fixing a constant frequency and measuring S_{11} scattering parameter as a function of the position **x, y** on the sample, CW images can be obtained. On Fig. 9, one can observe real and imaginary parts of the S_{11} images.

Due to the oscillatory behaviour of the CW V(z) curve (fig. 8), CW images are extremely sensitive to the topology of the sample. As real and imaginary components contain different information, a phase sensitive detection will allow easier separation of the different sources of contrast.

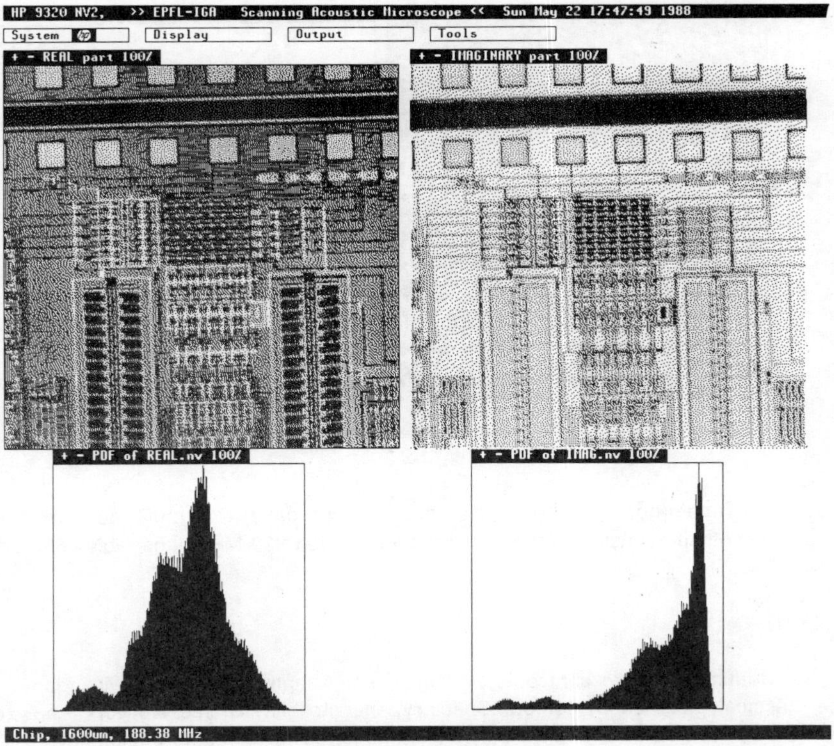

Fig. 9. Real and imaginary part CW images with corresponding intensity histograms underneath.

Equivalent pulse echo response

Our Network Analyser is equipped with a time domain option which is able to calculate the equivalent pulse echo response of the system from its frequency response.

Having the good signal to noise ratio and the possibility of easily adapting the frequency span to the transducer's bandwith, we can choose very short equivalent pulse width.

In Fig. 10, one can observe time domain separation of the specular reflection and signals delayed by the propagation of leaky SAW (top left). The sample is very close to the acoustic lens on the top of the graph. Increasing the distance z between sample and lens, the two pulses begin to interfere. That can be understood as a pulse-echo V(z) curve seen from the top (note intensity oscillations in the middle). SAW pulse disappears completely when focal point is on the surface (z = 0), and maximum amplitude is reached. From the fig. 10, one can calculate SAW velocity.

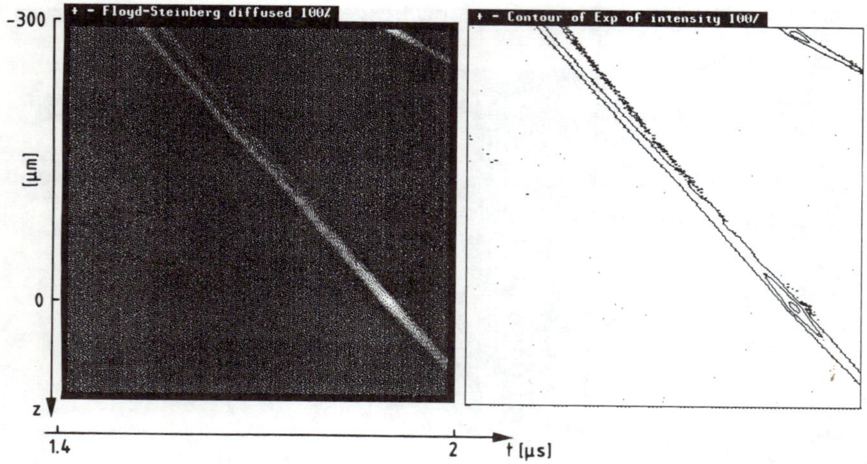

Fig. 10. Time (horizontally) - distance z (vertically) - amplitude (intensity) measurements in acoustic microscope. Frequency span 220 MHz, lens radius 500 µm.

CONCLUSIONS

Continuous wave ultrasonic system presents many advantages with respect to the classical pulse-echo method. Specially, the flexibility of the Network Analyzers allows us to realise several ultrasonic experiments, using the same equipment.

REFERENCES

1. Briggs A., "An introduction to Scanning Acoustic Microscopy", Oxford University Press. Royal Microscopic Society, 1985.

2. Kushibiki J. and Chubachi N., "Material Characterization by line focus beam Acoustic Microscope", IEEE Trans. SU 32 (2), 189 (1985).

3. Kushibiki J., Ishikawa T. and Chubachi N., "Precise measurement of film thickness by line-focus-beam acoustic microscope". Ultrasonics 87, Butterworth, Guilford (1987).

4. Bolef D.I. and Miller J.G., "High frequency continuous wave ultrasonics", Physical Acoustics, Vol. VIII, Ed. W.P. Mason and R.N. Thurston, Axademic Press (1971).

5. Kulik A., Bidaux J.E., "CAMELEONDE - Continuous Wave Automatic Ultrasonic Measuring System", J. de Physique, 48, C8-335 (1987).

6. Kulik A., Gremaud G., Sathish S., "Continuous Wave Reflection Scanning Acoustic Microscope (SAMCRUW)", presented at 17th International Symposium on Acoustical Imaging, Sendai, Japan, May 31-June 2, (1988).

7. Bidaux J.E., "Etude par méthodes acoustiques de la transition de phase hcp-fcc du Cobalt", Ph.D. Thesis, EPFL, Lausanne (1988).

HETERODYNE LASER PROBE AND APPLICATIONS TO SAW, THERMOACOUSTIC MICROSCOPY, THERMOELASTIC IMAGING

B. Cretin*, D. Hauden*

Laboratoire de Physique et Métrologie des Oscillateurs du C.N.R.S.
associé à l'Université de Franche-Comté-Besançon
32, avenue de l'Observatoire - 25000 Besançon - France

Abstract

After recalling the different ways to detect elastic waves in solids by special transducers, we present optical probes using a heterodyne technique in interferometric set ups. Sensitivities and resolutions are discussed before examining several applications. At high frequency, heterodyne laser probes are used to measure amplitude and phase of the SAW propagating in nonlinear mode at surface of Quartz and Gallium Arsenide wafers. Low frequency applications need special electronics to improve the signal/noise ratio (PLL and AGC). Measurements of silicon plate resonance modes or standing waves in metallic plates are shown and discussed. Coupled with an optical excitation provided by an intensity modulated laser, the probe is used as detector in a thermoacoustic microscope. Results on volume and multilayered samples, metallic or dielectric, are presented as the form of amplitude and phase images. The last domain of application of our laser probes are the thermoelastic imaging of an electronic circuit. The probe detects the synchronous displacements due to thermoelastic effect in the conductors or semiconductors strongly attached with the substrate. Finally, limits and new improvements are discussed to extend the field of applications to rough surface samples.

1 - INTRODUCTION

The acousto-optic interaction was first studied by Brillouin (1922) [1] who analyzed the light coherent scattering by plane acoustic waves. Then, using it in acousto-optic devices, it enables to improve signal processing method [2-3]. First, bulk acoustic waves (BAW)

* Professors at ENSMM - Besançon

interacting with laser beam were studied for BAW detection and acousto-optic devices (modulators, coders, ...). After 1968, acousto-optic interactions were applied to Rayleigh's waves by using different ways : optical array [4], holography [5], laser speckle [6], Fabry-Pérot interferences [7] to obtain the total field of the acoustic wave. The laser beam diffracted by the Rayleigh's wave enables to analyze the nonlinear propagation at high frequencies (up to the fifth harmonic) [8]. Heterodyne interferometric laser probes [9-11] conjugate together advantages of the diffraction light sensor and a very high sensitivity since its resolution is few 10^{-14} m. Moreover, one can obtain a measurement of amplitude and phase of the surface wave. This kind of probe can be used also as sensor in different other applications : thermoacousticmicroscopy [12] and thermoelastic imaging coupled if necessary with functional test of electronic circuit using the same sensor. Modulated light or electrical current provide periodic displacement within the scanned sample and the probe measures the coherent elastic displacements.

In this paper, we will describe the process on the different effects used to detect elastic waves, or thermal diffusion, or thermoelastic displacements. Each time, examples of applications will be shown as the forms of graphs and pictures.

2 - HETERODYNE INTERFEROMETRIC LASER PROBES

All types of heterodyne interferometers are based on the phase modulation due to the path length modulation at the elastic wave frequency (or the mechanical displacement frequency) carried by a RF subcarrier introduced by means of an acousto-optic modulator-deflector acting on the light beam (or the optical fiber). One can distinguish three different types of heterodyne interferometers : symmetrical probe, asymmetrical probe and fiber optics probe.

The symmetrical probe principle is shown on Fig. 1 which shows two symmetrical brackets, one to measure the displacements, the other is used as reference. The signal is carried at the double of theBragg cell frequency ($2f_B$).

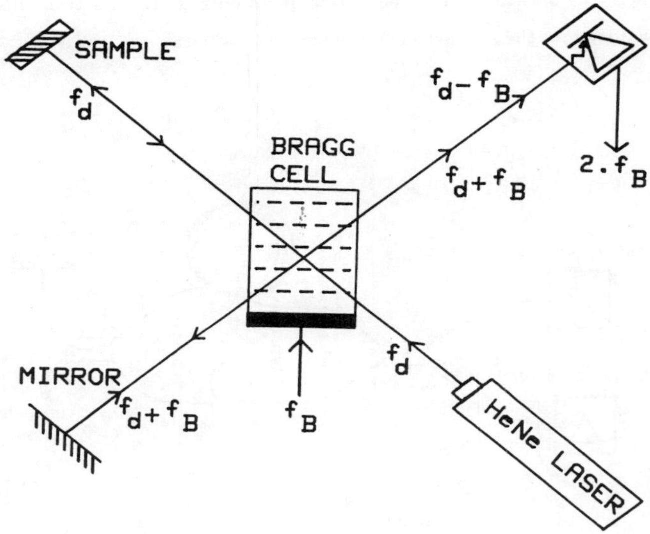

Fig. 1 : Diagram of a classical heterodyne interferometer

The asymmetrical probe is presented on the schematic n° 2. The signal can be carried either at theBragg cell frequency (f_B), or at the double frequency ($2f_B$). The efficiency was checked to have approximately the same value.

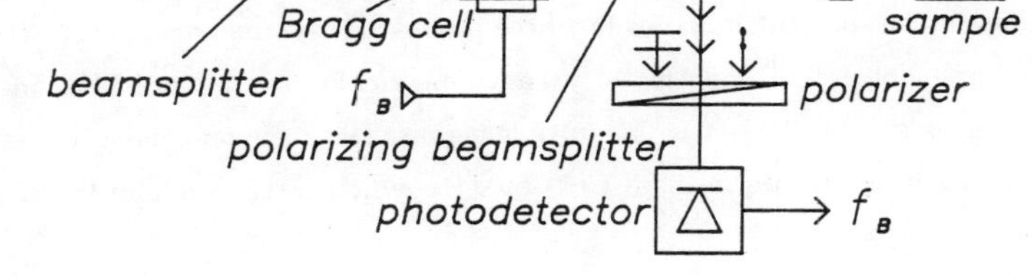

Fig. 2 : Compact asymmetrical probe diagram

The principle of the optical fiber probe [10] is given on Fig. 3. This probe needs a special modulator to introduce the subcarrier frequency.

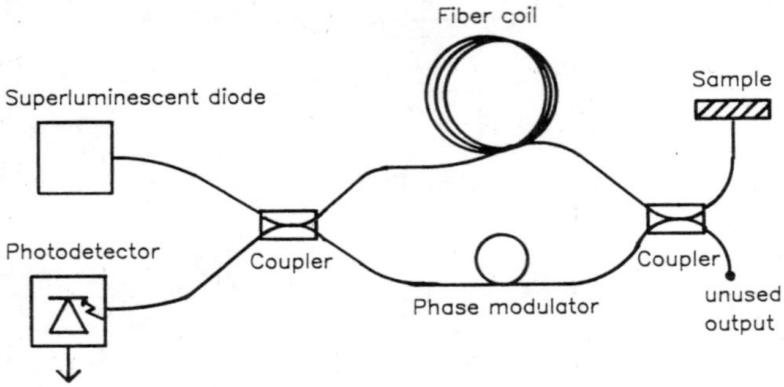

Fig. 3 : All fiber probe diagram

The interest for this probe is the compact sizes we can reach while conserving the performances for low frequency applications.

The sensitivity to detect high frequency elastic waves is function of the ratio between the light spot and the acoustic wavelength. The resolution and therefore, finally, the ultimate accuracy of the measurements is limited both by the own noise of the optical part and by the electronic noise of the phase demodulating system : mechanical vibrations, phase fluctuations of the laser beam, photodetector noise, spurious multiple beams.

An analysis of the phase noise spectral density shows the different contributions. Fig. 4 presents phase noise spectral densities for three probes : (1) symmetrical probe (2) compact symmetrical probe built by Cilas Alcatel, (3) asymmetrical probe. Up to 1 kHz, one can see effects of vibrations, light fluctuations, multiple beams. The electronic noise appears above 1 kHz. Then the probe resolution is 3.10^{-14} m/$\sqrt{\text{Hz}}$ which corresponds to the theoretical noise level.

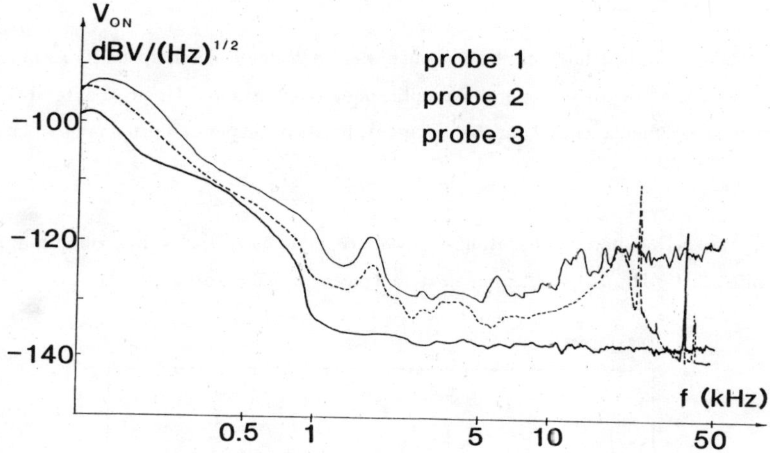

Fig. 4 : Density of phase fluctuations versus frequency

To measure low frequency displacements, a matched Phase Locked Loop is added in the electronic chain which decreases strongly the noise level below 1 kHz (Fig. 5).

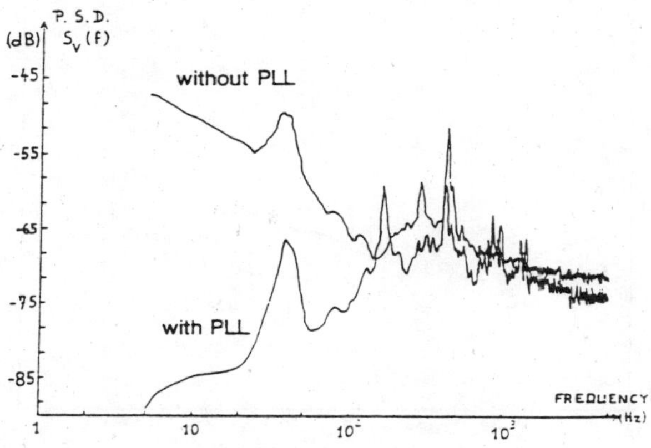

Fig. 5 : Power spectral densities of phase fluctuations with and without PLL

3. HIGH FREQUENCY APPLICATIONS - NONLINEARITIES IN SAW DEVICES

The main field of applications for these probes was SAW devices and specially to measure velocity, diffraction, beam-steering and displacement cartography. One can extend that to nonlinear measurements as 2nd and 3rd harmonic levels or dispersion properties in layered sample.

Fig. 6a shows the harmonic generation of SAW propagating in the X-direction on a Y-cut quartz plate at 91 MHz. The excitation electrical power was 0.8 watt.

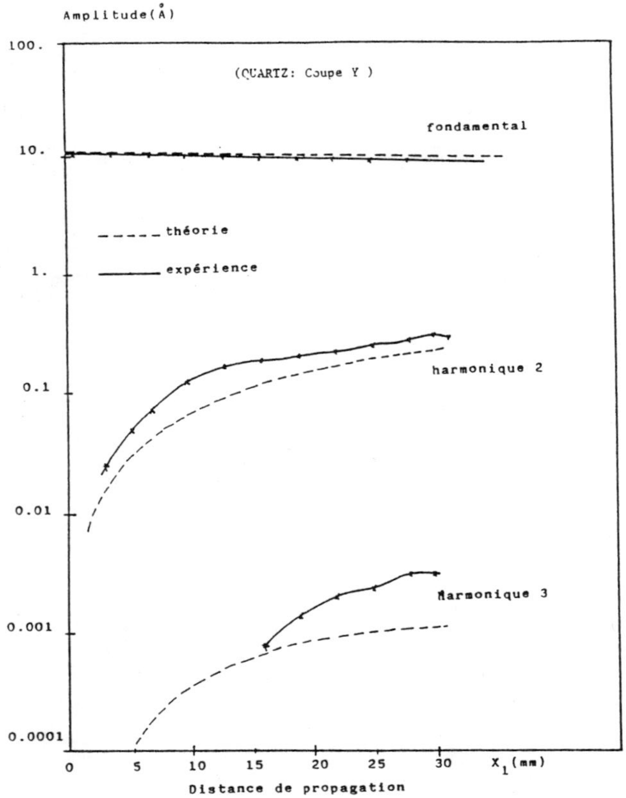

Fig. 6a : Harmonic generation of a Y,X quartz cut (f_a = 91.2 MHz, P_{ex} : 0.8 Watt)

Table I compares experimental results [13] with theoretical ones obtained by M. Planat [14] for the second harmonic level to the fundamental level ratio.

SAW-cuts	Experimental results		Theoretical results Planat [14]
	Alippi [13]	Hauden	
Y-cut	0.035 (79 MHz)	0.028 (91 MHz)	0.021
X-cut	0.068 (79 MHz)	0.058 (95 MHz)	0.06

Results obtained in the case of GaAs substrate (001), X-propagation are presented on Fig. 6b (frequency = 83 MHz, P_{ext} = 0.2 W).

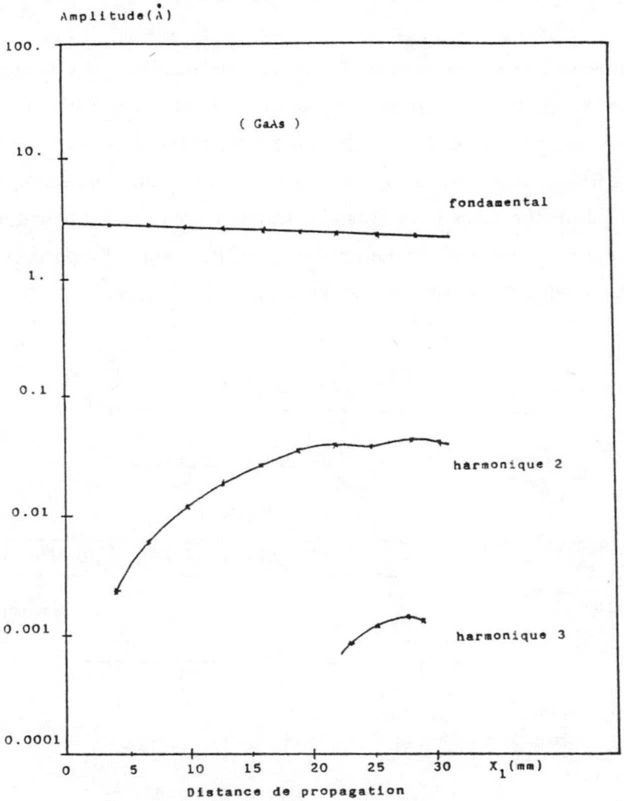

Fig. 6b : Harmonic generation of a (100) GaAs cut, X propagation direction (f_a = 83 MHz, P_{ex} : 0.2 Watt)

The phase difference between the fundamental wave and the second harmonic wave that its instantaneous phase is translated to the fundamental frequency enables to evaluate the dispersion due to a metallic layer plotted on the propagation surface.

$$\Delta\phi = \omega_o\left(\frac{1}{V_R(2kh)} - \frac{1}{V_R(kh)}\right).x_1$$

x_1 is the propagation distance.

The dispersion curve leads to measure the layer thickness directly from the phase difference value. The accuracy is about 10 % for layer thickness above 500 Å.

4 - LOW FREQUENCY ACOUSTIC APPLICATIONS

Using the symmetrical laser probe to measure acoustic waves at low frequency needs improvements in the electronic chain to increase the signal to noise ratio. That was made on this type of probe and extended to the others. We present here two examples with plate waves. The first case consists to detect standing waves in an aluminum plate where a slot was machined on the faces [15] (Fig. 7). Lamb waves are launched with a PZT transducer and optically detected. Standing waves are detected between the slot and the transducer and travelling waves are generated from the cut off modes.

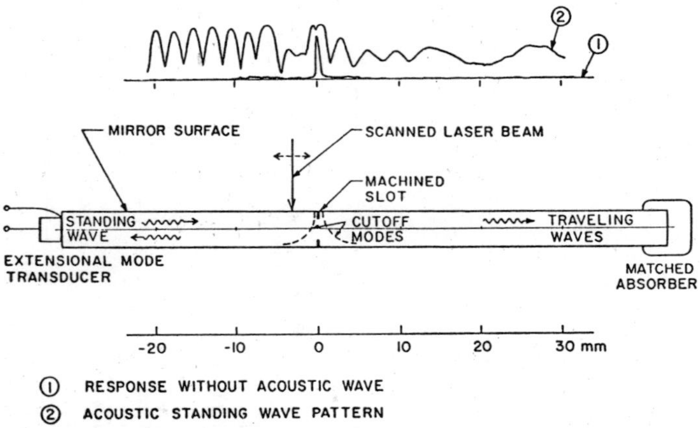

Fig. 7 : Standing wave measurement

Resonance modes in silicon plates excited by light and detected by laser probe are shown on Fig. 8. This figure represents the vibration level versus the modulation frequency of the excitation laser beam. A strong flexural mode appears at 9 kHz on the silicon plate (25 mm diameter, 370 μm thickness) clamped in three points [16].

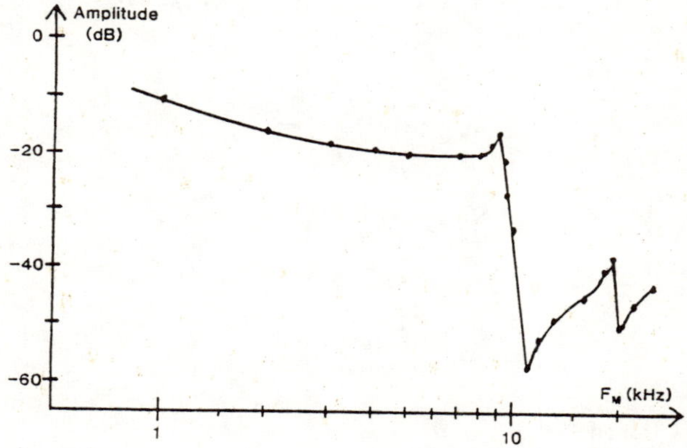

Fig. 8 : Thermoacoustic signal amplitude versus frequency for a 370 μm thick silicon plate

5 - THERMOACOUSTIC MICROSCOPY

The physical process occuring in thermoacoustic generation consists in two energy conversions :

a) modulated light energy is absorbed by the sample surface,

b) thermal diffusion creates mechanical expansion at the modulation frequency at the hit point,

c) elastic waves are generated from the hit point acting as an elastic wave source.

Then, two ways are available to detect optically the thermomechanical properties : reflected signal on the excited surface or transmitted signal on the opposite surface.

The general problem is modelized from the basic equations of thermal diffusion and elastic constitutive equations and by fixing boundary conditions for temperature and displacements

$$\sum_{i,j} K_{ij} \frac{\partial^2 T}{\partial x_i \partial x_j} - \rho C_V \frac{\partial T}{\partial t} + H = T \sum_{ij} \lambda_{ij} \frac{\partial e_{ij}}{\partial t} \tag{1}$$

with

$$\lambda_{ij} = \sum_{k,\ell} C_{ijk\ell} \alpha_{k\ell} \quad ; \quad \alpha_{k\ell} = \left(\frac{\partial e_{k\ell}}{\partial T} \right)_{\sigma = cte} \tag{2}$$

$$\rho \frac{\partial^2 u_i}{\partial t^2} = \sum_j \frac{\partial \sigma_{ij}}{\partial x_j} \tag{3}$$

$$\sigma_{ij} = \sum_{k,\ell} C_{ijk\ell} \delta_{k\ell} - \lambda_{ij} (T - T_c) \tag{4}$$

$$\delta_{k\ell} = \frac{1}{2} [u_{i,j} + u_{j,i}] \tag{5}$$

Solution given for 1-D model is

$$u(x,t) \simeq B(\omega) [C(\omega) e^{-jkx} - De^{-\mu x}] e^{j\omega t} \tag{6}$$

with

$$k = \omega \sqrt{\rho / C}$$

B(ω), C(ω) and D depend also on thermoacoustic and thermoelastic properties of the medium.

$$\mu = (1 + j) \sqrt{(\omega \rho C)/(2K)}$$

μ characterizes the penetration depth of the thermal diffusion.

If we represent these results as plane thermal waves properties, it leads that equivalent thermal wavelength and penetration depth decrease as $\sqrt{f}$.

Reflection can appear on thermal discontinuities and in terms of reflection coefficients and normalized thermal impedance

$$r = \frac{1 - Z_t}{1 + Z_t} \quad \text{with} \quad Z_t = \frac{K_2}{K_1} \sqrt{a_1 / a_2}$$

K_1 and K_2 are respectively thermal conductivity of mediums 1 and 2, a_1 and a_2 are thermal diffusivity constants.

3D-modelization is studied taking into account the experimental set up parameters.

Fig. 9 is the schematic of the microscope principle operating in transmission mode. The main characteristics are the following :

a) excitation laser can be He-Ne 25 mW laser or Nd Yag (0.5 to 2 W) laser. Light is modulated in the bandwidth DC to 2 MHz.

b) Heterodyne laser probe has a resolution of 3.10^{-14} m at 10 kHz and a good linearity up to 200 Å in the bandwidth 650 Hz - 2 MHz.These performances are obtained by using PLL detection and AGC to control the averaged signal amplitude. Phase and amplitude images are rebuilt in a computer after the sample has been scanned by means of stepping motors, synchroneously driven with the detection.

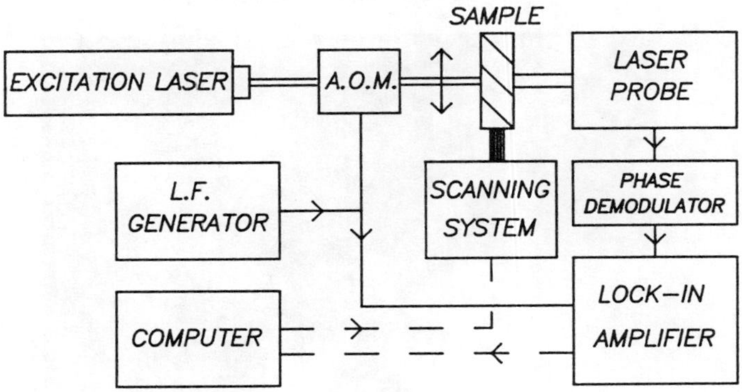

Fig. 9 : Principle of thermoacoustic microscope operating in transmission mode

Different samples have been tested : some are bulk metallic pieces with defects, others have layers on metallic blocks or others are dielectric substrates [17].

The first sample (photo 1) is a duralumin piece drilled with two holes under the surface. Diameter is 1 mm and depth is respectively 150 μm and 350 μm from the surface. Photo 2 shows amplitude image where appear grooves to give the hole positions and holes themselves. Modulation frequency is 5 kHz and the image resolution is 64 × 64 pixels. Each pixel corresponds to 75 μm.

Photo 1 : Duralumin drilled sample

Photo 2 : Amplitude image of the sample in the grooves region

The second example is obtained in the case of defects (letters) engraved in stainless steel sample, polished and filled up with epoxy, polished again and overlayered with aluminium sheet (20 µm-thick). Phase and amplitude images obtained at 15 kHz show clearly letters "MO" on 100 × 100 pixels image with 5 × 5 mm size. (Photo 3).

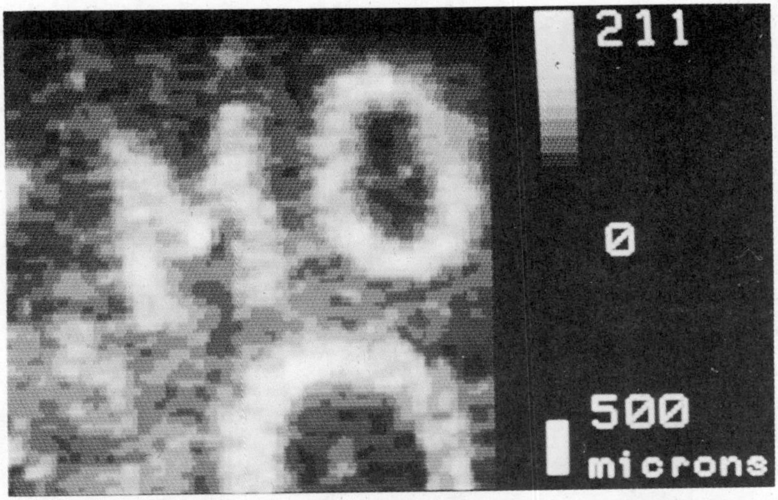

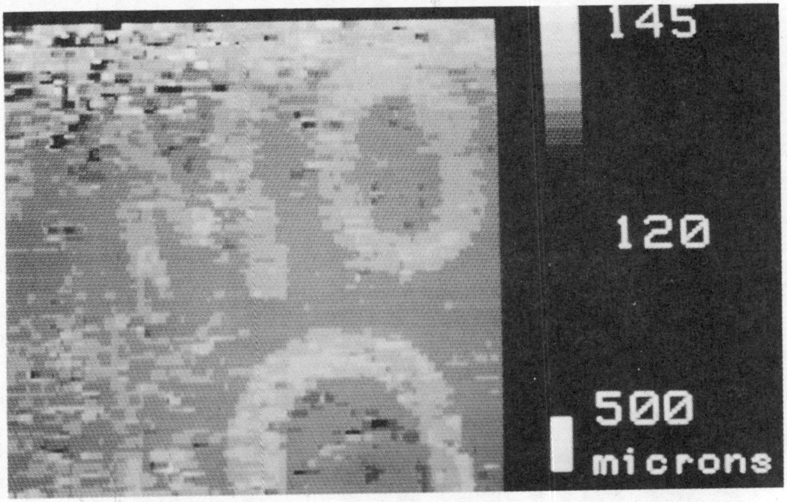

Photo 3 : Amplitude and phase images of the overlayered letters "LPMO"

3a : amplitude image - 3b : phase image

Optical wave guide image in amplitude (Photo 4) is also made at the process-step between titanium diffusion and electrodes photolithography processes. Then the sample was overlayered by a 1 000 Å thick aluminium film.

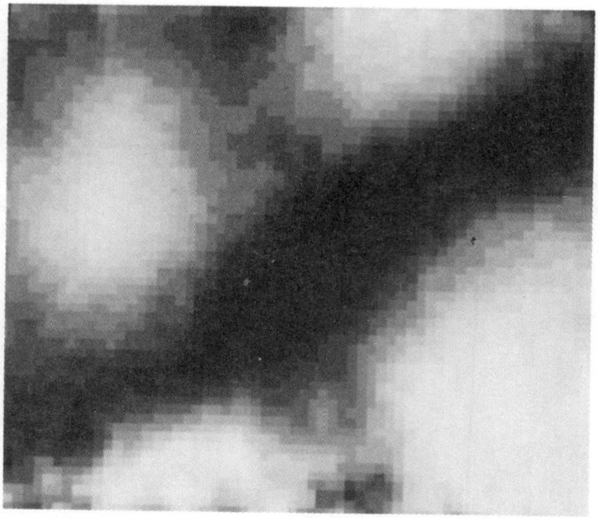

Photo 4 : Optical wave guide amplitude image

Photo 5 shows finger-print images obtained on the same sample before cleaning up it. The wave guide image is superimposed with the finger-prints response.

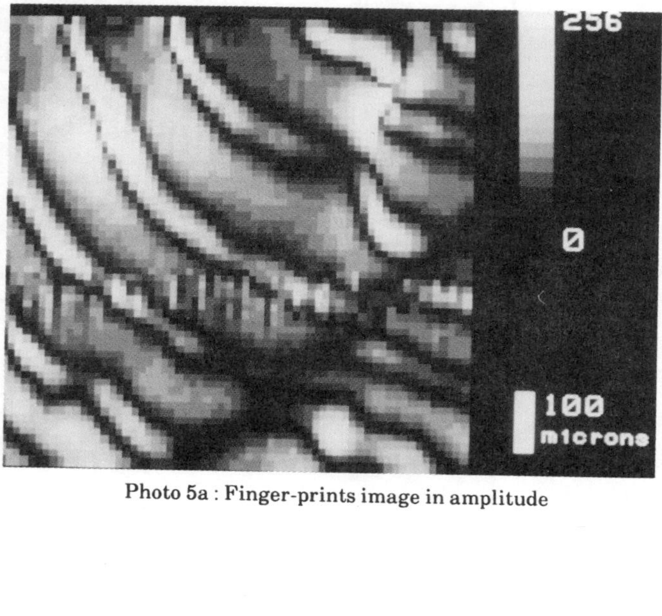

Photo 5a : Finger-prints image in amplitude

Photo 5b : Finger-prints image in phase

All images shown here are obtained in transmission configuration for the microscope. Images (phase and amplitude) may be plotted in the reflection mode. The detected signal is about 100 times higher than previously, but informations are a little bit different because in transmission the signal is carried by acoustic wave and in reflection it is given by the direct thermal expansion.

6- JOULE DISPLACEMENT MICROSCOPY (OR THERMOELASTIC MICROSCOPY)

The method consists in detecting, by means of the probe, pulsed displacements induced by thermal expansion (Joule effect) due to pulsed current in electric conductors or semiconductors (metallic layer, resistive layer, transistors or diodes, integrated circuits).

If we consider a 1-D model like on Fig. 10, the coupled thermal diffusion equation and the thermoelastic equation lead to a mechanical displacement given by the relation [18]

$$D \simeq \frac{\alpha_c P}{\rho_c C_c \omega} |X||Y|$$

α_c : expansion coefficient, ρ_c : specific mass, C_c : calorific capacity
and P is the peak power density per unit area in the conductor. |X| is the thermal transfert conductor to substrate and |Y| is the insulating substrate expansion function. But |X| and |Y| are generally close to unity.

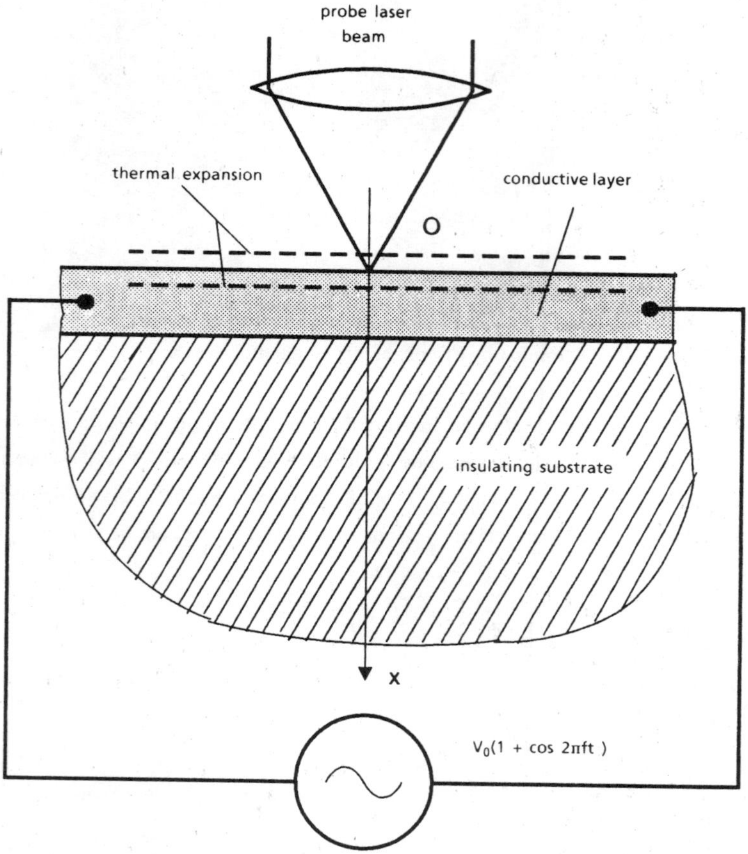

Fig. 10 : Conductive layer on insulating substrate model

Photo 6a represents the part of the sample having Cupper strips on epoxy substrate. The amplitude image recorded at 6 kHz for a pulsed current equal to 1.5 A is shown on Fig. 6b. This image is 64 × 64 pixels corresponding to 1 × 1 mm.

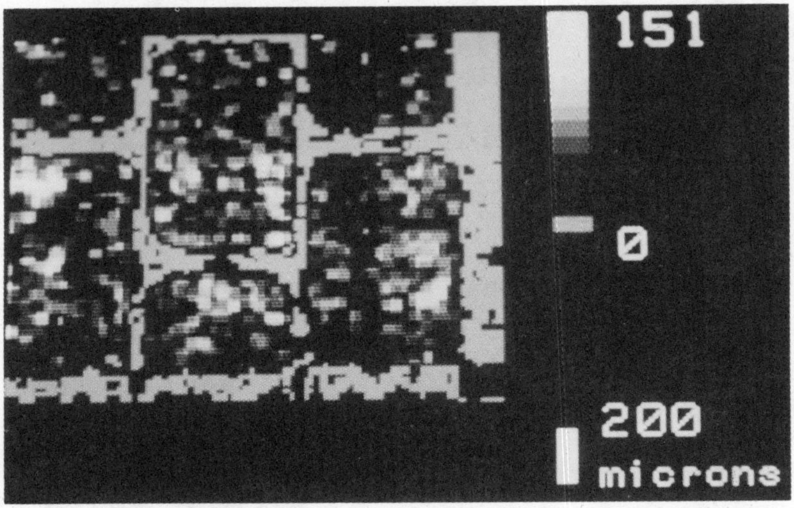

Photo 6a : View of the sample : cupper strips on epoxy substrate

Photo 6b : Amplitude image at 6 kHz

The further case is this one of an open case transistor 2N2219 acting in switch mode at the frequency of 5 kHz. Photo 7a is the transistor optical image. Photo 7b and 7c are amplitude and phase images when the dissipated mean power is 100 mW. Images size is 100 $\times$100 pixels over 400 $\times$ 400 µm. Maximum displacement amplitude is about 0,5 Å.

Photo 7a : Optical image of a 2N2219 transistor

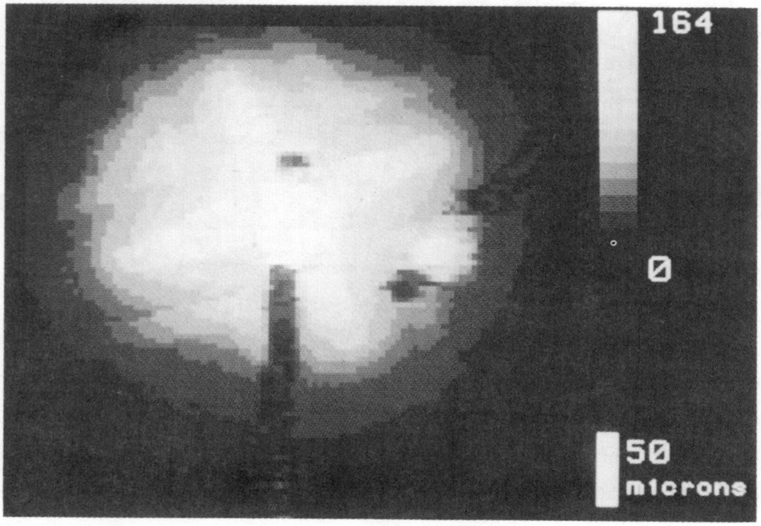

Photo 7b : Amplitude image at 5 kHz

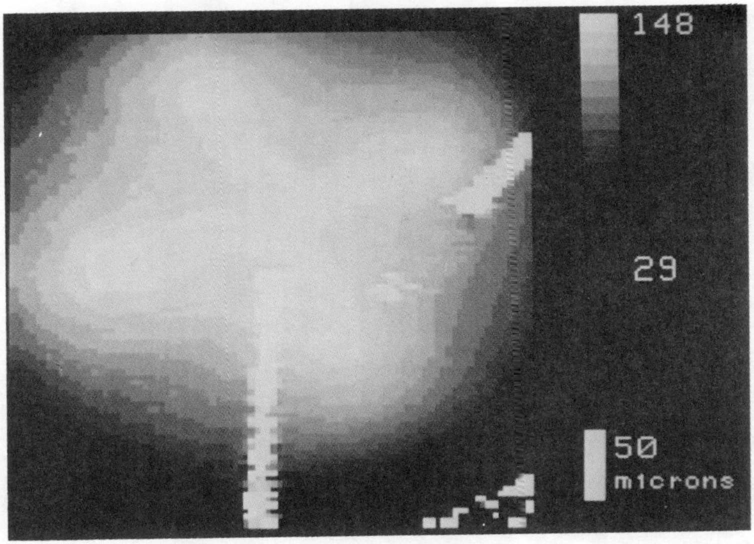

Photo 7c : Phase image

The same method is also avalaible to record a thermoelastic chart of an integrated circuit (analog circuit for instance) having powered components. Photo 8a shows the design of IC part with resistors, low signal bipolar transistor and medium current bipolar transistor. At a frequency of 3 kHz, 100 × 100 pixels images are obtained at 2 magnifications : figure 8b where the sizes are 1 × 1 mm and figure 8c where the sizes are .3 × .3 mm.

Thermal energy is dissipated in the zone of power transistors where thermoelastic effects are stronger compared to other parts of the integrated circuit.

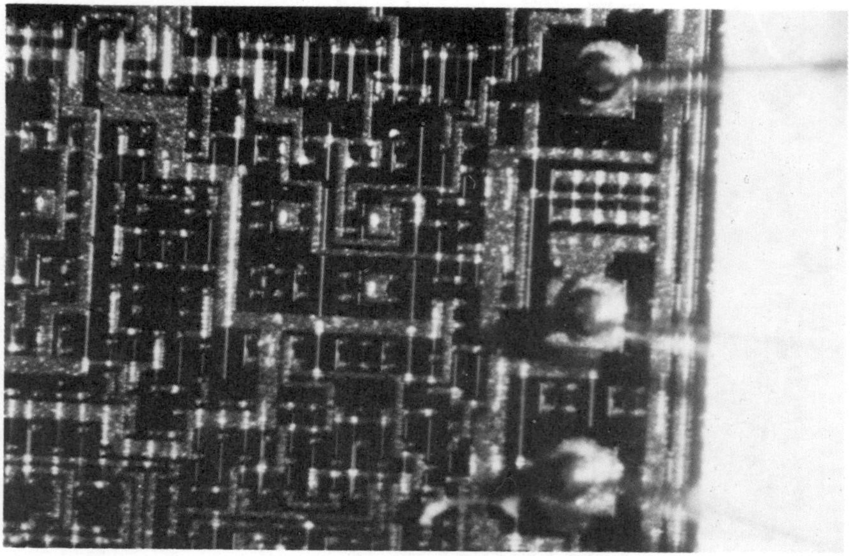

Photo 8a : Analog gate array integrated circuit with two power transistors

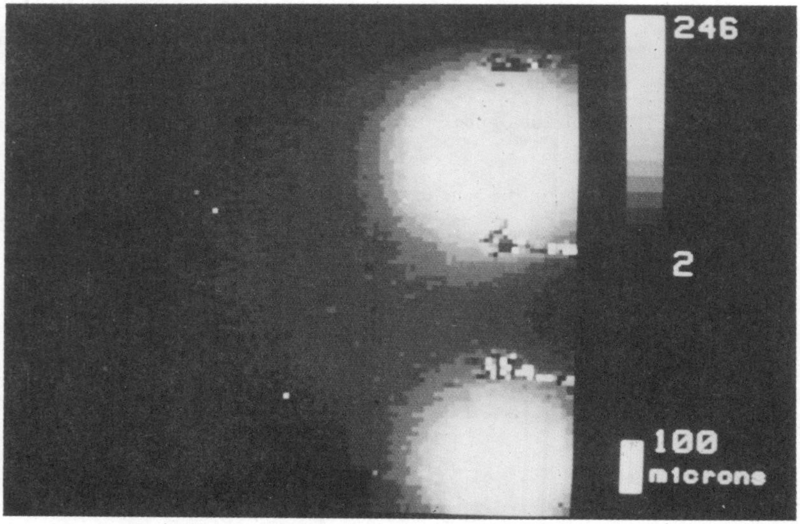

Photo 8b : Amplitude image of the IC

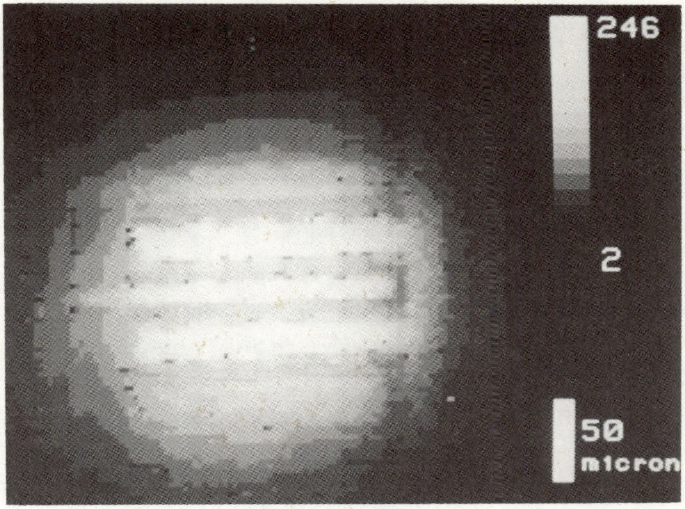

Photo 8c : 3 × magnifications of the power transistor in the IC

CONCLUSION

Interferometric heterodyne laser probes are useful to detect and measure vibrations at high and low frequency in many applications. However, in the last frequency range where large new field of application exists (N.D.T., thermoelastic imaging, near field microscopy, ...) some limitations must be studied and improvements proposed. One type of limit due to the noise level must be reduced by spatial integration, narrow bandwidth electronics or cross-correlation of compressed pulses. The other limitation is due to the surface roughness and the optical reflectivity which can be controled by an automatic intensity controller and a speckle reduction system.

REFERENCES

[1] Brillouin, L., C.R. Acad. Sci., 158, 1331 (1914).

[2] Korpel, A., "Acousto-optic : a review of fundamental", Proc. IEEE, 69 (1) (1981).

[3] Rhode, W.T., "Acousto-optic signal processing : convolution and correlations", Proc.IEEE, 69 (1) (1981)..

[4] Konotjku, K.A., Black, T.D., "Analysis and application of stationary reference operating method for optical detection of SAW", J. Appl. Phys., 52 (1) (1981).

[5] Chiba, T. and Togani, Y., "Optical observation of acoustic wave and electrical characteristics in LiTaO3 SAW devices", IEEE Trans. Son. Ultrason., SU-29 (1), 58 (1982).

[6] Dandliker, R. and Willemin, J.F., "Measuring microvibrations by heterodyne speckle interferometry", Optics Lett., 6 (4), 165 (1981).

[7] Lawson, R.L., Hunsinger, B.J., Cho, F.Y., "Visualization of acoustic surface waves in an experimental Fabry-Perot interferometer", Appl. Opt., 9 (12), 2805 (1970).

[8] Lean, E.G., Tseng, C.C., "Nonlinear effects in SAW", J. Appl. Phys., 41 (10), 3912 (1970).

[9] Delarue, R.M., Humphryes, R.F., Pason, I.M., Ash, E.A., "Acoustic surface wave amplitude and phase measurements using laser probes", Proc. IEEE, 119, 117 (1972).

[10] Bowers, J.E., "Fiber optical sensor for acoustic surface waves", Appl. Phys. Lett., 41 (3), 231 (1982).

[11] Royer, D., Dieulesaint, E., Martin, Y., "Improved version of a polarized beam heterodyne interferometer", Ultrasonics Symposium Proc. 1985.

[12] Cretin, B., Xie, W.X., Wang, S., Hauden, D., "Heterodyne interferometers : practical limitations and improvements", Optics Communications, 65 (3) (1988).

[13] Alippi, A., Palma, A., Palmieri, L., Socino, G., "Phase and amplitude relations between fundamental and second harmonic acoustic surface waves on SiO2 and LiNbO3", J. Appl. Phys., 48, 2182 (1977).

[14] Planat, M., Thèse d'Etat, Besançon (1984).

[15] Hauden, D., Hassani, A., Planat, M., "Characterization of nonlinear propagation of Rayleigh waves", 1984 Ultrasonics Symposium Proc.

[16] Cretin, B., Hauden, D., "Thermoacoustic imaging using a laser probe", Acoustical Imaging, 15, 423 (1986).

[17] Cretin, B., Hauden, D., "Thermoacoustic microscopy using optical excitation and detection", SPIE, vol. 809 : Scanning Imaging Technology (1987).

[18] Martin Y., Wickramasinghe, H.K., "Study of dynamic current distribution in logic circuits by Joule displacements microscopy", Electronics Letters, 18 (16), (1982).

ACOUSTIC IMAGING AND SUPERRESOLUTION

H. W. Jones and H. W. Kwan
Acoustic Institute, Technical Unversity of Nova Scotia
Halifax, Nova Scotia, Canada.

ABSTRACT

The principles of producing diffraction patterns which are narrower than the classic limit has been discussed in the literature, by R. K. Luneburg and G. Toraldo di Francia among others. The theory which was developed is outlined and the limitations stated in the literature are discussed and compared with the physical circumstances relating to ultrasound. Experiments on a superresolving lens are outlined and commented upon.

INTRODUCTION

The subject of the resolving limits of optical instruments has a long history but much of the work relates to methods of increasing the relative aperture by geometrical design, change of wavelength and similar methods[1]. Similarly, methods of using the diffraction patterns to obtain information beyond the normal limits, e.g. the Michelson interferometer[2], have been developed over the years. Other methods have been developed to deal with distortion created in images[3] or special problems related to particular optical difficulties[4] so that better resolved images might be obtained.

Ultrasound presents a special opportunity which is not available in the majority of optical systems. It is possible to use the pressure or velocity information directly and process the received signals in real time in many cases. For example, it is easy to obtain a hologram by "adding the reference beam" in the electronic processing after the signal is received. Further, the use of the pressure or velocity signal, instead of the energy signal, allows a reconstruction of the hologram which has less noise than would otherwise be the case. These opportunities are offset to a considerable degree by the problems relating to the relatively small apertures available in ultrasound optical systems because of the long wavelengths which we are obliged to use in many circumstances.

This focuses our attention on diffraction effects. The problems in the area are well understood but there is much work to be done, at least, in the practical application, of existing theory to ultrasonic imaging and probably to other areas of acoustical signal processing. We will show that it is possible to obtain superresolved images by the use of acoustic lenses and we will describe such images which we have obtained, if somewhat fortuitously, during our experimental work.

THEORY

The control of diffraction by the shading of an aperture appears to be an idea which dates back to the turn of the century with the work of Straubel[5]. The work excited little interest for many years until it was the subject of the Luneburg lectures[6] and a little later by Toraldo di Francia[7]. This work was expanded by others and later in this section, some of the work of Barakat[8] will be listed.

Luneburg considered the case of a circular aperture in which the transparancy over the aperture was varied in some symmetrical fashion about the optical axis. In this case it follows that:

$$F(r) = \int^{\rho} \rho A(\rho)\, J_0(\rho r)\, d\rho \qquad (1)$$

where ρ is radius from optic axis in exit pupil, A is transparency, ρr is consistent with equation (2). In the investigation which follows Luneburg established the following:

1) among all diffraction patterns (1) of equal total energy the highest central maximum is obtained by the normal pattern i.e. for $A(\rho)$ = a constant, where $A(\rho)$ is the transparency of the aperture, and
2) it is impossible to increase the resolving power with regard to periodic structures by coating the aperture of an optical instrument.

During the investigation of this problem Luneburg demonstrated that the integral equation (1) could be manipulated to give diffraction patterns with different half power widths provided always that the condition (1) above was satisfied. As an example of this situation[1], figure 1 shows the change in the diffraction pattern for a lens with shading, all be it that the shading is simply blocking the central part of the lens.

Toraldo di Francia[7] took a particularly direct approach to the design of apodization; he started with the equation:

$$A(x) = J_0(x) \tag{2}$$

where $x = \pi D \sin \theta/\lambda$, D is diameter of aperture, θ is angle to point in focal plane, and λ is wavelength. He noted that for an aperture consisting of a series of concentric rings of area A_1, A_2 etc. that the amplitude of the diffraction pattern is given by:

$$A(x) = A_1 J_0(n_1 x) + A_2 J_0(n_2 x) + \text{ etc} \tag{3}$$

where n_1, n_2 is some number which may be determined arbitrarily in relation to A_1, A_2 etc. It was then noted that the zeros of the diffraction pattern can be chosen arbitrarily to suite any given condition and the following example was given (see figure 2). If we choose to make:
$A(x) = 0$, at $x = 2.4, 3.83, 5.52$ and 7.02, when $A(0) = 1$, then we find that
$A_1 = 12.43$, $A_2 = -34.11$, $A_3 = 40.03$, $A_4 = -25.31$, and $A_5 = 7.96$.
If we do not like the sidelobes we are getting from the choices of A_n we can manipulate the aperture to reduce any side lobe as much as we wish. The price, as Toraldo di Francia points out, is however, that increasing resolution "means that a smaller and smaller percentage of the luminous flux passing the pupil is utilized in the central disc". This was that which had been observed earlier by Luneburg.

Barakat[8] considered the apodization problem and the following is a summary of his analysis. First he addressed the problem of maximising the central lobe of the diffraction pattern subject to the first zero crossing moving inward from the Airy radius i.e. $v = 3.832$ to a radius β.

The pupil function is

$$\int^1 | T(\beta,r) |^2 \, rdr = 1 \tag{4}$$

(i.e. normalising the pupil function,

$$T = \frac{\text{amplitude distribution over exit pupil}}{\text{amplitude distribution over entrance pupil}} \quad \text{, i.e. } T = T(\beta,r) \text{).}$$

The amplitude is given by (c. f. Luneburg)

$$D(\beta,\beta) = \int^{1} T(\beta,r)\, J_0(\beta r)\, r dr = 0 \tag{5}$$

$$\text{given} \quad |\, D[0,\beta]\,|^2 = \text{maxim.} \tag{6}$$

He shows that 4, 5 and 6 lead to

$$T(\beta_1 r) = B(\beta) - C(\beta) J_0(\beta r) \tag{7}$$

where $C(\beta) = Q(b)\, B(b)$,

$Q(\beta) = (1/\beta)\, [2J_1(\beta)] \,/\, [J_0^{\,2}(\beta) + J_1^{\,2}(\beta)]$

and $B(\beta) = \{(1/2) - (1/2)\, [\, (2J_1(\beta))/\beta]^{\,2} / \, [J_0^{\,2}(\beta) + J_1^{\,2}(\beta)]\}^{\,(-1/2)}$

Figure 3 shows numerical values for T versus r. r is normalised to unity and the first Airy minimum of the diffraction pattern (ie. $\beta/3.832 = \beta'$). Negative values for T are simply a phase change of π. Figure 4 shows the shape of the diffraction pattern for $\beta' = .8$ and .9. Barakat then examined other problems relating to superresolved circular and slit apertures, the reader is referred to his paper for this discussion.

EXPERIMENTAL WORK

Kwan[9] performed a series of experiments on a well designed acoustic lens made of polystyrene (see figure 5). First he designed a lens by a development of a numerical program written by Jones & Kesner[10]. In particular he designed a lens which had abberations which would have produced a circle of least confusion in the image plane of 1/4 of the diameter of the first Airy ring. Table 1 lists the value of the third order abberation constants for that lens. The operating frequency used in the experiments was 1 MHz. The illuminating transducer effectively shaded the entrance pupil slightly see figure 6. This shading would have the effect of broadening the first maxima of the diffraction pattern. The interfacial effects of the lens were understood to some degree from earlier work, see figure 7. The interfacial effects shade the lens willy nilly. The experiments were performed in a tank (see figure 8), which used digital machine tool measuring scales so that the positional accuracy was determined to about 0.01 mm; this accuracy compares with a sound wavelength of approximately 1.5mm. Plots of the phase and amplitude of the received signal could be obtained in various planes, after refraction, by the use of a small transducer. Figure 9 shows such plots. The phase distribution across a Gaussian sphere situated some 5 cm behind the lens is shown in figure 10.

Figure 11 shows plots of the diffraction pattern in the focal plane for objects on and off axis. This data shows that superresolved diffraction patterns occur over the range 1 - 10° off axis and over this range the resolution in terms of β' being from 0.87 to 1.0. For comparison Figure 12 shows the approximate pupil function compared with the calculated functions of Barakat.

The pupil function which was present was not designed into the lens, it occurred fortuitously because of phase and amplitude changes associated with the transmission through the lens. It was, and is our intention to present results of other shading experiments later. However it is noted at this time that changes of shading can be effected simply by changing the pulse length of the transmitted sound. The reason for this is explained by reference to the dependence of the interfacial effects on pulse length[11], we shall comment in detail on such effects later.

CONCLUSION

This paper outlines the theory of superresolution and describes experimental work on the subject. It is shown that superresolution can be obtained in practice, provided sufficient care is taken in designing and manufacturing ultrasound lenses.

REFERENCES

1 Born, M. and Wolf, E., "Principles of Optics, Pergamon Press, N.Y. 1980.

2 Michelson, A.A., Phil. Mag., 30, (5), 1890.

3 Bates, R.H.T., "Astronomical Speckle Imaging", Physics Reports, 90, No.4, 203-297, 1982.

4 Gabor, D., Nature, 161, 777, 1948.

5 Straubel, C.R., Physik Z., 4, (16), 74, 1902.

6 Luneburg, R.K., "Mathematical Theory of Optics", University of California Phess, L.A., 1966.

7 G. Toraldo di Francia, "Super-Gain Antennas and Optical Resolving Power", Suppl. Nuovo Cimento, 9, (3), 426-438, 1952.

8 Barakat, R., "Solution of the Luneberg Apodization Problems", J. Opt. Soc. Am., 52, (3), 264-275, 1962.

9 Kwan, H.W., "A Study of Acoustic Lenses", Ph.D. Thesis, Dalhousie University, Halifax, Canada, 1988.

10 Jones, H.W., and Kesner, J.W., "A Program for the Numerical Evaluation of Acoustic Lenses (ACOLEN)", Research Report 73-1M7-DPSUB-RI, Westinghouse Research Labs., 1973.

11 Jones, H.W., and Kwan, H.W., "Ultrasonic Lenses for Imaging", Ultrasonics, 23, (2), 63-70, 1985.

Table 1. Aberration Function Constants.

3rd order abberation constants (ref 1, page 213 Eqn. (7))				
B	C	D	E	F
-0.0557	-47.9	-3070	817	-2.12

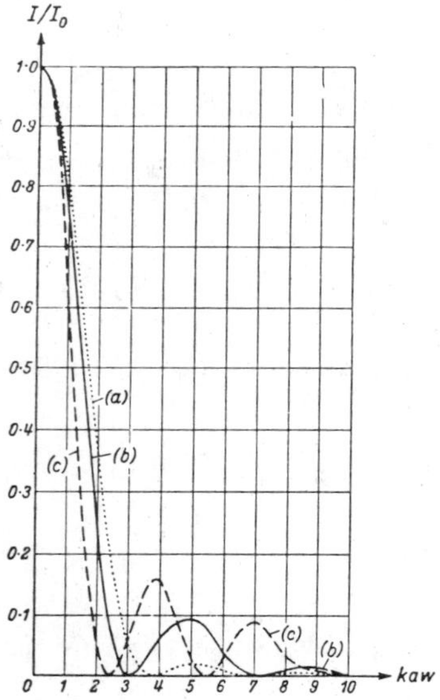

Figure 1. Illustrating the effect of central obstruction on the resolution. Normalized intensity curves for Fraunhofer patterns of (a) Circular aperture, (b) Annular aperture with r'= 1/2, and (c) Annular aperture with r' → 1. (after Born and Wolf).

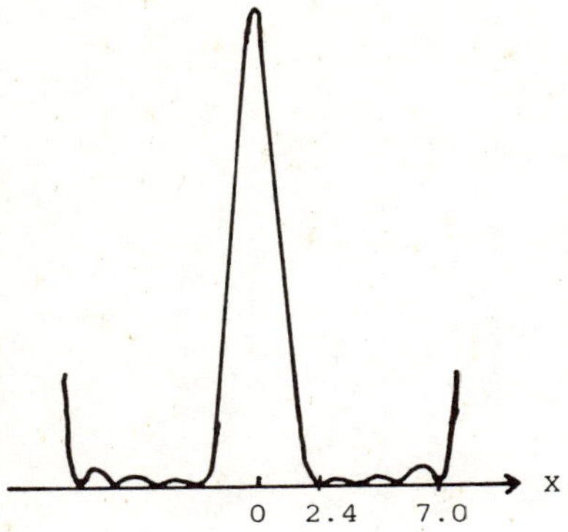

Figure 2. Diffraction pattern obtained with five ring apertures and coefficients. (after Toraldo di Francia)

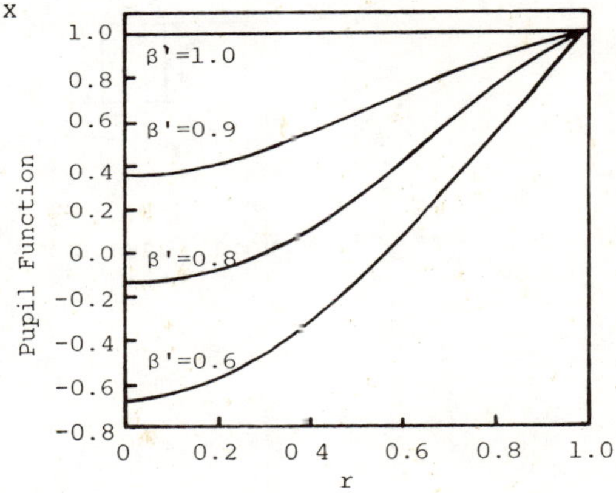

Figure 3. Normalized pupil function for circular aperture at selected values of β'. (after R. Barakat).

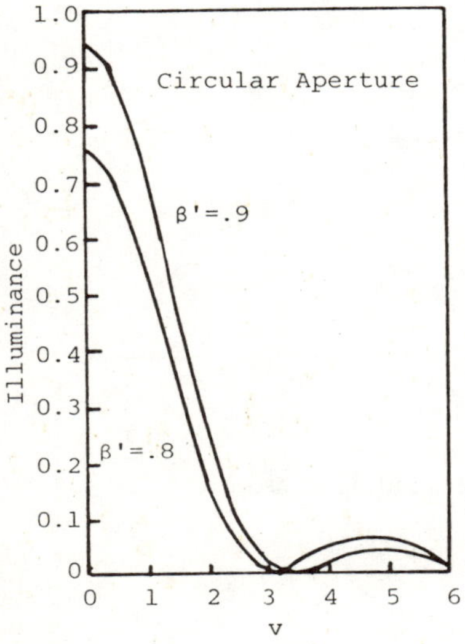

Figure 4. Illuminance in the Fraunhofer receiving plane for circular aperture. (after R. Barakat).

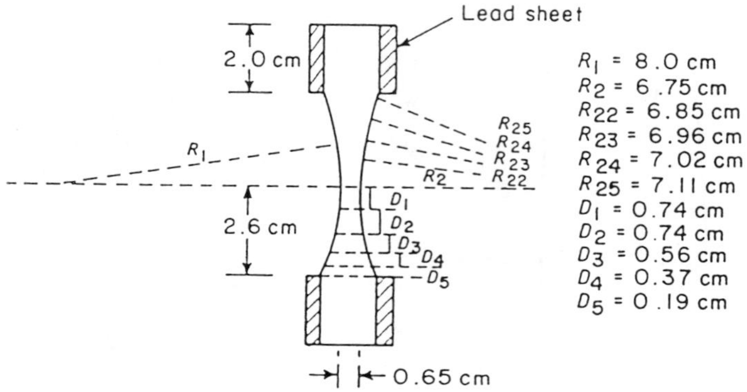

Figure 5. Dimensions of lens.

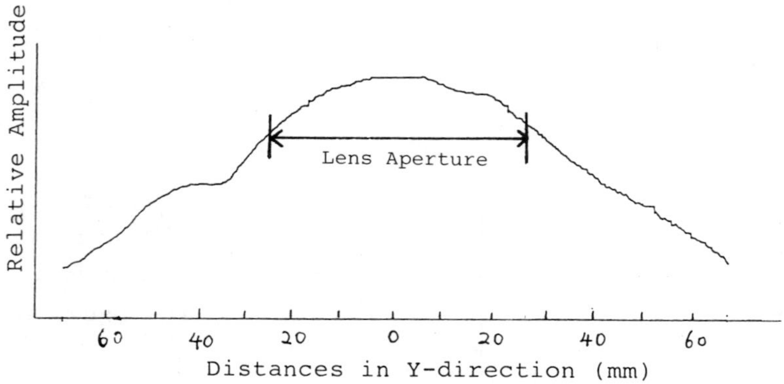

Figure 6. The amplitude pattern of the transmitting transducer.

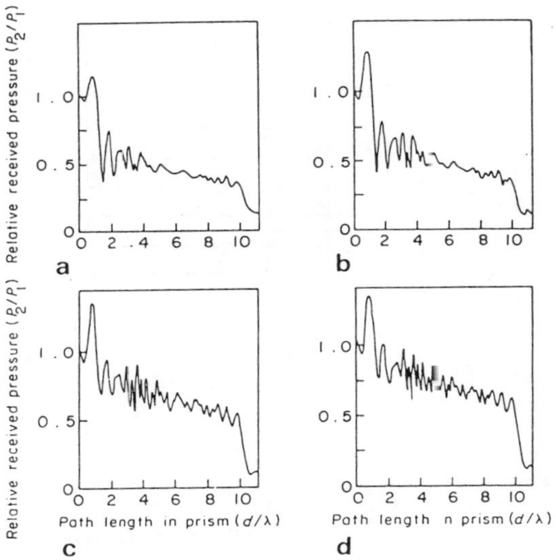

Figure 7. Plot of relative received pressure against path length in a 20 degree lucite prism for f = 646 kHz, angle of incidence 10 degrees: (a) pulse length = 5 wavelengths; (b) pulse length = 10 wavelengths; (c) pulse length = 20 wavelengths; (d) pulse length = 30 wavelengths.

Figure 8. General view of the tank.

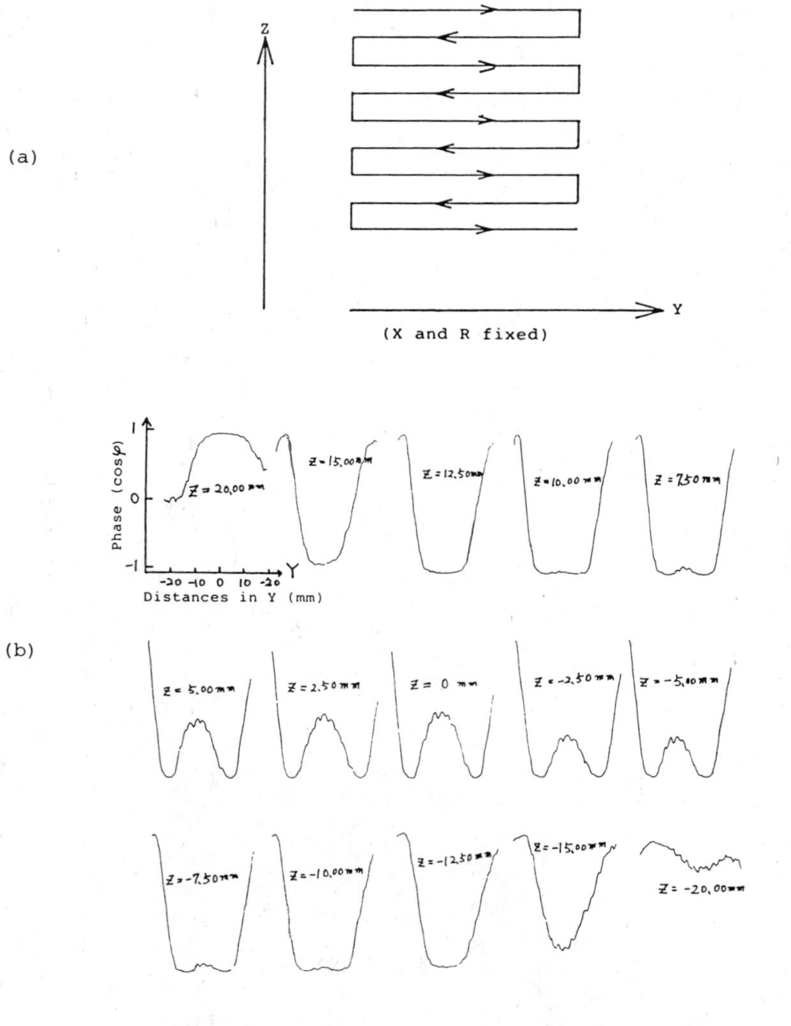

Figure 9. (a) Scanning pattern of the receiving transducer. (b) The measured variations in relative phase (cos ϕ) of image on the Y–Z plane at a distance of 53.22 mm from lens.

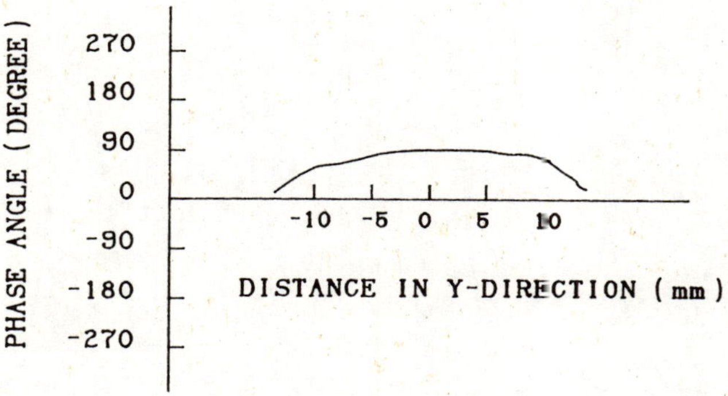

Figure 10. The variation of phase across a Gaussian sphere at a distance of 53.22 mm behind the lens.

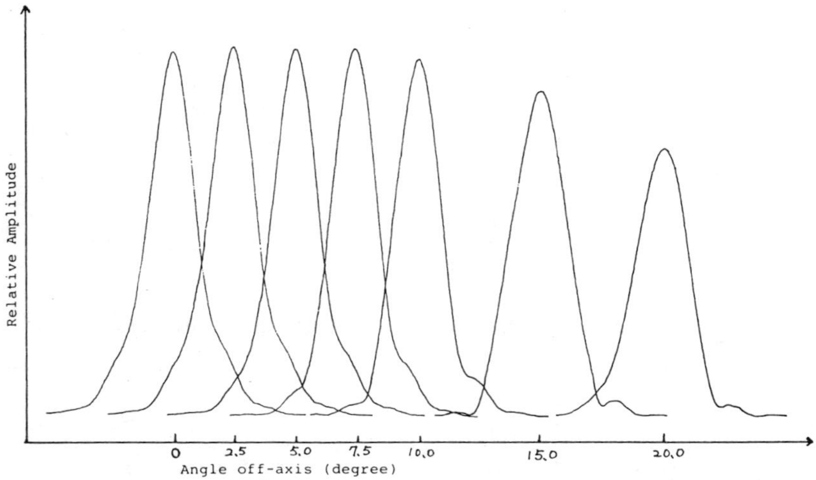

Figure 11. The measured pressure amplitude at different angles off the axis by rotating lens.

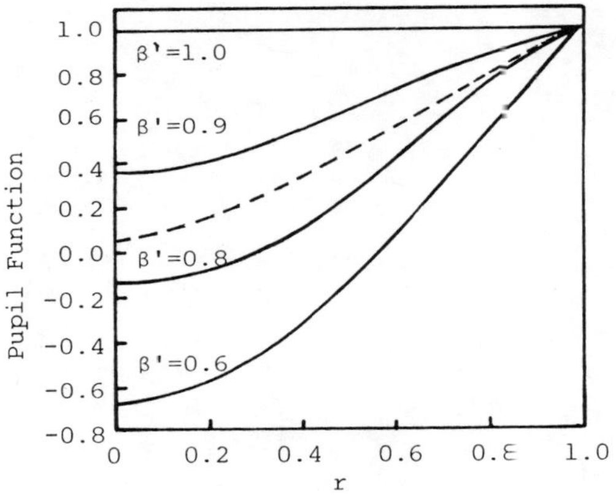

Figure 12. Dashed line showing the approximate value of β' from the lens experiments.

Application of Scanning Electron Acoustic Microscopy (SEAM) to Gallium Arsenide Technology

K.Kaufmann and L.J.Balk

Universität Duisburg Fachgebiet Werkstoffe der Elektrotechnik
Leiter: Prof. Dr. Ing. E. Kubalek; Sonderforschungsbereich 254
Kommandantenstr.60 D 4100 Duisburg 1

Abstract

SEAM has become a useful tool for microcharacterization of many materials, e.g. metals, biological specimen, and semiconductors. The technique is based on the generation of elastic waves inside the specimen due to an intensity modulated electron beam of high primary energy. These elastic waves are detected by attaching a transducer to the specimen, the most common one being a PZT ceramic. In the case of a harmonic or a square wave beam modulation both magnitude and phase of the correspondingly achieved electrical signal depend on various instrumentation parameters, such as energy and current of the primary electron beam, the chosen modulation frequency, and, most important, on the local material parameters of the specimen.

After a principal introduction into this method, examples of SEAM application to GaAs are discussed, like the detection of pn–junctions or doping inhomogenities. The spatial resolution of the technique is better than 1μm for this material. Further, recent results will be presented showing the dependence of the SEAM magnitude on some fundamental parameters of the electron beam for GaAs.

1. Introduction

SEAM is an operation mode within scanning electron microscopy (SEM) which is based on the generation of ultrasound inside the specimen. This mode has been introduced in the year 1980 by Brandis and Rosencwaig /3/ and by Cargill /4/. By means of SEAM it is possible to get informations on thermoelastic, structural, and electronic specimen features, what is of great use in semiconductor technology, especially as for semiconducting materials and devices good spatial resolutions and high contrasts can be attained. Further there are high SEAM signals due to the simultaneous appearence of various sound generation mechanisms, the most important one in the case of Gallium Arsenide (GaAs) are assumed to be the coupling of excess carriers with the piezoelectric properties of this material. This paper presents initial measurements towards a quantitative evaluation of the sound generation mechanisms in GaAs.

2. Signal Generation and Contrast Mechanisms

2.1 Electron Beam – Specimen Interaction

Until now there have been suggested several mechanisms for the generation of SEAM signals, and first descriptions have been formulated to enhance the understanding of the sound production /2,3/. But these models are only qualitative, and it is not possible to predict the magnitude of the electron–acoustic signal at the transducer output. One

reason for that is the complicated nature of the interaction of the electron beam with the specimen.
Because of the low frequencies of the SEAM signals (10 to 250 kHz) used in most experiments, corresponding to an acoustic wavelength in the order of 0.5 to 0.01 meters, the contrast generation mechanism, the spatial resolution, and to some extend the magnitude do not depend on the propagation of sound but on the sound generation.
The generation of sound mainly occurs close to the electron beam energy dissipation volume in the specimen.
When bursting into the specimen the primary electrons are scattered in different directions by elastic and inelastic collisions within the dissipation volume until all energy is lost. This strongly limits the attainable spatial (along the investigated specimen surface) and the axial (along the probe direction) resolutions. The possible penetration depth R of the electron beam energy can be computed by a formula proposed by Kanaya and Okayama /5/, as shown in Fig. 3. The diameter of the energy dissipation volume is in the order of the penetration depth. Thus using a lower beam energy may increase the attainable resolution.
The generation of sound is caused by the direct or indirect interaction of the electron beam with the material. The most important interaction of interest considering semiconductors is the generation of excess–charge carriers along the electrons path during slowing down. The exited electrons get the band–gap energy and an extra amount of kinetic energy. The kinetic energy is transfered to the lattice causing a slight increase of the local specimen temperature. In the case of a direct band–gap semiconductor as for GaAs, the band–gap energy is partially radiated as IR–light by the recombination of the electron–hole pairs. Furthermore, a local electric field is caused by the excess carriers, which may interact with the material. Finally, the charge and the momentum of the electron beam are to be considered.

2.2 SEAM Coupling Mechanisms

Here a short survey of the models for sound generation being of importance to semiconductors is given.

2.2.1 Thermal Generation of Acoustic Waves

This model is based on the exitation of elastic waves by a periodically heating of the specimen due to electron beam impact. Starting from the dissipation volume the heat diffuses away as a heat wave according to the temperature gradient. As this wave is strongly damped, its penetration is stopped within only one thermal diffusion length d_{th} which is given by /6/

$$d_{th} = (2^*K \,/\, \omega^*\rho^*c_{th})^{0.5}$$

K – thermal conductivity ω – frequency ρ – specific density c_{th} – specific heat.
If the material under test possesses a non–vanishing thermal expansion tensor, the periodic temperature variation leads to periodic elastic expansions and contractions of the lattice causing acoustic waves to propagate through the specimen. If the signal generation is mainly determined by this thermal wave coupling, the spatial resolution is limited by the diffusion volume of the heat wave; due to the frequency dependence of the thermal diffusion length d_{th} the resolution increases with increasing frequency. The contrast, which may be a change in either magnitude or phase, then results from local alterations of material parameters, such as heat conductivity, thermal expansion coefficient, heat capacity, and elasticity tensor.

2.2.2 Generation of Acoustic Waves by Electric Fields

In semiconductors the generation of sound is probably strongly influenced by alterations of the charge distribution, that is the appearence of local electric fields or field gradients within the material, in contrast to e. g. metals. The electric fields interact with the specimen because of the materials piezoelectricity or the electrical polarizibility. GaAs is piezoelectric due to the mixed (ionic/covalent) binding between both elements. The coupling between the electric fields and the permanent dipole moments is a linear effect, that means the driving force is proportional to the electric field. The electric fields may occur because of the input of the primary electrons charge, the output of the secondary electrons, and – likely most important – the different distribution of the exited electrons and holes due to their different diffusion constants and recombination rates. The electrons possess a higher mobility than the holes and their diffusion is restricted by the built up electric field. By this the SEAM signal is sensitive to the electrical properties of the semiconductor. Thus SEAM contrast arises, if the concentration of recombination centers and traps varies or if the local resistivity of the material changes. Device related structures as pn–junctions are visible, too.

3. Experimental Arrangement

3.1 The SEAM set–up

In the following, the SEAM set–up used for the presented measurements is described. Fig. 1 shows a block diagram of the arrangement of the necessary instrumentation. The main component is a slightly modified SEM. A chopping system is used to generate an intensity modulated electron beam of the frequency f. It consists of a pair of condenser plates being aligned parallel to the electron beam and supplied by a square wave generator. Typically the beam energy is varied in the range from 3 to 30 keV with beam currents in the range from 0.1 nA to 0.1 μA. The detection unit, containing the specimen and the transducer, is mounted on the specimen stage of the SEM. A description is given in the next section.
The electrical signal of the transducer, which is in the order of microvolts, is supplied to a lock–in amplifier, which is tuned to the chopping frequency f and phase–locked to it by the reference signal of the square wave generator controlling the beam chopping assembly. The used lock–in amlifier also makes it possible to detect the second harmonic (2f) of the electrical signal. By this the lock–in amplifier may deliver three informations about the initial signal: the (noise reduced) harmonic amplitude, the second harmonic amplitude, and the phase shift against the reference signal. All these quantities depend both on local specimen parameters and on various instrumentation parameters.
In order to process the SEAM signal, the lock–in amplifiers output can be supplied to the display of the SEM, to an image processing system or to an x–y–recorder.
An alternative to this experiment is the possibility to carry out time–resolved SEAM experiments by replacing the lock–in amplifier by a boxcar integrator.

3.2 Detection of the SEAM Signal by Piezoelectric Transducers

The elastic waves, which are generated inside the specimen and which travel through the specimen, are detected by a piezoelectric transducer. For this purpose the specimen is fastened on the top of a piezoelectric lead zirconate titanate (PZT) ceramic by Electrodag, a high conductivity silver paint. A good acoustic and electrical contact is given. The PZT–disc is acoustical backed by an attached brass cylinder.
To avoid intrusion of SEAM signals by other signals within the instrumentation,

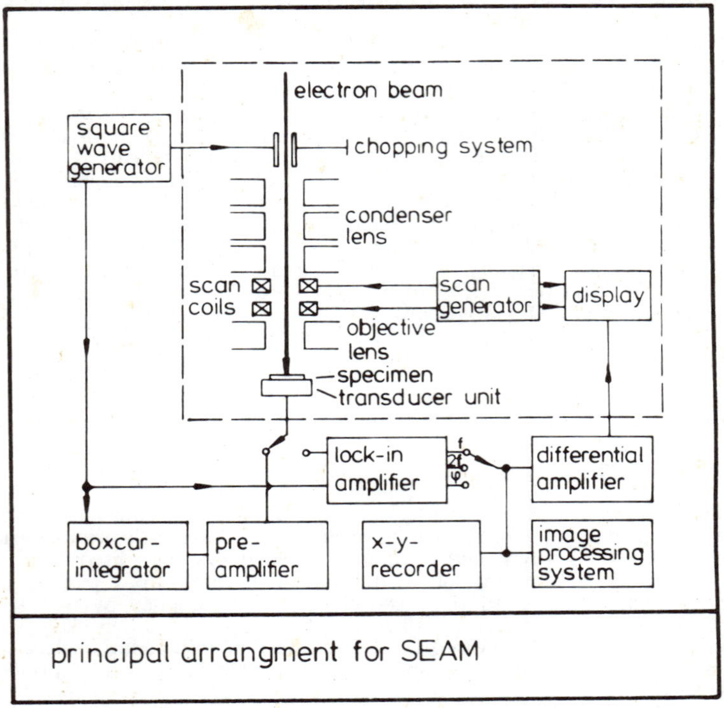

Fig. 1

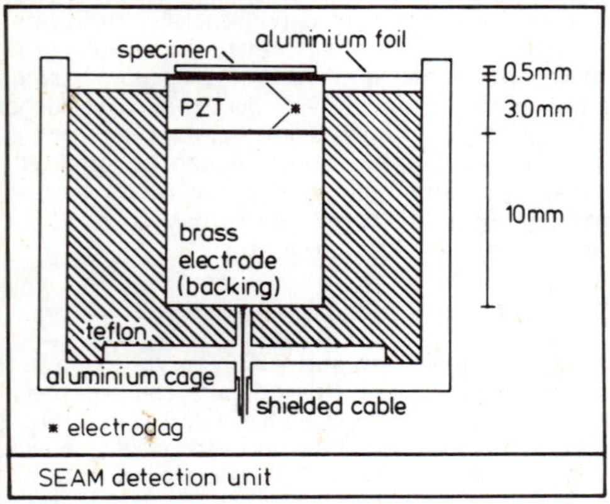

Fig. 2

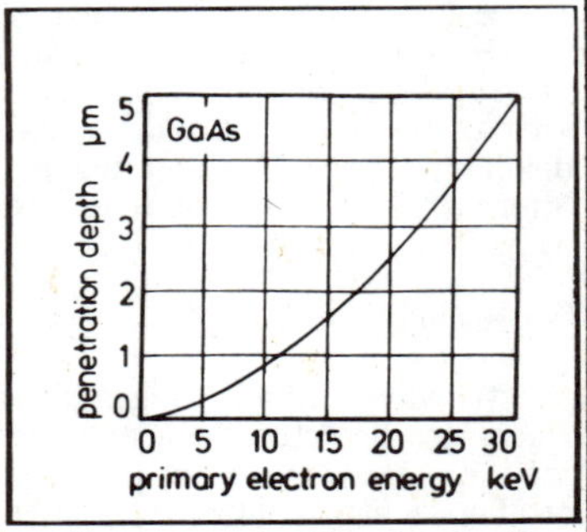

Fig. 3 penetration depth of primary electrons vs. energy for GaAs especially by other electron beam caused signals of the same frequency as the SEAM signal, a good electrical shielding of the PZT and all cables and connectors is necessary. The PZT is encapsulated within a metal cage; a alumimium foil is used at the top between the PZT and the specimen. A drawing of the detection unit is given in Fig. 2. An essential property of this arrangement is the strong frequency dependence of the electrical SEAM signal influenced by the resonance frequencies of specimen, PZT, and backing system. A frequency scan of the amplitude respectively of the phase is shown in section 4.1.

4. Application of SEAM on GaAs

Up to now SEAM theories have been developed to the point that a qualititative explanation of the reasons for a contrast appearing in a micrograph can be given in favourable cases. But no comprehensive quantitative theory has been developed, neither allowing prediction of contrasts due to the variation of specimen properties or determining functional dependences between signal levels and operation parameters. In the case of GaAs an exact theoretically treatment is difficult due to the complicated interaction of the electron beam with the material and the complicated electronical conditions in GaAs. In order to enable a consistent theory a lot of crucial experiments seems to be necessary. Therefore first measurements are presented in the following regarding the dependence of the SEAM–signal, as delivered by the described detection unit, on the parameters: electron beam current, energy, and chopping frequency.

4.1 Dependence of the SEAM Signal on Frequency

The electrical signal, delivered by the transducer, depends strongly on frequency. In Fig. 4 frequency scans of the SEAM amplitude and the phase over the frequency range from 10 to 250 kHz are shown. They are gained in the spot mode of the SEM for semiinsulating GaAs substrates. The resonances appear both,in the amlitude and phase curves at the same frequencies, the one with the greatest amplitude being at f = 234400 Hz.
In order to gain more information out of this frequency dependence, time–resolved boxcar integrator investigations have been carried out. The lower frequencies are all superposed by a frequency of f = 234400 Hz, as can be seen for f = 20000 Hz in Fig. 5.a. Assuming a velocity of sound of v_{brass} = 4650 m/s in brass the corresponding wavelength λ to f = 234400 Hz is λ = 0.02 m. This is just the double of the length of the brass backing of the transducer unit, leading to the conclusion that there exists for f = 234400 Hz a standing longitudinal wave in the brass backing with great elongations at the "open ends". This would result in a high electrical transducer output signal.
One part of the other resonances appearing in Fig. 4 , e. g. at f = 31600 Hz and f = 157000 Hz, probably can be explained by superpositions of different reflected acoustic waves in such a manner that the measurement by the lock–in amplifier results in a maximum signal. On the other hand, comparing measurements at f = 50000 Hz taken with the lock–in amplifier and with boxcar integrator experiments shows a low lock–in amplifier amplitude is recorded in spite of an actually high transducer signal due to an unproper beat frequency. Thus a great part of the frequency dependence of the SEAM amplitude can be explained by the technique of measurement.
But there is a frequency dependence of some specimen properties, too. As can be seen in Fig. 7, which shows the same specimen area at different chopping frequencies. The specimen is a mesa–etched GaAs layer structure. A 1.2 μm thick n–doped GaAs layer is grown on a semiinsulating GaAs substrate. In Fig. 7a, f = 94650 Hz, the layer area possesses a darker contrast than the substrate, in Fig. 7b, f = 133340 Hz, the layer area

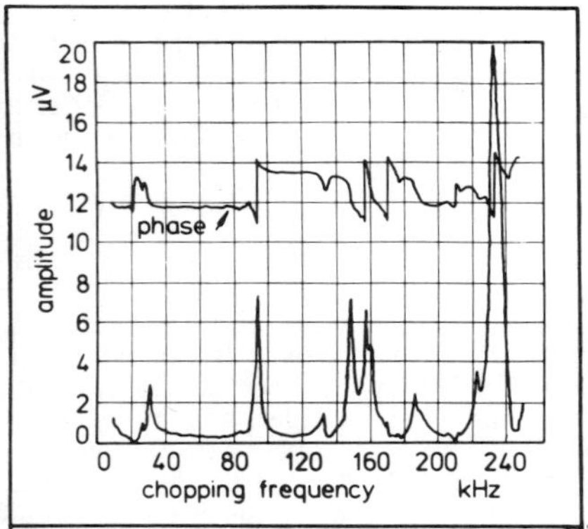

frequency scan of SEAM-amplitude and phase (phase without unit)
$E_{PE} = 20$ keV $I_{PE} = 5.8 \cdot 10^{-8}$ A

Fig. 4

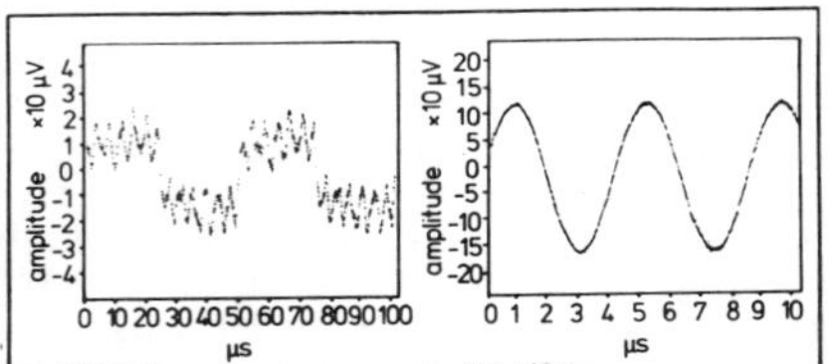

time-resolved electrical electron acoustic signal

Fig. 5

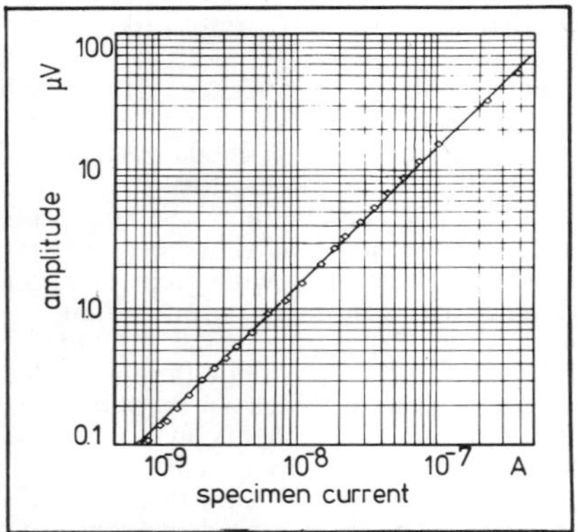

SEAM-amplitude vs. specimen current for $E_{PE} = 20$ keV

Fig. 6

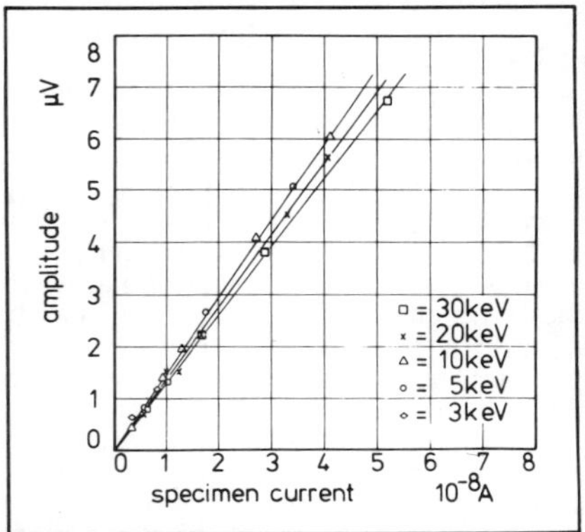

SEAM-amplitude vs. specimen current for different primary electron energies

Fig. 7

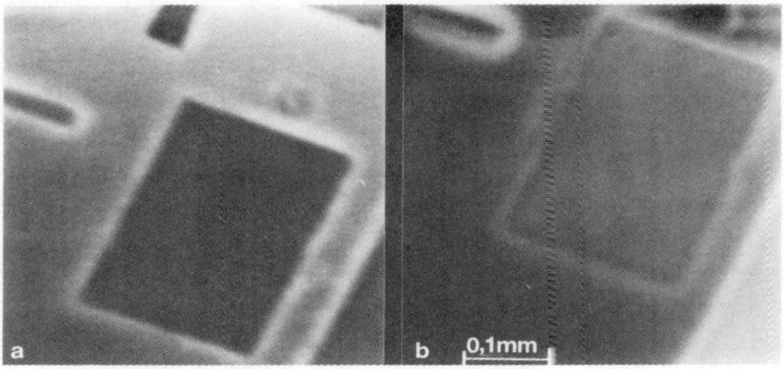

Fig. 7 SEAM micrographs of a mesa–etched GaAs–layer structure
E_{pe} = 5 keV $I_{pe} = 6.2 \cdot 10^{-10}$A a) at f=94650 Hz b) at f=133340 Hz

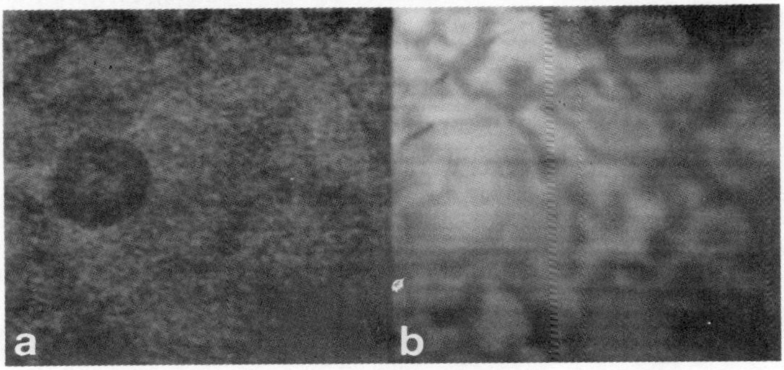

Fig. 8 SEAM micrographs of dislocation networks in a GaAs–layer structure
linear SEAM–amplitude images: a) E_{pe} = 10 keV b) E_{pe} = 30 keV

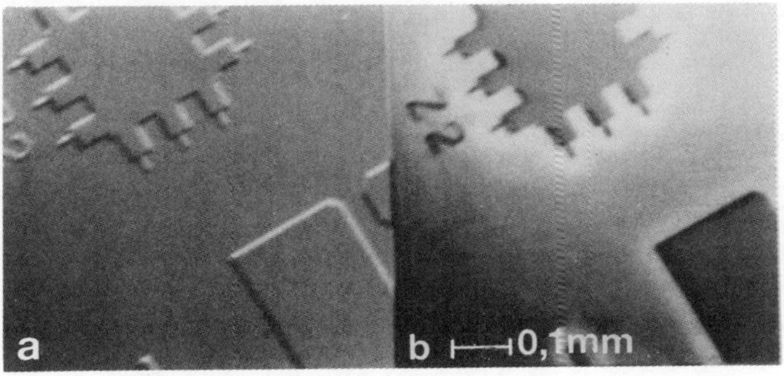

Fig. 9 micrographs of a mesa–etched GaAs–layer structure
E_{pe} = 20 keV $I_{pe} = 5.8 \cdot 10^{-8}$A a) SE–image b) linear SEAM–amplitude image

has a brighter contrast than the background, although the n–i–junction (that is the white border around the layer structures) has the same contrast in both images. An explanation of this effect cannot be given at present.

4.2 Dependence of the SEAM Signal on Current and Energy of the Primary Electron Beam for GaAs Substrate Material

In order to gain reliable informations on a device structure by an SEAM micrograph, it is essential to choose proper beam parameters. The first results regarding the investigations of GaAs substrates are:

a) The electrical SEAM amplitude depends linearly on the beam current at fixed chopping frequency and primary electron energy. Fig. 6 presents a curve of a measurement at a primary energy of $E_{pe} = 20$ keV, which clearly shows the linear dependence over a range from $I_{pe} = 8 \cdot 10^{-10}$ A to $I_{pe} = 3 \cdot 10^{-7}$ A. By this one can conclude that the sound generation is a linear effect within GaAs substrates. Furthermore this implies that the GaAs substrate is not in a degenerate state with respect to the generation of acoustic waves inspite of high electron beam currents being used. Further one can assume that the detection unit responds linearly to the acoustical SEAM magnitude.

b) The electrical SEAM amplitude decreases only very weakly with increasing primary electron energy, as can be seen in Fig. 7. The decreasing factor is in the order of 0 95 to 0.80 per 10 keV, this factor slightly depends on the chopping frequency. The decrease of the SEAM amplitude is at first sight surprising, since a higher power input should result in an increase of the signal. But another effect of increasing the primary electron energy is the increase of the penetration depth of the primary electrons. This means an increase of the energy dissipation volume, too. By that the impacted power density is also reduced, indicating that the sound generation mechanism is not only depend on the total power brought into the specimen, but more likely on the power distribution.

4.3 SEAM micrographs

Fig. 8a,b show SEAM micrographs of a GaAs layer structure (layer thickness: 0.8 μm) at different electron beam energies. In Fig. 8a the energy of the primary electrons is E = 10 keV, according to a penetration depth d of the electrons of d = 0.8 μm. In Fig. 8b the penetration depth of the electrons reaches d = 5 μm according to a electron beam energy of E = 30 keV. In Fig. 8b the dislocation networks in the GaAs substrate are clearly visible due to different density of recombination centers. In contrast in Fig. 8a the networks are visible only vaguely. Instead of this contaminations and scratches between the GaAs substrate and the layer are visible.
The secondary electron (SE) image and the linear SEAM amplitude image of a mesa–etched layer structure are shown in Fig. 9. In the SEAM micrograph other contrasts than in the SE–image, which mainly shows a surface topography contrast, are visible. The layer structures possess a darker contrast than the substrate and are surrounded by white borders, which appear because of the existence of a n–i–junction in this region.

5. Conclusions

SEAM allows to detect doping variations, alteriations in the concentration of recombination centers, and device related structures as pn– or ni–junctions in GaAs.

Towards an understanding of this method with regard to GaAs initial measurements have been carried out. The SEAM amplitude depends linearly on the electron beam current and decreases slightly with an increasing primary electron energy for GaAs substrates. Further investigations are necessary in order to understand this effect. A strong frequency dependence on the detection unit used is given. An improved detector design seems to be necessary to exclude superimposing of artificial contrast. For a qualitative analysis, however, the resonances can be used to get high SEAM signals and by this high signal–to–noise ratios.

6. Acknowledgements

The authors wish to thank Prof. Dr.–Ing. E. Kubalek for his encouragement and helpful discussions and the department of solid state electronic of the university of Duisburg for the preparation of specific specimens. The work was funded by the Deutsche Forschungsgemeinschaft.

7. References

/1/ Balk L J 1988 Advances in Electronics and Electron Physics vol. 71 pp 1–73
/2/ Balk L J and Kultscher N 1987 Inst.Phys.Conf.Ser.No.87: Section 11 pp 675–684
/3/ Brandis E and Rosencwaig A 1980 Appl. Phys. Lett. 37 pp 98–100
/4/ Cargill G S 1980 Nature 286 pp 691–693
/5/ Kanaya and Okayama 1977 J. Appl. Phys. 43 D5
/6/ Rosencwaig A 1982 Science 218 4569 pp 223–228

SIGNAL PROCESSING IN ACOUSTIC MICROSCOPY WITH APPLICATIONS TO NONDESTRUCTIVE EVALUATION

A. Morsch and W. Arnold

Fraunhofer Institute for Nondestructive Testing
Bldg.37, University, D-6600 Saarbrücken 11, FRG

ABSTRACT

We present our recent results to use signal processing techniques to extend the capability of Acoustic Microscopy in nondestructive evaluation. The V(z)-measurement technique for the evaluation of surface wave velocities with the Scanning Acoustic Microscope (SAM) was applied to nanocrystalline Pd-samples. Powerspectrum and cepstrum-analysis were used to derive the grain size from a SAM line-scan over a polycrystalline material. Scanning Laser Acoustic Microscope (SLAM) images which are acoustic holograms can be improved by backpropagation in the Fourier domain of the displacement field measured at the surface of the sample. We applied this technique to delineate the size of defects in advanced ceramics.

1. Introduction

Acoustic Microscopy uses for imaging the interaction of high-frequency acoustic waves with the sample material. The contrast of acoustic images is therefore mainly determined by the elastic properties of the object under investigation. Since acoustic waves can penetrate also opaque materials, it is possible to image internal structures of the sample with acoustic microscopes. These qualities made acoustic microscopy an important tool in nondestructive testing and material characterization. Especially two types of acoustic microscopes gained significance in practice: the Scanning Laser Acoustic Microscope (SLAM) and the Scanning Acoustic Microscope (SAM).

The SLAM uses for imaging a plane ultrasonic wave travelling through the sample. By scattering and absorption, the internal structures of the sample modify the sound field according to their elastic and geometric properties. The surface elongations caused by the sound field are transmitted to a light-reflecting layer that is coupled acoustically to the sample surface by water and are detected in amplitude and phase with a knife-edge scheme /1/. The real and the imaginary part of the sound field can be obtained seperately. Together they form an acoustic hologram /2/.

In SAM, a pulsed focused ultrasonic beam generated by the transducer of an acoustic lens and coupled with water to the sample, is scanned over the sample /3/. The sound field reflected by the sample is modulated in amplitude and phase according to the elastic properties and the topography of the surface. The lens recollects the returning waves and the transducer converts them back into electrical signals. The intensity of the received signals is displayed on a monitor synchronously to the scanning of the lens. If the focused ultrasonic beam contains rays whose incident angle to the surface is larger than the critical angle, surface waves are excited which in turn radiate back energy to the lens and thus strongly influence the contrast mechanism of the microscope. They are scattered by discontinuities of the sample surface such as cracks and grain boundaries. This is the reason why these surface features can be imaged although their size can be smaller than the resolution limit of the lens.

Signal processing and evaluation techniques can support the use of Acoustic Microscopy in nondestructive characterization of materials. Here we present some of our recent results of efforts in this area.

2. Characterization of the Elastic Behavior of Nanocrystalline Materials

Nanocristalline materials are a new kind of polycristalline solids whose grain-sizes comprise the range from 1 to 100 nm. The volume portion of the grain-boundaries in these materials is approximately as large as that of the cristalline phase. X-ray scattering experiments show that the structure of the grain boundaries resembles that of a gas /4/.

In a nonscanning mode the SAM can be used for material characterization. The output voltage of the lens V(z) as a function of distance z of the lens to the sample displays periodicities which depend on the elastic properties of the sample /5/. Because of this, V(z)-curves are also called Acoustical Material Signatures (AMS). The periodicities are caused by the interference of the leaky wave modes excited by rays in the acoustic lens under oblique incidence and the central ray. The periods Δz are related to the phase velocities c_{sw} of these modes, travelling parallel to the surface of the sample:

$$c_{sw} = \frac{c_{sw}}{\sin\Theta_c} \qquad (1)$$

with $\Theta_c = \arccos(1 - c/2\nu\,\Delta z)$

Here, c is the sound velocity in the coupling medium, and ν is the ultrasonic frequency. On an isotropic surface of a sample that can be considered as semiinfinite, two leaky waves can exist: leaky Rayleigh-waves and leaky skimming compressional waves. Usually, the periodicity produced by the Rayleigh-mode is used for material characterization because it is much more pronounced than the one produced by the compressional mode. If, however, the opening half-angle of the lens is smaller than the critical angle for the Rayleigh-mode, this mode is not excited. This is the case for materials with a low longitudinal wave velocity such as the nanocristalline material Palladium.

We made V(z)-measurements at frequencies from 1-2 GHz on the surface of nanocristalline Pd-samples /6/, and we interprete the observed periodicity as caused by a skimming compressional wave. We found that the deduced phase velocity v (v = 3 mm/µs, depending on the size of the nanocrystals) is about 30% lower than the velocity of a compressional wave in the bulk as determined by time of flight measurements. Further details can be found elsewhere /7/.

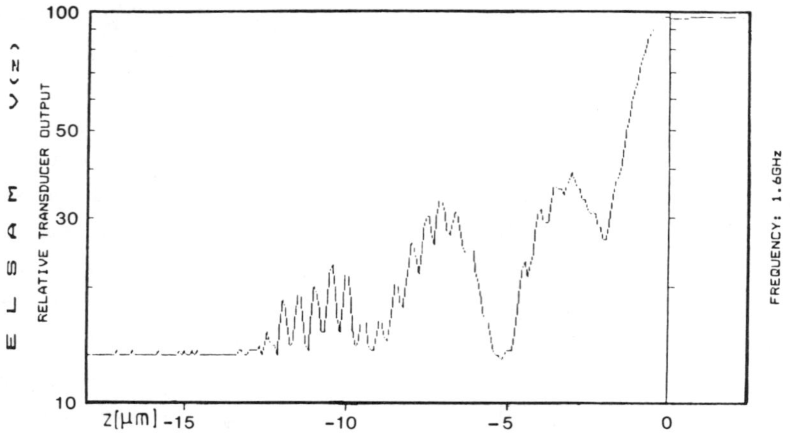

Figure 1: V(z)-curve of a nanocrystalline Pd-sample showing a well defined periodicity. The velocity of the surface wave excited by the acoustic lens can be calculated from the measured period Δz.

3. Grain-Size Measurements by Spectral Evaluation of Line-Scans

In SAM-images showing the surface of sufficiently flat polycristalline samples, grain contrast is caused by several factors /5/. These are: i) the reflection of surface waves at the grain-boundaries, ii) different reflection-functions of neighbouring grains due to their anisotropy, and iii) from the topography. The fact that in SAM the image information is provided as an electronic signal, which can be processed either digitally or in an analogue way, may form the base for an automated grain-size measurement technique.

3.1 Power-spectrum evaluation

The signal V(x) that is obtained with a line-scan in a SAM is schematically shown in Fig. 2. It already contains information about the grain-size. The contrast of the grain boundaries causes deviations in the lens-signal which repeat more or less periodically provided the scan-velocity is constant. Fig. 3a shows such a signal measured on a ferritic steel sample with a mean grain size a of 10 µm. The corresponding frequency $\eta = 1/a$ and its harmonics $n \cdot \eta$ ($n = 1,2..$) dominate in the spectrum of the signal, and the amplitude spectrum shows maxima at these frequencies. The spectrum can be measured with an analogue spectrum analyzer from the video-output of the instrument /8/. It is then straightforward to calculate the spatial frequencies η from the temporal frequencies ν by $\eta = \nu/c$, where c is the scanning velocity. They can also be discerned from a digitized image with the aid of a Fourier transformation on a computer. In both cases a sequence of maxima has to be recognized with a mean distance η_m related to the mean grain-size by $a_m = 1/\eta_m$.

Experience shows that the powerspectrum defined as the square of the absolute value of the amplitude spectrum is more adequate for practical evaluation because the maxima appear better resolved. Fig. 3b shows the powerspectrum of the V(x)-signal taken in the ferritic steel sample. A maximum appears at the frequency $\eta = 106\ mm^{-1}$ (arrow) from which the mean grain-size is discerned to be 9.4 µm. The example also demonstrates that the evaluation can be rather difficult because other maxima can occur in the neighbourhood of the main maximum.

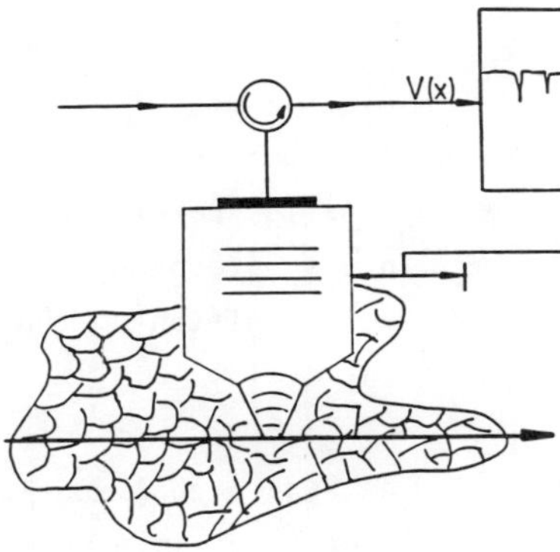

Figure 2: Principle of the measurement of a line-scan on a polycrystalline sample. The output voltage V(X) of the lens is processed.

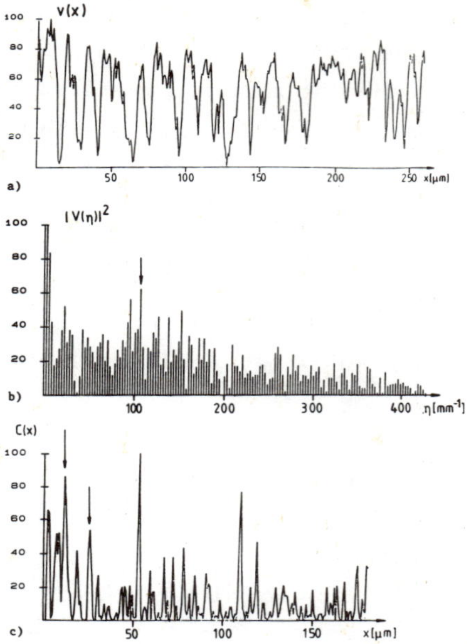

Figure 3: Grain-size determination on a ferritic steel sample
a) linescan, b) spectrum, c) cepstrum

3.2 Cepstrum evaluation

The information obtained in the spectrum can be further enhanced by calculating the cepstrum defined as the powerspectrum of the logarithmic powerspectrum resulting in a suppression of the unperiodic noise /9/. This greatly facilitates the recognition of the signals from which the mean grain-size are delineated.

Fig. 3c shows the cepstrum of the V(x)-signal mentioned above. A first well resolved maximum can be observed at 11.6 μm and another at a distance of 13.6 μm from the first (arrows). The average value of 12.5 μm corresponds to the mean grain-size. This signal-processing technique was applied to several different samples and the results were compared with the ones obtained from metallography. Although good agreement can be obtained in general, they also show that considerable deviations can occur when the sample surface, apart from the grain structure, contains many scratches or an uneven topography.

4. Scanning Laser Acoustic Microscopy

4.1 Holographic Backpropagation of defect images

Besides the simple detection of defects in a given sample by SLAM, their characterization may be of a certain interest. In general this is a complex problem because the SLAM provides two-dimensional images of a three-dimensional defect geometry deblurred by diffraction effects, so that the defect appears much larger than it is in reality. It is possible, however, to detect both the amplitude and phase of the acoustic field in a SLAM image rendering possible the reconstruction of the defect by acoustic holography-techniques /2/.

4.2 Acoustic holography and backpropagation

Acoustic holography is a two-step process. Firstly, the amplitude and phase of the acoustic field emanating from an insonified object is detected by a suitable electronic set-up at a plane adjacent to its surface. Secondly, from this field-data Φ, the defect can be reconstructed with a resolution of one wavelength.

The relation between the fields $\Phi(x,y;0)$ and $\Phi(x,y;\Delta z)$ in two parallel planes in the sample under investigation, separated by a distance Δz, is given by a linear filtering process:

$$\tilde{\Phi}(k_x,k_y;\Delta z) = \tilde{\Phi}(k_x,k_y;0) \cdot \exp[i\ k_z\ \Delta z] \tag{2}$$

with: $k_z = \sqrt{k^2-k_x^2-k_z^2}$ and $\tilde{\Phi}$ the corresponding fields in k-space.

k_x and k_y can be interpreted as the x- and y-component of the wave-vector of a plane wave with amplitude $\tilde{\Phi}$.

Thus, the field distributions between the detection plane and the object can be obtained by calculating the two-dimensional Fourier-transform of the field Φ, multiplying it with the filter function $\exp[-ik_z\Delta z]$ and Fourier-backtransforming it into spatial coordinates. This process is called backpropagation. Since outside $k_x{}^2+k_y{}^2 = k^2$ the filter function is steeply increasing /8/, the filtering is restricted to the inner of the circle k^2 so that noise is reduced. The S/N ratio in the backpropagated image can be further enhanced by deconvoluting the field-data with the transfer function of the laser detection scheme employed in the SLAM, resulting in a total improvement of the S/N ratio of ~ 10 dB compared to the original image /8/.

4.3 Reconstruction example

Fig. 4a shows the image of an iron-inclusion in a SiSiC-bending-bar obtained with SLAM at a ultrasonic frequency of 100 MHz. The diffraction of the sound waves at the inclusion causes a concentric ring-pattern. By subsequently calculating the field distributions in planes in increasing depth, an image with an apparently optimal defect contrast is obtained (Fig. 4b). In the example presented here, this is the case for Δz = 1500 μm. The extension in x-direction of the defect is 240 μm. From a radiographic micrograph, we evaluated a value of 250 μm and a depth of 1600 μm agreeing well with the holographic data. The holographic image is, however, deblurred in the y-direction. To our experience this effect is caused by an inhomogeneous insonification of the defect.

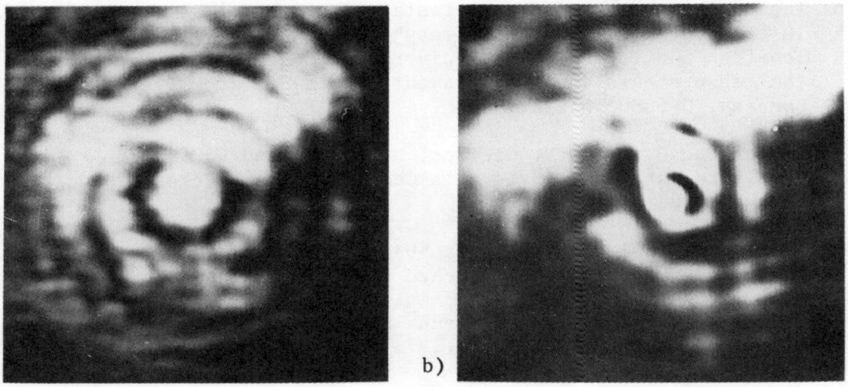

a) b)

Figure 4: a) SLAM-image of a Fe-inclusion in SiSiC as obtained at the output detector (ultrasonic frequency: 100 MHz, field of view 2800 x 2800 μm²) b) reconstructed image, the Fe-inclusion becomes clearly visible; field of view: 2150 x 2150 μm².

5. Summary

Summarizing we should like to stress that acoustic microscopy is of great value in non-destructive material characterization. By combining signal processing techniques with SAM and SLAM, the interpretation and evaluation of acoustic images is enhanced.

7. References

/1/ L. W. Kessler, Proc. IEEE, 67, 526 (1979)

/2/ Z. Lin, H. Lee, G. Wade, M. G. Oravecz, and L. W. Kessler, IEEE Trans., UFFC-34, 293 (1987)

/3/ R. A. Lemons and C. F. Quate, Phys. Acoust. 14, 1 (1979)

/4/ R. Birringer, H. Gleiter, P. Marquardt, and H. P. Klein, Phys. Lett. 102A, 365 (1984)

/5/ see for example: A. G. D. Briggs, "An Introduction to Scanning Acoustic Microscopy", Microscopy Handbooks 12 (1985), Oxford University Press

/6/ The experiments on the manocristalline Pd-samples were performed in collaboration with H.Gleiter "Abteilung für Werkstoffwissenschaften und Technologie", Universität Saarbrücken. Furthermore, the measurements were carried out with a SAM from the Wild-Leitz Company, D-6330 Wetzlar, FRG

/7/ A. Morsch, D. Korn, H. Gleiter, M. Hoppe, and W. Arnold Proc. 3rd Int. Symp. Materials Charact., Saarbrücken 1988, to be published

/8/ A. Morsch, Diploma-Thesis, Fachbereich Physik, University of Saarbrücken and IzfP Saarbrücken (1988) unpublished

/9/ G. Ehrenstrasser, "Stochastische Signale und ihre Anwendung", UTB Taschenbuch, Heidelberg (1973)

II. CONTRIBUTED PAPERS

B. Transducers and materials

ANALYSIS OF NEARLY PERIODIC IDTs BY MEANS OF SCATTERING MATRIX FOR PERIODIC METAL STRIPS

E. Danicki, Institute of Fundamental Technological Research
Polish Academy of Sciences, 21 Swietokrzyska Str.,00-049 Warsaw, PL

ABSTRACT

A model of an interdigital transducer of surface acoustic wave is introduced on the strength of the scattering matrix for single metal strip of the periodic system of strips deposited on a piezoelectric substrate surface. The model describes properties of IDT within a wide frequency band and takes into account $\Delta v/v$ SAW reflections from transducer electrodes. The analysis of narrow-band dispersive delay lines performed on a strength of the model shows that both the SAW reflections from the transducer electrodes and the influence of the input transducer admittance may equally heavily distort the filter response. The presented method for analyzing inhomogeneous systems may have some other applications, e.g. in a theory of superlattices.

1. INTRODUCTION

There can be as many as some few thousands fingers in dispersive interdigital transducers. This is a reason that the analysis of such long transducers is extremely difficult and there are not many papers in literature presenting full numerical analysis of DDLs, including their input/output admittances [1]. Even in a simpler cases, the correct model of IDT may be useful in some modern measurements concerning physical acoustics and SAW devices [2].

To perform such a full analysis, one can apply Green's function model [3], but this model is very difficult in application for long IDTs because it leads to very large system of simultaneous equations (each transducer finger is represented in the model by at least few unknowns describing charge distribution over the finger).

This paper presents a new model for analysis of long IDT, the fingers of which are nearly periodic, or locally periodic, exactly in the same meaning as it takes place in dispersive IDTs. The foundations of the model are given by the rigorous theory [4] developed for periodic metal strips, the main results of which is the scattering matrix for each transducer finger separately [5].

There is close correspondence between the proposed model and Milsom-Morgan's model [3], in fact they are identical for periodic fingers, however the presented model does not lead to simultaneous equations of such huge dimensions as former model does. It is because some relations are derived analytically in the presented model and one need not to calculate charge distribution over the transducer electrodes explicitly.

There are three steps in the presentation of the model below. The first one gives short view of the theory of periodic metal strips (bulk waves are excluded from considerations throughout the paper), the second one introduces the model explicitly for the system of "locally periodic" strips. Some numerical results are presented of analysis of dispersive delay lines. The last section of the paper presents some preliminary results on improvement of the model by more exact description of the performance of the transducer edge electrodes [6].

2. THEORY OF PERIODIC METAL STRIPS

2.1 Propagation Of SAW On Piezoelectrics

Consider a harmonic wave on the surface of a piezoelectric substrate with angular frequency ω and wave-number r in a form

$$exp(j\omega t - jrx) \qquad (1)$$

After Ingebrigtsen [3] we apply an approximated form of the effective surface permittivity of the substrate

$$\varepsilon(r) = \Delta D_\perp/E_{||} = -j\varepsilon_o\varepsilon_e \frac{r}{\sqrt{r^2}} \frac{r^2-k_v^2}{r^2-k_o^2} \quad ; \quad E_{||} = jr\Phi \tag{2}$$

where $\Delta D_\perp$ and Φ are complex amplitudes of waves of surface charge density and electric potential on the substrate surface, correspondingly, k_v and k_o are the wave-numbers of SAW propagating on free or metallized substrate surface, and ε_e is determined by dielectric constants of the substrate [5].

The above effective surface permittivity fully describes the property of the piezoelectric halfspace with stress-free surface. he SAW wave-number r can vary from k_v to k_o depending on electric conditions applied to the substrate surface. It can be shown that the magnitude of SAW Poynting vector depends on the complex amplitude of the electric field on the substrate surface and on the wave-number of SAW as follows

$$\Pi = \tfrac{1}{2}|E_{||} / (r-k_o)|^2 \tag{3}$$

It is convenient to introduce the quantity of SAW amplitude *a* , defined in ordinary way as follows

$$\Pi = \tfrac{1}{2}|a|^2 \tag{3'}$$

By comparison with (3) one obtains the relation between electric field amplitude and SAW amplitude as follows

$$a = E_{||} / (r-k_o) \tag{3''}$$

Let's consider now the eigenvalue bourdary problem of SAW propagation in the system of periodic metal strips on a piezoelectric halfspace. We need to find such an electric charge distribution $\Delta D_\perp$ on the metal strips and such a potential Φ between them which fulfill the following boundary conditions

$$\begin{aligned} &\Delta D_\perp = 0 \quad \text{on the surface between strips} \\ &\Phi = const \ (\text{or} = 0) \quad \text{on metal strips} \end{aligned} \tag{4}$$

To solve the above problem one can apply Milsom's spatial representation of Green's function [3], being the inverse Fourier transformation of the above-mentioned effective surface permittivity. The problem can be then led to a certain singular integral equation for charge distribution on the strips. This equation can be solved numerically by the method of collocation, for instance.

We apply the spectral method introduced by Ingebrigtsen and others. Accordingly to Floquet theorem we have an expansion for electric field and charge distribution in the form [4]

$$E_{||} = E_n e^{-j(r+nK)x} \qquad ; \qquad K=2\pi/\Lambda \qquad (5)$$
$$\Delta D_{\perp} = D_n e^{-j(r+nK)x}$$

(there are summations over n), where Λ is the period of strips.

The method gives the solutions for E_n and D_n satisfying (4) but the most important results for us now are the explicit formulae for amplitudes of current I flowing to the electrodes and their potential V, which can be written as follows [5]

$$I(r) = j\omega\varepsilon_o\varepsilon_e 2\alpha_o I_r$$
$$V(r) = \alpha_o V_r / sin(\pi r/K) \qquad (6)$$

I_r and V_r are some explicitly given functions of r, depending on strip dimensions as well as on the properties of the substrate, α_o is an arbitrary constant.

There are two different wave-numbers of SAW for two substantially different electric conditions applied to the strips, namely for free or short-circuited electrodes. In the last case the dispersion relation is $V(r_o)=0$ that gives the rigorous relation for r_o if r<K (it's the case considered here)

$$r_o = K/2 + \frac{1}{v} \left[(\omega-\omega_1)(\omega-\omega_3)\right]^{1/2} \qquad (7)$$

where
$$\omega_3 = \omega_1[1 - (\Delta v/v)s(1-s)]$$
$$\omega_1 = Kv/2 \qquad (8)$$
$$s = \tfrac{1}{2}\omega/\omega_1 \quad ; \quad (s < 1)$$

The same rigorous theory gives the relation between backward SAW amplitude a_{-1} and forward SAW amplitude a_0 as follows

$$a_{-1}/a_0 = \gamma = (1-Z)/(1+Z) \qquad (9)$$

where
$$Z = [(\omega - \omega_1)/(\omega - \omega_3)]^{1/2} \qquad (10)$$

One notes the similarity of the above relation to the relation describing the reflection of wave that takes place at a junction of two different waveguides, the first having wave impedance equal to 1 and the second one having wave impedance Z. This is not only the case (Appendix), the relation (9) was verified experimentally for the case of SAW, simply by measurement of SAW reflection coefficient from a system of short-circuited metal strips [5].

As it results from (10) Z is a function of ω and is a complex valued quantity for stop-band frequencies between ω_3 and ω_1. It appears, that exactly the same dependence of wave impedance can be introduced for any periodically inhomogeneous transmission lines (Appendix). What's more, if the nonhomogeneities of the transmission line is not periodic, one can divide the line into sections, each one having its own wave impedance Z that is dependent on ω and on its own ω_1 and ω_3 as given by (10). Moreover there are wave reflection and transmission phenomena between the neighboring sections, the coefficients of which are given by well known relations including their wave-impedances

$$T_{LR} = 2\sqrt{Z_L Z_R}\,/(Z_L+Z_R)$$
$$\Gamma_{LR} = (Z_R-Z_L)/(Z_L+Z_R) \qquad (11)$$

in spite of the dependence of Z on frequency. These relations remain correct even for the case where one or both Z_L and Z_R are complex-valued. In the above formula Z_L and Z_R means the wave impedance of left and right neighboring section and the wave is assumed propagating from left to right.

Having above in mind we make important generalization of the theoretical results obtained for periodic metal strips. Namely we expect that in the case of nearly periodic system of metal strips one can introduce the sections, each one including at least one metal strip, and that these sections can be characterized by their own wave impedancies Z, and their own stop-band frequencies ω_1 and ω_3. These quantities should be considered as local and varying along the system of metal strips accordingly to their "local period".

Similarly to the case of an inhomogeneous transmission line, there are transmission and reflection of SAW between these sections, the coefficients of which are given by the above-mentioned relations (11). Additionally if a wave covers the distance Λ between one edge of the section to the other edge, its complex amplitude is phase-shifted by the angle (r_0 depends on the local period Λ, as given in (7))

$$\tau = r_0\Lambda \tag{12}$$

2.2 Transadmittance Of Periodic Electrodes

Let's return to the relation (6) describing the potential of metal strip as a function of arbitrary wave-number r. This wave-number can vary in the area of one Brillouin's zone, say (-K/2, K/2). Taking inverse Fourier transformation for discrete functions as given by

$$V_n = \int_{-K/2}^{K/2} V(r)e^{-jrn\Lambda}\, dr/K \tag{13}$$

one obtains the electric potential of the given n-th electrode.

We consider the periodic system of metal strips in which all strips are grounded with exception of some of them, which have potentials V_n. It means that the left-hand side of (13) is known. Then, performing Fourier transformation of (13) one can determine α_0 as a function of r (see (6)) what makes function I(r) fully determined, its Fourier transformation gives

$$\int_{-K/2}^{K/2} I(r)e^{-jrm\Lambda}\, dr/K = I_m \tag{14}$$

After some manipulations this results in the following rigorous formula for current flowing to the m-th electrode as dependent on the potential of n-th electrode of the system

$$I_m = y_{mn}V_n \tag{15}$$

where, for electrodes of $\Lambda/2$ width we have

$$y_{mn} = -j\omega C_{|m-n|} + Y\, exp(-j\,|m-n|\,\tau)$$

$$C_k = \frac{1}{\pi}\,\varepsilon_o\varepsilon_e/(k^2-1/4) \quad ; \qquad Y = E_f/Z \tag{15'}$$

$$E_f = \omega\varepsilon_o\varepsilon_e 4\pi\frac{\Delta v}{v}\, s(1-s)sin\pi s \quad ; \quad \tau,\ Z,\ s \text{ as given above in (8-12)}$$

It is worth to note that we obtained these relations without explicit calculation of charge distribution over the electrodes. Also note that the wave impedance Z appeared once more above.

Neglect temporarily the fact that the transmittance y_{mn} is singular at ω_1 as Z^{-1} is. In the other words consider angular frequency ω being far away from very narrow stop-band area (ω_3, ω_1). One can make a comparison of the electric power delivered to the electrode supplied with the potential V and the power carried away by SAW in both left and right directions from the electrode. This gives the amplitude of SAW generated by the supplied electrode as follows

$$a = V \sqrt{(Y/2)} \tag{16}$$

Consequently, when a SAW with amplitude a_L^+ propagates from left to right under the supplied electrode, it changes its amplitude to the value

$$a_R^+ = a_L^+ + V \sqrt{(Y/2)} \tag{16'}$$

after passing the electrode. Similar considerations yield the value of the current flowing to the grounded electrode that is caused by SAW propagating under it and having amplitude a

$$I = a \sqrt{(2Y)} \tag{16''}$$

One may ask if these relations are still correct in the stopband, where Y is imaginary-valued quantity. The answer is yes, and it has also been verified experimentally. In fact, the well known distortion of the frequency response of simple interdigital transducer working at its fundamental frequency is strictly connected with the above-mentioned imaginary-valued quantities Y and r_0. This was a subject of papers [7], where especially arranged structure of metal strips were investigated both theoretically and experimentally. The results agreed quite well and this proved not only the relations (15-16), but also the reasoning concerning wave-impedance Z, as well.

3. A MODEL OF NEARLY PERIODIC IDT

3.1 General Model For Transducer Fingers

Let's summarize the results obtained for periodic metal strips and presented till now

item 1 - we found the explicit relation for a change of SAW amplitude under the supplied electrode with potential V,

item 2 - we have the explicit relation for current flowing to the grounded electrode caused by SAW propagating under it

(in terms of familiar δ-function model we have determined both SAW and current sources connected with given finger),

item 3 - we introduced the wave-impedance Z that determines at least the reflection coefficient of SAW from the transition point between the area covered by the periodic metal strips and the free area of a substrate surface.

Now consider nearly periodic electrodes. One can expect that both charge distribution on the given electrode and the potential distribution between this electrode and the neighboring ones are not change substantially because of slight shift of very distant electrodes from their periodic positions. Then, the above-mentioned SAW and current sources are expected to be good approximations for the case of nearly periodic metal strips. It is worth to note that the above reasoning allows to apply at least first few terms of mutual electrode capacitance given by (15') for estimation of the nonperiodic transducer capacitance.

From the other side, if wave-impedance Z which is characteristic for one section of periodic strips describes well the reflection of SAW from the free substrate surface, it certainly should describe SAW reflection from the another section having different Z, as well.

All the above allows to introduce a model for one electrode of interdigital transducer (assumed to be nearly periodic) as shown in Fig.1. There are marked SAWs propagating right and left, the amplitudes of which are changed when passing the voltage-supplied electrodes. Mutually, the propagating SAWs give contributions to the electrode currents (mutual capacitances are not shown in the figure). There are also the transmission and reflection phenomena between the neighboring one-electrode sections, and phase-shift τ of a wave that covers the distance from one edge of the section to the other.

One can prove that the energy conservation law is fulfilled in each section separately.

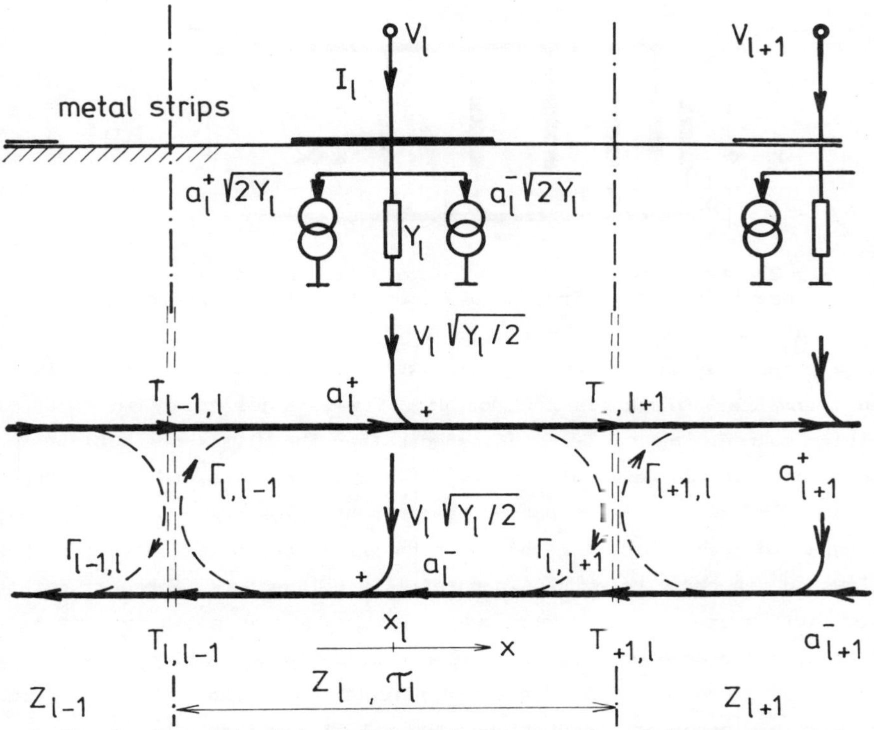

Fig. 1. A model of transducer electrodes where all fundamental phenomena taking part in propagation, generation and detection of SAW are depicted

All parameters introduced into the model have clear physical meanings and they are explicitly related to the substrate and metal strip parameters on a ground of SAW theory. The model can be easily applied for numerical calculations giving a number of linear equations, the number of an order of the number of electrodes in the transducer. The equations can be solved on a nearly recurrence way.

To obtain a description of the transducer as a whole, one should take into account that

- a free substrate surface on both side of the transducer has wave impedance equal to 1,
- some electrodes are connected to one transducer bar and some to the other (or others in the case of unidirectional transducer, for instance), giving contributions to the current of one transducer bar or

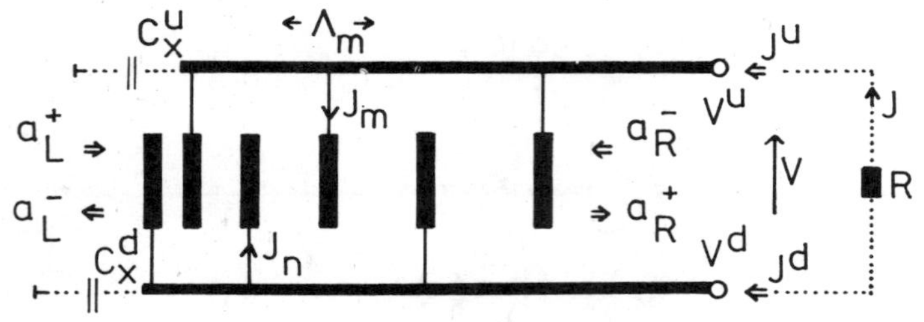

Fig.2. Description of transducer as a whole, including connections of the fingers to one or the other transducer bar.

of the another,

- the potentials of transducer bars, and what follows the potentials of the transducer electrodes are not directly established by an external voltage source. The source establishes rather the difference between the bar potentials. To calculate the potentials of the bars one has to apply Kirchhoff's law that the summa of the currents flowing to both bars must be equal to zero. This, together with the given bar potential difference allows to determine both unknown potentials of the bars. Performing this calculations one should account additionally the parasitic capacitances of the transducer electrodes to the ground, depicted in Fig.2 as C_x.

It is convenient for further applications of the results to find (by suitable arranged computations) the relations like this

$$\begin{bmatrix} a_R^+ \\ a_L^- \\ J \end{bmatrix} = \begin{bmatrix} S_{11} & S_{12} & \frac{1}{2} S_{31} \\ S_{12}^* & S_{22} & \frac{1}{2} S_{32} \\ S_{31} & S_{32} & Y_T \end{bmatrix} \begin{bmatrix} a_L^+ \\ a_R^- \\ V \end{bmatrix} \tag{17}$$

that is similar to the scattering relation for a transducer as a whole (the denotations are shown in Fig.2). Y_T is the admittance of the transducer, this and the other matrix components are to be calculated on a basis of the model for a specified transducer.

3.2 Analysis Of Dispersive Filters

The above model was applied for numerical analysis of two dispersive delay lines composed of two IDTs, one dispersive and one wideband having a few split fingers (both IDTs aperture widths were 4mm). Up-chirp and down-chirp filters differs only by the side on which the wideband IDT is placed with respect to the dispersive one. The assumed parameters were - center frequency 30MHz, passband 4MHz, impulse response 8μs. Calculated were dispersive IDT admittance and both impulse and frequency responses of the filters for two values of electric loadings of the dispersive transducer, namely 1Ω or 50Ω. This gives a picture how much the impulse and frequency responses might be affected by the electric loading of the filter. The compressed pulse was also calculated for these two cases of loadings.

Fig.3a) presents a kind of ideal characteristics of the dispersive filters the dispersive transducer of which have periodic electrodes with apodization [8]. To avoid spurious effects quartz YX was applied as the filter substrate as well as split fingers in dispersive IDT. In spite of a certain rounding of the time response of the filters that was caused by a passband characteristic of the wideband transducer, the frequency response of each of the filter is nearly ideal and so is the compressed pulse.

Next figure (Fig.3b) presents results of analysis of up- and down-chirp filters having dispersive transducers working on their fundamental frequencies (there are two strips on a wavelength). The substrate was assumed lithium niobate YZ. One can expect some distortions affecting filter responses due to SAW Δv/v reflections from transducer fingers. Indeed, both time and frequency responses are heavily distorted, what was expected but we also see that the responses are significantly different for two cases of filter loadings, 1 and 50Ω. In a consequence there are significantly different shapes of compressed pulses in these two cases. This makes evidence how important full analysis of filter might be, including calculation of filter input/output admittances.

It is not the aim of this paper to give any conclusions concerning the properties of dispersive filters. We have seen just an example how important calculations may be performed using the above proposed model of interdigital transducers.

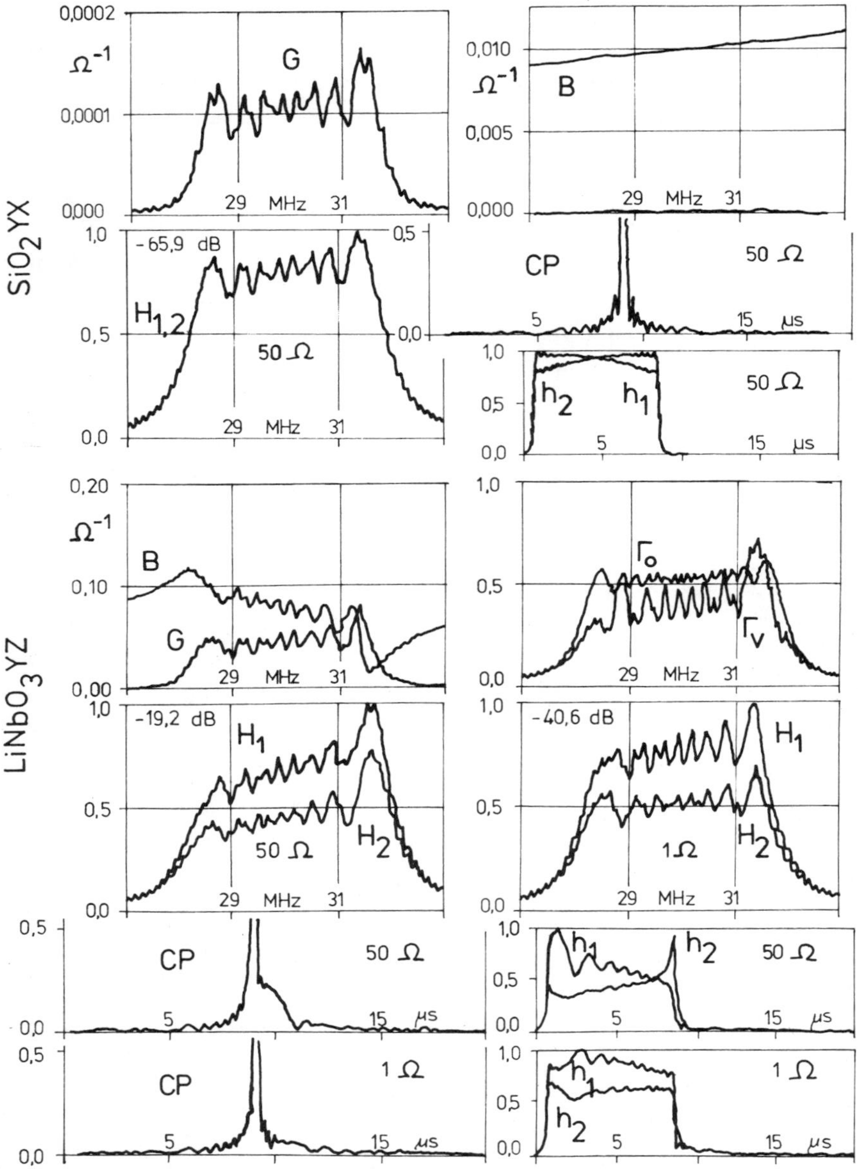

Fig.3. Time (I) and frequency (H, in linear scale, IL is depicted in [dB] above) responses of DDLs. Index 1 - down-chirp, 2 - up-chirp, CP - compressed pulse, Y_T=G+jB.

4. IMPROVEMENTS OF THE MODEL

One important feature of the presented model is that it is ready for further modification allowing one to account some additional phenomena taking place in SAW devices. One such important phenomenon is the reflection of SAW connected not only with $\Delta v/v$, what was accounted in the model, but with mechanic properties of the electrodes as well.

SAW propagation under heavy and elastic electrodes was a subject of many papers, for example [9], where SAW reflection coefficient from one electrode deposited on anisotropic and piezoelectric substrate was derived explicitly. That can easily be applied in the model, simply by suitable correction of coefficients T and Γ. Both the diffraction and scattering of SAW into bulk waves can also be accounted.

Finally, the slightly different performance of the edge electrodes of a transducer can also be accounted, This may be significant improvement of the model for transducers having only a few fingers, or wideband transducers. Some preliminary results have been obtained on this matter which are presented below.

Consider a system of seven electrodes uniformly spaced (Λ=.2mm, strip width = .1mm) and deposited on $LiNbO_3$YZ. Rigorous analysis [10] similar to that above for periodic strips gave the results for electrode radiation transadmittance as a function of frequency and the electrode position in the system

$$\textit{radiation part of } I_m = \omega\varepsilon_o{}^e{}_e 4\pi B_{mn}\, exp(-j|m-n|k_v\Lambda) \qquad (17)$$

Some results of computations are shown in Fig.4. Marked are frequencies changing from 4MHz to 13MHz by every $\frac{1}{2}$MHz (fundamental frequency of the system is 8.4MHz). One curve presents the admittance of the edge electrode of the system, the other curve shows the transadmittance between the edge electrode and the second one from the transducer edge and so on. We see the differences between the values of the transadmittances of the edge electrode and these of electrodes placed deeply inside the system. To help appreciate these results it may be noted that the above transadmittances correspond closely to the so-called "element factor" [3].

What concern the diffraction, several preliminry results were also

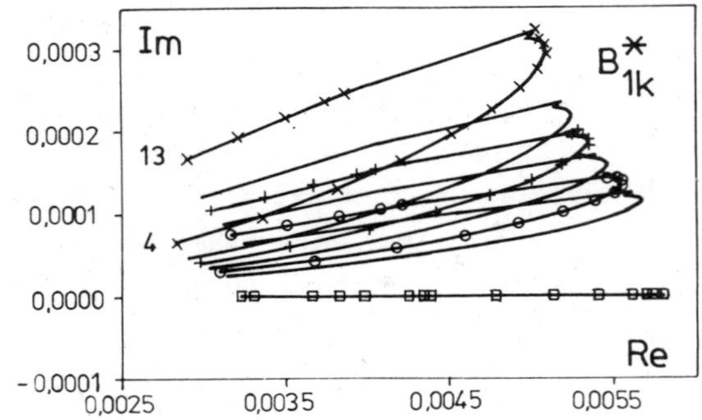

Fig. 4. The efficiency of electromechanical transformation of metal strips shown as the complex phasors on a complex plane.

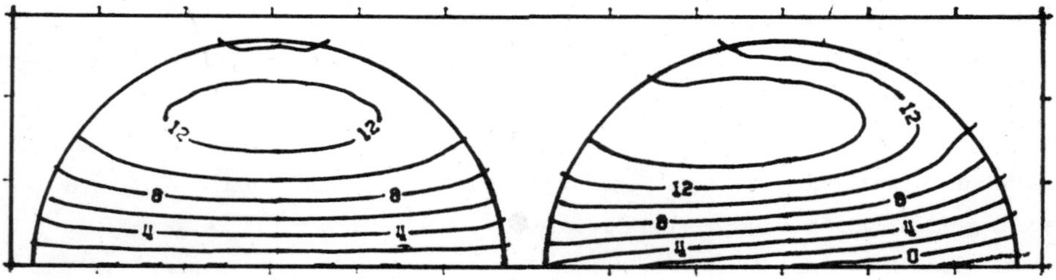

Fig. 5. A map of $Im\{\rho\}$ (18) for isotropic (a) and for equivalent anisotropic substrate (b) (the equivalence resulting from transformation of 1:4 elliptic disk - the long axis of which was turned 47° from X - to a circular one on $LiNbO_3$ YZ)

obtained. It was considered an elliptic metal disk on a piezoelectric anisotropic substrate that could model a single transducer electrode. The scattering of SAW on the disk was a subject of rigorous theory, as well as the disk capacitance and a magnitude of electric current flowing to the grounded disk, generated by the scattered SAW. The problem is described by singular integral equation the solution of which is represented in polar coordinates for equivalent circular disk by [11]

$$\sigma = \rho/\sqrt{(1-r^2)} \quad ; \quad \rho = \rho_{mn} r^n exp(jm\vartheta) \quad ; \; m \le n \tag{18}$$

Fig. 5 shows the map of ρ obtained for isotropic and anisotropic disk. The next step of the research will be the modelling of interaction between two different elliptic disks, one radiating, and the other detecting SAW.

APPENDIX

Consider a transmission line composed of three sections of different but homogeneous lines having the impedances *R* and *R'*. The relation between voltages and currents on the line ports can be written in the following form [12]

$$\begin{bmatrix} i_{n+1} \\ e_{n+1} \end{bmatrix} = \begin{bmatrix} \cos(\tau) & -j\sin(\tau)/Z \\ -jZ\sin(\tau) & \cos(\tau) \end{bmatrix} \begin{bmatrix} i_n \\ e_n \end{bmatrix}$$

where Z has similar dependence on frequency as given by (10).

Similarly one may derive the relation for τ (9). Note that the wave impedance Z may have a complex value. Next consider the chain (transmission line) composed of different two-port networks (with different parameters *R* , *R'* etc). The network with the number n is described by the values τ_n, Z_n instead of τ, Z, respectively. Let's introduce the wave a^+ propagating to the right and the wave a^- propagating to the left. These waves are defined for each network of the chain in the following standard way

$$i_n = [a^+ exp(j\tau_n/2) - a^- exp(-j\tau_n/2)]/\sqrt{Z}$$

$$e_n = [a^+ exp(j\tau_n/2) + a^- exp(-j\tau_n/2)]\sqrt{Z}$$

The relations between the amplitudes of the waves a^+ and the waves a^- in the successive networks of the non-homogeneous transmission line under consideration can be written in the form (the equality of currents and voltages on the boundaries of the adjoining networks should be taken into account)

$$a_n^+ = T_{n-1,n}\, a_{n-1}^+ exp[-j(\tau_{n-1}+\tau_n)/2] + \Gamma_{n,n-1}\, a_n^- exp(-j\tau_n)$$

and similarly for the wave a_n^- , where T and Γ are the transmission and reflection coefficients of wave given in (11).

In conclusion the wave reflection coefficient at junction of the homogeneous line of the impedance *R*=1, and the semi-infinite line periodically inhomogeneous is (1-Z)/(1+Z).

REFERENCES

[1] Lewis B., Electrode Reflections, Directionality, and Passband Ripple in Wideband SAW Chirp Filter, IEEE Trans., **SU-32**, 3(1985), 409, 422

[2] Alippi A., Palma A., Socino G., Verona E., Phase Contrast Optical Filtering For Mapping SAW Second Harmonic Field, IEEE Ultrasonic Symposium, 1980

[3] Morgan D.P., Surface Wave Devices For Signal Processing, Elsevier, NY, 1985

[4] Blotekjaer K., Ingebrigtsen K.A., Skeie H., A Method For Analysing Waves in Structures Consisting of Metal Strips in Dispersive Media, IEEE Trans., **ED-20**, 12(1973), 1123-46

[5] Danicki E., Unified Theory of IDT and SAW Reflectors, J.Tech.Phys., **21**, 3(1980), 387-403

[6] Gafka D., doctoral thesis, IFTR, Warsaw, 1988

[7] Danicki E., Bartnicki S., A Method For SAW Reflection Coefficient Determination, IEEE Ultras. Symp., 1979

[8] Danicki E., Filipiak J., Kawalec A., SAW dispersive delay line utilizing apodised IDT with periodic electrodes, Electr. Let., **22**, 19(1986), 976-7

[9] Danicki E., Perturbation Theory of SAW Reflection From a Periodic Structure With Arbitrary Angle of Incience, Arch.Mech., **22**, 5-6(1984), 623-638

[10] Danicki E., Gafka D., a paper in preparation

[11] Danicki E., a paper in preparation

[12] Danicki E., Analysis of Narrowband Dispersive Interdigital Transducers, Arch.Acoust., in print

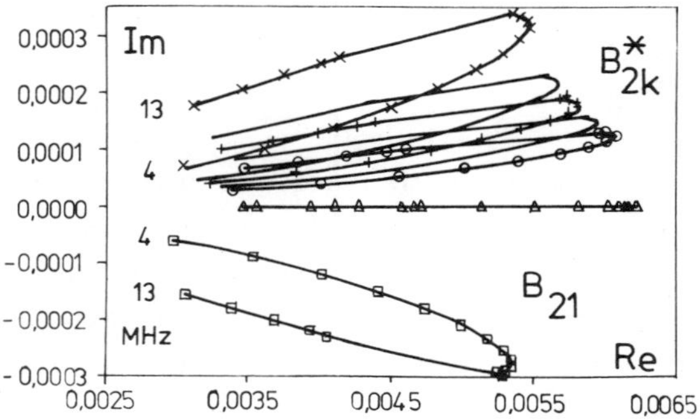

RESONANCES OF FINITE COMPOSITE PLATES

F. Craciun

CNR, Istituto di Acustica "O.M.Corbino"

Via Cassia 1216, I-00189 Roma, Italy

Abstract

The resonance spectrum of finite composite plates made of alternating elements of piezoelectric ceramic and epoxy resin is calculated by means of a new approach which gives the transmission through finite plates below their thickness resonances, taking into account both the modifications of sound velocities in the plate and the finite number of elements. The results are compared with the electrically excited acoustical resonances of samples with different number of elements and good agreement is obtained. New resonances, attributed to the end element, are observed in the stopbands when the surface element has lower impedance.

1 Introduction

Acoustic mode resonances in infinite periodical structures due to Bragg scattering from the interface between two different alternating media are related to the stopband edges in the dispersion relation at the center and boundary of folded Brillouin zone[1,2].

Acoustic wave propagation in infinite composite plates has been previously studied with the coupled mode approximation[3,4] and Kronig-Penney model[5,6]. It has been shown that below the first thickness resonance of both materials a single symmetrical plate mode is allowed to propagate in the structure[6]. The dispersion curves have been calculated in this frequency range and the predicted stopband edges have been found in good agreement with the experimental resonances[6].

In a real periodical structure with a finite number of periods full translational symmetry is lost but distinct dips in the transmission for acoustic waves with frequencies in the stopbands are obtained[7].

In this paper a review of some recent results obtained in the study of resonance spectra of periodical composite plates[6,8]is presented. The approach for calculating the transmission function is based on a recursive method[8,9] and takes into account both the modifications of the sound velocity in the plate and the finite number of elements. The results are compared with Kronig-Penney approach predictions and close similarity in the dispersion of finite and infinite plates is observed. New resonances are predicted and experimentally observed for a finite plate which terminates with the softer medium[9], similarly with the surface states of finite semiconductor superlattices[10,11]. A more extensive treatment will be given in a following paper[9].

2 Method

We consider the propagation of a single acoustic mode along a plate of thickness h composed of alternating elements of two different materials, a piezoelectric ceramic and an isotropic resin. Indeed, only the first symmetrical Lamb mode s_o will be excited with real wavevector for symmetrical excitation in the frequency range below the threshold of the first thickness mode of both material plates. In this case the defining parameters of the structure are the geometrical ones (the widths

d_1 and d_2 of the ceramic and resin elements, respectively) and the reflectivity coefficient r of the s_o mode at the interface that would separate two semiinfinite plates of the given materials. In order to estimate the last parameter, one needs to know the s_o velocities in both media. It has been shown[6] that in the low frequency range the s_o phase velocities are approximately constant and consequently effective values v_1 and v_2 can be successfully used in each material for the frequency range below the thickness s_1 mode.

The frequency response of the plate composed of n elements corresponds to the transmittance of the plate itself to the s_o Lamb mode, calculated in the frequency range of interest. From the boundary conditions relating the incoming with the outgoing waves at the $(n+1)$ interfaces, a system of $2(n+1)$ equations in the $2(n+1)$ unknown amplitudes is obtained, which is then recursively solved[9]. The transmittance function of the plate is given by the ratio between the last outgoing wave amplitude and the first incoming one.

Fig. 1b represents the allowed and prohibited regions for acoustic transmission for a finite plate made of 21 elements (11 ceramic and 10 resin). A Kronig-Penney model has been applied to calculate the dispersion curves of the infinite plate shown in Fig. 1a. The dimensions of elements are the same in both cases ($d_1 = 0.9mm$ and $d_2 = 1.1mm$). A perfect agreement is noted between the transmittance results and the Kronig-Penney model predictions, despite the relatively low number of elements, due to the particular combination of two materials with great mismatch in their acoustical impedances. Figs. 1c and 1d represent the transmittance of a single element of resin and ceramic respectively. It can be observed that the individual peaks in the allowed regions (Fig. 1b) are clustered around each resonance of the single element.

The region bounded by the dashed line in Fig. 1 is enlarged in Fig. 2, where the mini-stopband edges in the calculated transmittance of the finite plate (Fig. 2b) are assigned to the wavevector values corresponding to the confining length

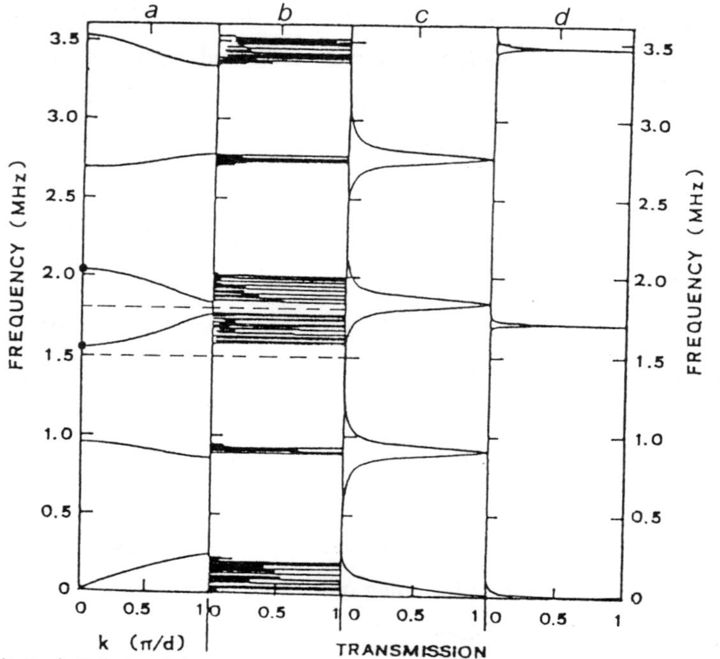

Fig.1 a) Calculated dispersion curves of the infinite periodical composite plate ($d_1 = 0.9mm, d_2 = 1.1mm, v_1 = 3200m/s, v_2 = 1800m/s, r = 0.841, t = 0.540$). b) Calculated transmission for a composite plate with 21 elements. c) and d) Calculated transmission for a single element of resin and ceramic, respectively. The parameters used in the calculus are the same as in a).

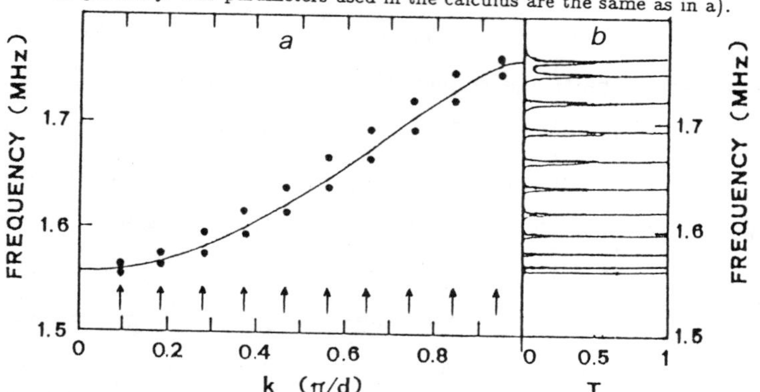

Fig.2 Mini-stopband edges (a) in the calculated transmission T (b) of the finite plate. The arrows mark the wavevector values corresponding to the confining length D of the entire plate.

$$k' = m\frac{\pi}{D}$$

where $D = n_1 d_1 + n_2 d_2$ and $n_{1,2}$ is the number of elements of one kind. The points marking the mini-stopbands follow very closely the dispersion curve of the infinite plate. As the number of elements increases, the peaks become denser and completely fill the allowed branch, thus sharply defining the band edges.

3 Experimental results and discussion

The samples used in this work were prepared by simply cutting periodical interruptions with parallel boundaries in a piezoelectric plate and filling them with epoxy resin that solidifies and gives a compact structure. The major surfaces (perpendicular to the piezoceramic polarization) are fully metallized and connected to a bridge meter for measurement of admittance function. It is worthwhile to note that, due to this geometry of excitation, only symmetrical waves may resonate within the plate.

In order to avoid sample reproducibility errors, the resonance spectra shown in Fig. 3 belong to the same sample, which has been progressively reduced as can be seen in the insets. The solid and dashed arrows mark the positions of the calculated peaks, and refer respectively to the symmetrical and antisymmetrical resonances with respect to the whole structures. A very good agreement between the predicted and experimental peaks is obtained, both for their number and their frequency.

The bottom spectrum, measured on the plate reduced to 13 elements, is very similar to that corresponding to the sample having the maximum number of elements (29) and to all the other samples with varying number of elements between these two values. This means that already in a structure of this size the k symmetry selection rule, which would strictly hold only for an infinite structure, is respected.

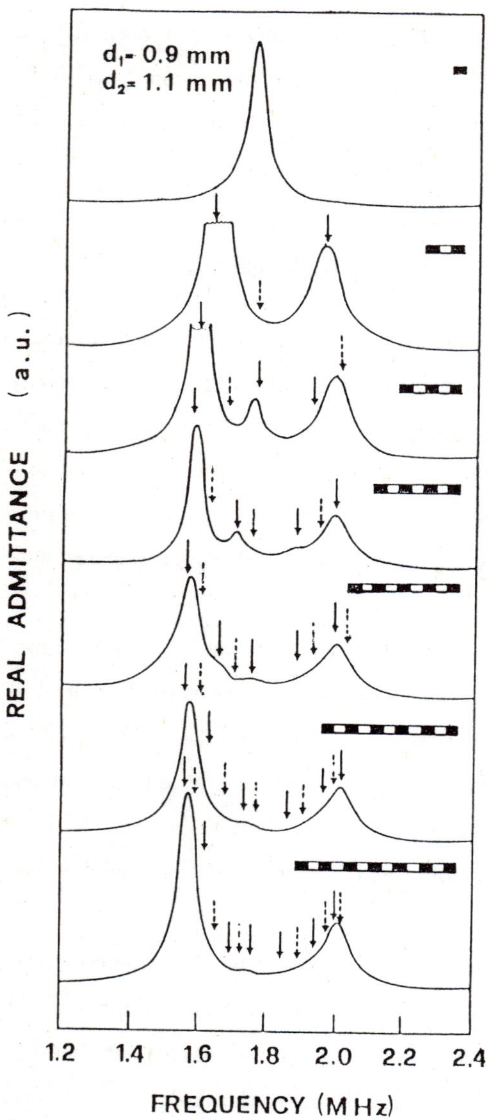

Fig.3 Real admittance function *vs.* frequency for composite samples with variable number of elements (see insets). The solid and dashed arrows mark the calculated symmetric and anti-symmetric resonances, respectively.

Indeed, what we expect to see are only the first symmetrical zone center stopband edges[6] (marked by dots in Fig. 1a). These values are rapidly attained by the mini-stopband edges, even for relatively small samples, due to the flatness of the dispersion curves produced by the great mismatch in the acoustical impedances of the two materials. By decreasing the number of elements, an increasing effect of the sample finiteness can be observed in the corresponding spectra: *i*) the resonances are broadened; *ii*) symmetric resonances due to mini-stopband edges are excited more strongly.

It is important to note that all the samples in the above figure end with elements of the harder material (ceramic), represented in the inset by dark segments. Indeed, the calculated transmission for a sample which ends with the softer element[9] presents new peaks placed in the stopbands (top of Fig. 4a); they are clearly resonant states of the soft end element. A good agreement between the calculated and experimental values of the new resonances is obtained (see the top spectrum in Fig. 4b); in the resonant spectrum of the sample without the soft surface element, presented for comparison in the bottom of Fig. 4b, these resonances are clearly missing.

4 Conclusions

In this work we have shown that the resonance spectrum of finite composite plates can be predicted by means of a new approach, which takes into account both the modifications of the sound velocities in the plate and the finite number of elements. The resonances of finite plates have been correlated with mini-stopband edges in the calculated transmission function and finite size effects have been experimentally put in evidence. New resonances, attributed to the composite surface element, are observed in the stopbands when the surface element has lower impedance. The excellent agreement between theory and experiments over a wide

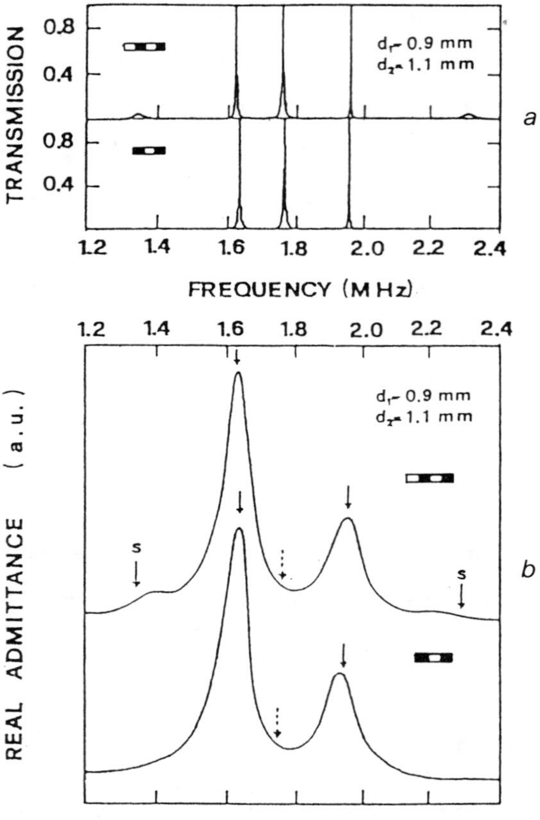

Fig.4 Calculated transmission (a) and experimental resonance spectra (b) for a sample which ends with resin element (top of the figures) and for a sample without the soft element (bottom of the figures).

set of structures gives confidence in the possibility of using the present approach to taylor the resonance spectra of composites in the frequency range below the thickness mode.

References

[1] S. M. Rytov, Sov. Phys. Acoust. **2**, 68(1956) [transl. from Akoust. Zh. **2**, 71(1956)].

[2] B. Jusserand, F. Alexandre, J. Dubard and D. Paquet, Phys. Rev. B **33**, 2897(1986).

[3] B. A. Auld, Y. A. Shui and Y. Wang, Journ. de Physique **45**, C5-159(1984).

[4] B. A. Auld and Y. Wang, Proc.1984 IEEE Ultrasonics Symp.,p.528.

[5] F. Craciun, L. Sorba, E. Molinari and M. Pappalardo, IEEE Trans. Ultrason., Ferroelectrics, Frequency Control (in press).

[6] A. Alippi, F. Craciun and E. Molinari, Appl. Phys. Lett. (in press).

[7] S. Tamura, D. C. Harley and J. P. Wolfe, Phys. Rev. B **38**, 1427(1988).

[8] A. Alippi, F. Craciun and E. Molinari, IEEE 1988 Ultrasonics Symp., to appear in Proc. 1988 IEEE Ultrasonics Symp.

[9] A. Alippi, E. Molinari and F. Craciun, to be published.

[10] E. Molinari and A. Fasolino, Superlattices and Microstructures **4**, 449(1988).

[11] L. Sorba, E. Molinari and A. Fasolino, Proc. ECOSS-10, 1988, to be published in Surface Science.

PVDF-TRANSDUCERS AND THEIR USE IN NDE OF ENGINEERING CERAMICS

S. Pangraz and W. Arnold

Fraunhofer-Institute for Nondestructive Testing
Bldg. 37, University, D-6600 Saarbrücken 11, FRG

ABSTRACT

Ultrasonic transducers made out of Polyvinylidene fluoride (PVDF) exhibit a broad bandwidth because the piezoelectric constant d_{33} can have a gradient inside the material. Furthermore, the bandwidth of PVDF-transducers is also determined by the electrical matching network employed. A broadband characteristic can be obtained by matching the impedance with an inductance and by transforming the value of the effective impedance to 50 Ω with a toroidal transformer. Experimentally, we achieved an insertion loss of about 20 dB for a 35 MHz-PVDF-transducer, its 3 dB bandwidth being about 17 MHz. For Polyvinylidene Trifluorethylene P(VDF-TrFE), a copolymer of PVDF, the insertion loss is about 10 dB rivaling the one obtainable with PZT-transducers. These transducers are well suited for defect-detection in ceramics. In ceramic components made out of SSi-SiC and Si_3N_4 with plane surfaces we have found defects with sizes down to 30 μm in a depth of 3 mm and with a signal/noise-ratio of 6 dB.

I. Introduction

In non-destructive testing by ultrasonic techniques, efficient piezoelectric transducers are a prerequisite for the detection of small defects in components. This holds good in particular for engineering ceramics where a small defect such as a pore or an inclusion in a critically loaded zone may cause failure of a given component because of the brittleness of the material /1,2/. We use PVDF-transducers for this task. They can be easily produced down to a thickness of approximately

5 µm. It is therefore possible to construct ultrasonic transducers from 10 to over 100 MHz out of this polymer. For certain applications the low acoustic impedance of PVDF /3/ is advantageous because it is close to that of water, and hence PVDF-transducers are well suited as probes in C-scan imaging using water as couplant. As it is known experimentally, PVDF exhibits a very broad bandwidth as ultrasonic transducer material. We believe that this large bandwidth is due to the inhomogeneity of the piezoelectric constant d_{33} after poling with electrical fields below 500 kV/cm. The PVDF-foil then oscillates not only in the $\lambda/2$-mode but also in the λ-mode, contrary to usual piezoelectric transducers. This behavior has been discovered previously /4/ and its full consequence has been discussed recently /5/.

Furthermore, the electrical impedance of PVDF is strongly dependent on frequency. In order to get a broad bandwidth, one has to match the transducer electrically. We have constructed a matching circuit which consists of an inductance for low frequencies and a toroidal transformer for impedance matching to the 50 Ω cable connecting the transducer to the rf-transmitter and receiver. With the aid of this kind of transducers we are now able to detect small defects in ceramics down to a size of 30 µm in a depth of 3 mm with a signal/noise-ratio of 6 dB.

II. Bandwidth of Piezoelectric PVDF-Transducers

PVDF is a polymer which can be polarized with the aid of an electrical field. Electrical fields below 500 kV/cm cause an inhomogeneous polarization in PVDF /4,6/ and also in its copolymer P(VDF-TrFE) /7/. For electrical fields larger than 1 MV/cm like corona poling, the polarization, however, becomes quite homogeneous /8/, although there are some contradictory reports /6/. Since the product of the gradient of the electrical field and the piezoelectric constant is the cause of ultrasonic generation, in inhomogeneously poled piezoelectrical materials ultrasound is not only generated at the surfaces but also inside the material as discussed previously /5/ and shortly summarized in what follows. It is assumed that the inhomogeneity inside the piezoelectrical material consists of two parts: an inhomogeneous one, characterized by its magnitude d_1 and an always present homogeneous part d_2. The inhomogeneous part is assumed to be exponentially decreasing with a decay constant α. This is depicted schematically in Fig. 1.

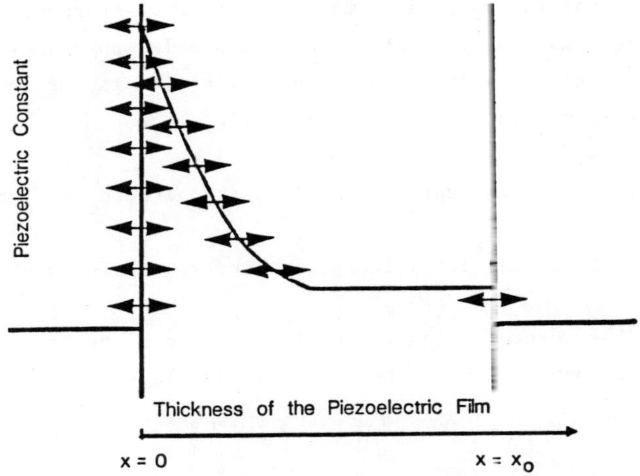

Fig. 1: Generation of ultrasound in an inhomogeneously polarized PVDF-film. The arrows indicate the location and direction of ultrasonic field generated.

The one-dimensional wave-equation for inhomogeneously polarized piezoelectrics can be written as:

$$\frac{\partial^2 u}{\partial x^2} - \frac{1}{c^2}\frac{\partial^2 u}{\partial t^2} =$$

$$\frac{\partial}{\partial x}\Big(H(x)(d_2 + d_1\exp(-\alpha x)) - H(x-x_0)(d_2 + d_1\exp(-\alpha x))\Big)E_0\exp(-i\omega t) \quad (1)$$

Here, u is the displacement, c is the sound velocity, H(x) is the Heaviside-function (step-function) describing the sharp change of the piezoelectric constant at the surfaces of the piezoelectric material, and E_0 is the magnitude of the time-varying field of frequency $\omega/2\pi$. The other variables are explained in Fig.1. Eq. (1) results from the combination of the equation of motion and the piezoelectric equation, and it can be solved with the aid of the Green's functions /5,9/. The right-hand side of Eq.(1) describes the source of ultrasound. In the case $d_1 \rightarrow 0$, or equivalently $1/\alpha \rightarrow 0$, $\alpha \rightarrow 0$, there is no gradient of the piezoelectric constant in the material under consideration, and Eq.(1) reduces to the expression already treated previously /10,11/. It was assumed that the materials adjoining the PVDF-transducer have the same acoustic impedance. This is an acceptable approximation if the adjoining material is water, because the reflection factor in this case is about 0.04. The solution of Eq. (1) is /9/:

$$u(x,t) = (i/k)E_0\exp(-i\omega t)\exp(ikx)[-d_2[1-\cos(ka))+(d_1/d_2)(1-\exp(-\alpha a)$$

$$\cos(ka))-d_1/(d_2((ka/\alpha a)^2)+1))(1-\exp(-\alpha a)(\cos(ka)-(ka/\alpha a)\sin(ka))]] \quad (2)$$

From this, the overall insertion loss Il can be calculated in a straightforward manner /5,9/. The result is displayed in Fig. 2.

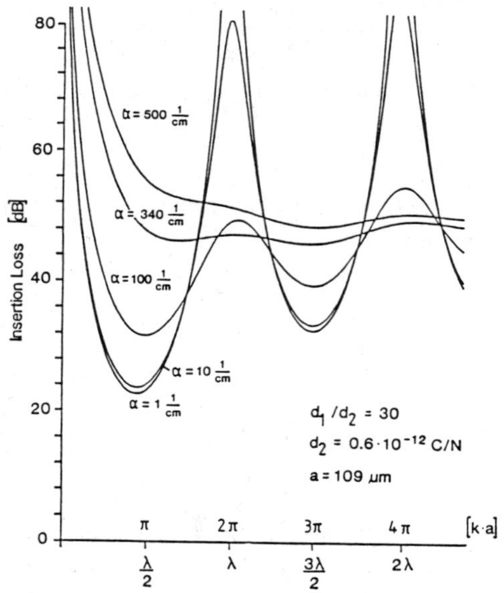

Fig. 2: Insertion loss of an inhomogeneously polarized PVDF-film as a function of the quantity ka and the parameter α. $1/\alpha$ is the penetration depth of the inhomogeneous polarization. For clarity, units of λ are also shown at the abscissa.

Il depends on the foil thickness a and the wave-vector k and with the decay constant α as a parameter. A ratio $d_1/d_2 = 30$ and $d_2 = 0.6 \cdot 10^{-12}$ C/N was assumed. These parameters were derived from experimentally determined values /6/. It can be seen that the transducer not only oscillates at $ka = (2n+1)\pi$ but also at $ka = 2\pi$ (n=1,2,3..), i.e. when a = $\lambda/2$ and a = λ. This becomes more pronounced as α increases and for α = 340 cm^{-1}, the transducers oscillates in the $\lambda/2$- and λ-mode with equal strength entailing an increased bandwidth It should therefore be possible to exploit this phenomenon in order to vary the bandwidth in a definite way. A large α can be produced with short poling times and poling fields below 500 kV/cm.

III. Electrical Matching Circuit

The electrical impedance of the PVDF-foil consists of the capacitance C_o of the foil and a series resistance R_E /12/, and the dominant resistance contributions are the dielectric losses. Since the electrical impedance at higher frequencies decreases rapidly, it is now required to keep the variation of the electrical impedance of the PVDF-foil at lower and higher frequencies as small as possible, thus maintaining a broad bandwidth. The electrical matching circuit (Fig. 3) employed by us consists of two parts: the first one matches the impedance at lower frequencies to the at higher frequencies. This can be done with an inductance L parallel to the capacitance of the foil. For an 35 MHz-transducer with a diameter of 10 mm, the crossover where the impedance is rapidly changing occurs at 10 MHz.

With a toroidal transformer /13/, this impedance is further matched to the impedance of the connecting cable, here 50 Ω. The gain in the radiated ultrasonic intensity is about 26 dB compared to the intensity without matching. The measured insertion loss Il by using an experimental arrangement exploiting the reciprocity theorem /14/ is about 20 dB. Additionally, if one uses the copolymer P(VDF-TrFE) instead of PVDF the insertion loss is reduced to about 10 dB. The reason for this is the higher crystallinity of P(VDF-TrFE) /15/ resulting in a larger electromechanical coupling-factor. Fig. 4 shows the insertion loss for two transducers made out of PVDF and P(VDF-TrFE) with an electrical matching circuit as described above. The 3 dB-bandwidth measured for both PVDF- (thickness 25 μm) and P(VDF-TrFE)-transducers (thickness 28 μm) was approximately 17 MHz for a center frequency of 35 MHz. The minimal insertion loss of these P(VDF-TrFE)-transducers becomes comparable to the one obtainable with PZT-transducers /9/.

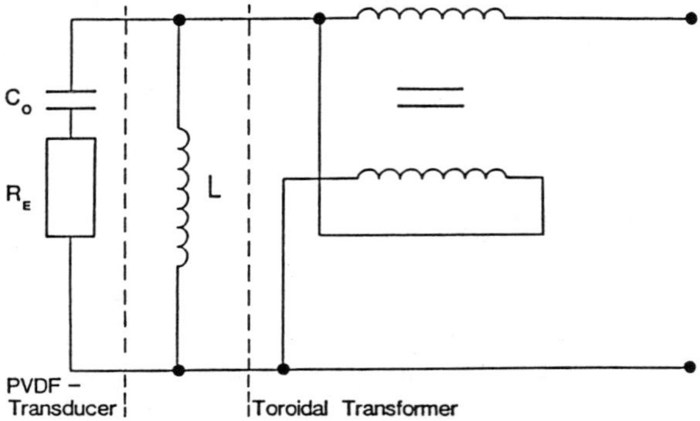

Fig. 3: Electrical matching circuit

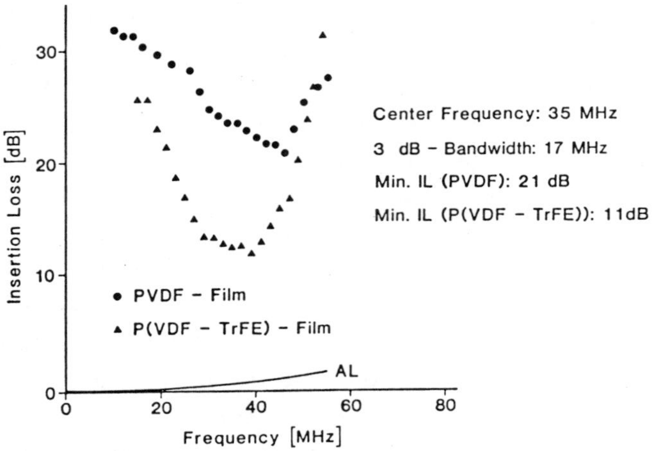

Fig. 4: Insertion loss of a matched PVDF- and P(VDF-TrFE)-transducer. For comparison the absorption loss AL in a 3 mm water-path is shown.

IV. Application of PVDF-Tranducers in NDE of Ceramic Components

In order to detect defects by ultrasound efficiently, it is necessary to choose the frequency such that the wavelength becomes comparable to the defect dimensions. This requires high-frequency ultrasonics due to the large sound velocity in ceramics. Additionally, in order to obtain sufficient spatial resolution, ultrasonic pulses shorter than 100 ns are needed. A large bandwidth of both the electronic signal amplifier and the transducer is then necessary.

In ultrasonic signal generation, the transducer is either excited by an exponentially decaying step-pulse (broadband excitation) or by a rf-carrier pulse (narrowband excitation). In our apparatus for narrow-band excitation this pulse is produced by pin-diode switches out of a cw-signal. The short tone-burst is then amplified by a power-amplifier up to 20 Watt rf-power at 50 Ω. The carrier frequency can be adjusted from 20 - 200 MHz and the pulse-width can be varied such that 1 - 100 rf-oscillations are included. Details of our electronic set-up using this so-called narrow-band excitation can be found elsewhere /17/. For broadband excitation we used a commercially available system /16/.

Depending on the ratio of the defect radius a to the employed ultrasonic wavelength λ, one can distinguish three regimes /18/: Rayleigh scattering ($a/\lambda \ll 1$), resonance scattering ($a/\lambda \sim 1$) and geometric scattering ($a/\lambda \gg 1$). If we employ 50 MHz ultrasonic waves, the ultrasonic wavelength is about 200 µm in ceramics, and we are therefore always in the Rayleigh-regime for defects smaller than 100 µm. Neglecting diffraction effects, the intensity I_{sc} backscattered from a given defect can be estimated to be:

$$I_{sc} = I_o (f(\pi))^2 \pi a^2 \exp(-\alpha_a r)\, d\Omega$$

$$\text{with } d\Omega = \Delta F / 4\pi r^2 \qquad (3)$$

Here, I_o is the ultrasonic intensity irradiated on the defect, $f(\pi)$ is the backscattered amplitude /19/, α_a is the attenuation in the material, ΔF is the area of the transducer, r is the depth of the defect, and $d\Omega$ is the solid angle subtended by the probe as seen from the defect. Because $f(\pi) \propto a^2$, $I_{sc}/I_o \propto a^6$! This means that very efficient transducers have to be used in order to get a large signal/noise (S/N) ratio for a given excitation voltage of the transducer. As outlined above, PVDF-transducers fill this condition. With these probes one can also easily focus the ultrasonic beam /12/, thus further increasing the incoming intensity at the defect. Table I shows our results for the S/N - ratio obtained in detection of various defects. The employed frequency was 50 MHz and the experiments were carried out in densely sintered and small-grained ceramics made out of SSiSiC and Si_3N_4 using focusing probes /16/. Hence, the influence of the attenuation was of minor importance on the measured S/N, although we had taken it into account.

Type of defect	Diameter(μm)	S/N (dB)	Depth of defect (mm)
Pores	500 100 50	47 29 14	2,9 2,8 2,7
Si-Inclusion	100 50	26 14	2,9 2,6
Iron-Inclusion	100 50	27 13	2,6 2,5
Tungsten-Carbide Inclusion	100	14	2,5

Table I

It is striking that the difference in S/N-ratio as measured for the various defects of the same size is so small, although the acoustic impedance differences to the host material are quite different. This reflects the fact that in the Rayleigh regime, the amplitude of the backscattered signal depends in a rather complicated way on the elastic parameters of the host material and the defect /19/. Due to Eq. (3) the dominant dependance for the available S/N ratio is, however, S/N ~ a^6 . The S/N-ratios obtained in table I were measured by exciting the transducers with a step voltage of 350 V exponentially decaying in 10 ns. By using rf-carrier excitation (20 W at 50 Ω) as discussed above, a further improvement of the S/N - ratio can be obtained (approx. 10 dB), albeit at the cost of axial resolution. This is demonstrated in Fig. 5 showing two A-scans obtained from the same defect with broadband and with narrowband excitation. As a rule of thumb, we can say that at present we can detect spherical defects as small as 30 μm in ceramic components in 3 mm depth with a S/N ratio of 6 dB by high-frequency ultrasound using PVDF-transducers. Still smaller defects may become detectable by using Synthetic Aperture Techniques /20/. Such systems are just being developed by us for the frequency range above 30 MHz.

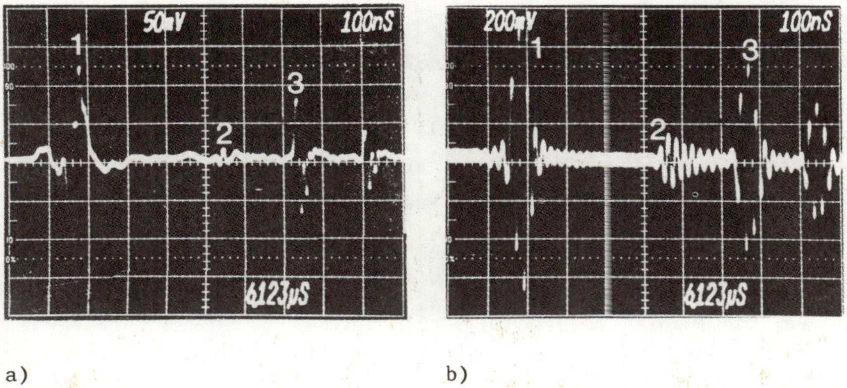

Fig. 5: A-scans obtained by a broadband system (a) and by a narrowband system (b) from the same defect in a SiC-ceramic. 1: Surface echo, 2: Echo from defect, 3: Backwall echo.

V. Summary

In this paper we have discussed the origin of the broad bandwidth of ultrasonic transducers made out of PVDF and its copolymer P(VDF-TrFE).

With a suitable electrical matching circuit, it is possible to reduce the insertion losses, and the bandwidth of the matched transducers becomes nearly the same as the one that is theoretically possible. These transducers are therefore well qualified for detection of small defects in high-strength ceramics in A- and C-scanning by using the immersion technique. At present we are able to detect defect sizes in the range of 30-100 μm in a depth of 3 mm and a signal/noise-ratio of 6 dB in ceramics.

Acknowledgement:

Part of this work was supported by the European Community (Material Technology Programme) and by the Materials Research Programme of the German Ministry of Research and Technology.

References:

/1/ Evans, A.G. in: Progress in Nitrogen Ceramics, Ed. Riley R.F., Marthinus Nijhoff Publ., Boston p.595-625 (1983)

/2/ Munz D., Rosenfelder O., Goebbels K., and Reiter H., in: Fracture Mechanics of Ceramics, Eds.: Bradt R.C., Evans A.G. Hasselmann D.P.H., and Lange F.F., Plenum Press, NY, 7, p. 265-283 (1986)

/3/ Ohigashi, H., J. Appl. Phys., 47, 949 (1976)

/4/ Sussner, H. and Dransfeld, K.,Coll. & Polym. Sci.,257, 591 (1979)

/5/ Pangraz S. and Arnold W., to be published in Ferroelectrics

/6/ Gerhard-Multhaupt, R., Sessler, G. M., West, J. E., Holdik, K., Haardt, M., and Eisenmenger, W., J. Appl. Phys., 55, 2769 (1984)

/7/ DeReggi, A. S. and Broadhurst M. G., Ferroelect., 73, 351 (1987)

/8/ Sessler, G. M. and Berraissoul, A., Ferroelect.76, 489 (1987)

/9/ Pangraz, S., Diplomarbeit, Fraunhofer-Institute, and Engineering Departement University of Saarbrücken (1988), unpublished

/10/ Jacobsen, E. H., J. Acoust. Soc. Am., 32, 949 (1960)

/11/ Tucker, J. W. and Rampton, V. W., "Microwave Ultrasonics in Solid State Physics", North-Holland Publ., Amsterdam, (1972)

/12/ Platte, M., Acustica, 54, 23, (1984)

/13/ Krauss, H. L. and Allen, Ch. W., Elect., Aug. 16th, p. 113 (1973)

/14/ Bömmel, H. E. and Dransfeld, K., Phys. Rev., 117, 1245 (1960)

/15/ Ohigashi, H., Koga, K., Suzuki, M., Nakanishi, T., Kimura, K., and Hashimoto, N., Ferroelect., 60, 263 (1984)

/16/ The measurements listed in Table I were obtained by broadband excitation with the USH 100 ultrasonic system and focussing probes manufactured by Krautkrämer-Branson Köln, FRG

/17/ Arnold W. and Simon H., IzfP-Rep. #860345-TW, unpublished (1986)

/18/ Ermolov I.N., Non-Destr. Test. 5, 87 (1972)

/19/ Hirsekorn S., IzfP-Report Nr 780157-TW, unpublished (1978)

/20/ Müller W., Schmitz V., and Schäfer G. , Nucl. Eng. Des. 94, 393 (1986)

MODE CONVERSION ULTRASONIC TESTING ON WOOD

V. BUCUR
Centre de Recherches Forestières de Nancy (I.N.R.A.)
CHAMPENOUX 54280 SEICHAMPS (France)

ABSTRACT : Ultrasonic nondestructive testing of wood based on procedures derived from mode-conversion phenomena are discussed in this paper. The detection, identification and characterisation of significant defects can be monitored indirectly through stiffness measurements. The procedure for the determination of the non-diagonal constants of stiffness matrix of wood using mode-conversion from bulk to surface waves is described. Defect detection by scattering based methods is commented in the second part of this note.

1 - INTRODUCTION

The ultrasonic nondestructive testing on wood products covers in considerable detail a variety of methods which may determine either the physical characteristics of materials and the presence of defects in wooden structural parts or assembles.

Procedures, based on waves propagation phenomena and mode-conversion, are reported in this paper. As part of this trend, procedures which will provide the complete quantitative nondestructive evaluation of the mechanical state of structures are tied together with stiffnesses change. Hence, the improved probability of detection, identification and characterization of significant defects can be monitored indirectly through stiffness measurement.

The first part of this note deals with the procedure for the determination of the off-diagonal constants of stiffness matrix (C_{ij}) of wood using mode-conversion from bulk to surface waves. The motivation of this approach come from the statement that by mode-conversion testing considerable gain can be obtained in the methodological improvement of all technical constants measurements from a very restricted number of experimental values.

The second part of this note is devoted to defect detection by scattering based methods that use travel times and employ a matrix of transducers sensitive to different types of waves. Some of the possible future contributions with mode-conversion based nondestructive technique in the fiels of quantitative defect characterization on wood are discussed.

2 - OFF-DIAGONAL TERMS OF STIFFNESS MATRIX OF WOOD AND MODE-CONVERSION TECHNIQUE

The common acceptation of an orthrotropic model for wood elastic behaviour investigation means that nine independent stiffness constants are required for complete characterization of one species : 6 diagonal and 3 off-diagonal terms. The diagonal terms can easily be obtained from in axis measurements.

The starting point of mode-conversion technique developed for C_{ij} terms determination, was the assumption that surface waves velocity investigation could give direct access to those off-diagonal terms only through velocity measurements along wood main axes of symmetry. In the present approach, mode conversion technique from bulk to surface wave was applied in contact testing. We seek to extract information from several components of the wavefields, induced by the modulation of ultrasonic propagating waves by the anatomical structure of the specimen under test. The theoretical background has been developed against the diverse literature (1, 2, 3, 4, 5) and only selected aspects are discussed later.

2.1 - Theoretical considerations

The relationships between the terms of stiffness matrix and the ultrasonic velocities are given by Christoffel equations.

To avoid unnecessary duplication, material on basic wave propagation phenomena has been excluded from the present text. However it is worthwile recalling that the expressions for C_{ij} off-diagonal terms involves off-axis bulk velocity measurements. Those velocities are in fact strictly dependent on the direction of propagation with the respect to wood main axes of symmetry. As a consequence, some optimisation procedure must imperatively be introduced.

In the case of surface waves the relationships between off-diagonal terms of stiffness matrix and ultrasonic velocities can be summarised as follow :

$$C_{ij} = f(\rho ; V_{surface} ; C_{ii} ; n_i) \quad (1)$$

where ρ is the density, $V_{surface}$ = surface velocity ; C_{ii} = diagonal terms of stiffness matrix ; n_i = components of propagation vector.

The equation (1) is thus of greater interest as far as direct determination of off-diagonal terms is concerned, since under specific in axis propagation conditions the equation governing surface waves propagation involves a single off-diagonal term. In other words we have to solve the following equation :

$$M.C_{ij}^4 + N.C_{ij}^2 + P = 0 \quad (2)$$

where M, N, P are real coefficients involving diagonal stiffness terms and measured in axis surface waves velocities (5).

Obviously, the agreement between C_{ij} terms computed from bulk and surface velocities can be used as an argument for this methodology approach.

An other possible solution of checking the validity of our approach is to compare the measured surface waves velocities with those deduced from equation (3).

$$(1-d)q^3 - (1-e+2f)q^2 + f(f+2)q - f^2 = 0 \quad (3)$$

where q depends on the surface waves velocity (V). The coefficients of this equation depend on all 9 stiffness terms and were first established by (2).

Table 1 summarizes the possibilities of measurements related to all off-diagonal constants using mode-conversion technique on an orthotropic solid. As we have seen previously, the motivation of the development of this technique arises from the necessity for a nondestructive approach to the determination of all technical constants on living trees, for both mechanical behaviour estimation and defect detection purposes.

TABLE 1 : Off-diagonal constants possibilities of measurements and corresponding velocities using mode conversion technique

Free surface	Off-diagonal constants	Velocities longitudinal waves	Velocities shear waves	Type of specimen
12	C_{13}	V_{11}	V_{55} ; V_{66}	sample or block
	C_{23}	V_{22}	V_{44} ; V_{66}	sample or block
13	C_{12}	V_{11}	V_{66} ; V_{55}	standing tree or log
	C_{32}	V_{33}	V_{44} ; V_{55}	standing tree or log
23	C_{21}	V_{22}	V_{66} ; V_{44}	log
	C_{31}	V_{33}	V_{55} ; V_{44}	log

Surface velocity measurements are thus expected to lighten the procedure. Within fairly well defined practical limits, the mechanical characterization and therefore defect detection on standing trees can already by answered if an appropriate signal processing would remove the ambiguity that occurs when trying to distinguish between the longitudinal and shear bulk waves components and surface waves components.

It is imperative to underline that mode-conversion technique arisses as a result of structure induced scattering mechanism. At the first view the occurence of energy transfers from one to different waves types can be apprehended as adding a degree of complexity to velocity readings. On the other hand, although including mode-conversion processes complicates the theoretical treatment of the scattering problem, but it offers the possibility of extracting additional information on the structure influence on the mechanical properties of wood material.

2.2 - Experimental approach

The experiments have been undertaken on two types of specimens : multifaced disks (0 35 mm x 20 mm) for bulk velocity measurements and block specimens (600 x 180 x 90 mm) for surface velocity measurements. The measurements were performed on beech (Fagus sylvatica L.) and spruce (Picea abies (L) Karst).

Ultrasonic measurements were restricted to time domain, using transmitting pulse technique by direct contact. In our particular case, surface waves were produced by a bulk to surface mode conversion, by fitting the emitter on the edge of the block cut at 45°. This particular method for generating surface waves was selected bearing in mind a possible development towards total waves technique. In our particular case simultaneous readings of both compressional and transversal bulk waves added to surface waves component are expected to reduce C_{ij} determination to a routine procedure. This explain why all experimental work was performed on a mode-conversion basis, without involving specific surface waves transducers.

2.3 - Results and discussion

Experimental results concerned with insuring the validity of the method applied for surface waves velocity measurements are given in table 2.

TABLE 2 : Off-diagonal terms of stiffness matrix (x 10^8 N/m2) determined from bulk and surface waves

Species	Density (kg/m3)	Off-diagonal terms (x 10^8 N/m2) bulk waves			surface waves		
		C_{12}	C_{13}	C_{23}	C_{12}	C_{13}	C_{23}
Beech	674	30,36	16,89	7,42	34,42	16,16	10,67
Spruce	485	2,84	2,31	7,60	3,78	3,58	7,88

It is now pertinent to ask whether the off-diagonal terms measured from surface waves are in agreement with those calculated from bulk waves. Both series of bulk and surface waves off-diagonal terms are given in previous tables. As a matter of fact good agreement between those terms is observed, as well as the agreement between the Young's moduli calculated from both methods (5, 6).

Comparing velocities measured on blocks to those computed from equation (3) provides another way of assessing the validity of the proposed methodology (table 3).

TABLE 3 : Comparison between measured and computed surface velocities (m/s)

Values	Species	Surface velocities (m/s) in planes 12	13	23
Computed	Beech	1455	1238	900
	Spruce	1294	1194	385
Measured	Beech	1404...1460	1260...1249	880...920
	Spruce	1173...1300	1113...1210	380...440

The rather small discrepances to be observed between the maximum and minimum values of surface velocities for beech can be easily explained from considerations of microstructure. Beech is one of the diffuse porous hardwood species, presenting a smooth transition from earlywood to latewood in annual ring. This species is closer to the homogeneous anisotropic medium assumed in the development of the continuum theory which ignores the ring structure of wood, than spruce. In spruce the transition between earlywood and latewood is very sharp. This layered structure induces dispersive propagation of the probe pulse and reduces the accuracy on velocity measurements. Ultrasonic wave is particularly dispersive in 1,2 and 2,3 planes. Further development of the acoustic stimulation technique on solid wood material will therefore require careful selection of the transducers and a signal processing stage for measurements.

3 - MODE-CONVERSION DUE TO ULTRASONIC WAWE WOOD STRUCTURE INTERACTION

The importance of microstructure in controlling mechanical properties of solids is, of course, well established. The ultrasonic nondestructive methods are particularly useful to explain the links between morphological parameters and mechanical properties of material.

Like most biological materials, wood is heterogeneous. However a non random organization of anatomical elements suggests to consider it as homogeneous anisotropic medium at macroscopic level for overall acoustic characterization. This assumption allows the use of the theory of wave propagation in orthotropic solids for the investigation of wood anisotropy. However the ultrasonic waves propagating in wood have a high potential for extracting information related to its structure that acts as a filter. An accurate estimation of the mechanical behaviour of solid wood requires simultaneous views on structure and propagation phenomena. An excellent review of the literature on the methodology of elastic constants measurements and wood microstructure modelling is reported by McINTYRE and WOODHOUSE (7).

The interaction between ultrasonic waves (bulk and surface waves) and wood anisotropy, (induced by the specific disposition of anatomical elements during the life of the tree), are commented in (8, 9, 10).

A very concise approach to waveguide effect and dispersive propagation due to the interaction of elastic waves with the medium microstructure will be commented further, for the dispersive propagation in RT plane. The annual ring structure (with latewood having almost solid thickwalled cells and with earlywood having thinwalled cells) induces wave guide effect in both longitudinal and transversal propagating waves. The propagation normal to the layering shows band effects (11). Up to this point the wavelength was assumed to be much larger than the material microscopic dimensions. However, as soon as matches the internal dimensions of the substrate, complicate periodic structure and waveguide effects can occur. The continuum theory which ignores the ring structure of wood can be use safely only if the wavelength is long compared with the ring spacing with the proportion of the latewood in the ring and fiber's dimensions. The diffuse porous hardwoods with a very small proportion of the latewood in the annual ring seems to be much closer to the hypothesis of continuum theory than the softwoods and oak. Similarities between the behaviour in acoustic field of oak and softwoods can also be deduced if we compare the corresponding acoustic emission responses under four-points bending test (12).

4 - DEFECT DETECTION

As defined in the literature (13, 14) the various stages that are usually taken into consideration during ultrasonic inspection are : detection, localization, characterization, and decision as to whether a defect is important enough to take action on it. Biological material like wood gives a large amount of scattering phenomena, induce by anatomical structure which obscure the image of the real defects. Ultrasonic techniques can, in general be divided into two families :

- the scattering-based methods that use travel times, frequency, amplitude ratio, waveformshape, ... Most of those methods are performed using waves in frequency range between 50 kHz and 2 MHz.
- imaging techniques which seek to provide a picture of discontinuity.

Knots detection and location on planks (18) are studied througth the variations of some local physical and mechanical wood properties. Stiffnesses mapping shows the location and the size of knots in all three main anisotropic directions considered simultaneously (figure 1).

Defect detection and location in particle boards are commented in (19). To verify the initial quality of the pannel the transit time was measured. The technique used still has the potential to be successfully implemented, using a peculiar pattern recognition technique.

5 - CONCLUSION

With recent advances in transducers, the production of an inspection configuration for any type of component, requires a good fundamental understanding of waves propagation phenomena into the member under test. The influence of several factors as frequency

FIGURE 1 : Stiffness mapping show the location and the size of knots (18)

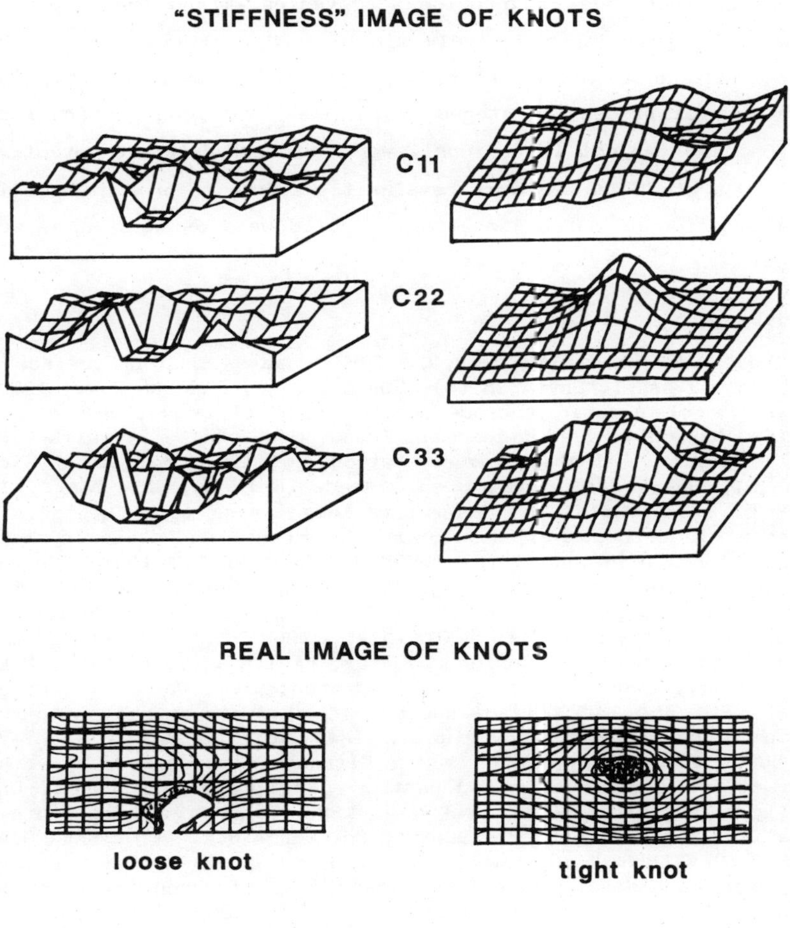

range, geometric beam spreading, pulse/receiver contact pressure, acoustic coupling, ... can be controlled in practice, and simultaneous measurements of several transient waveform parameters and velocity can be performed. Criteria need to be established, which involves changes in the measurements and which reliably indicates the position of the defects. The multiparameter technique as mode-conversion technique can provide a convenient means for inspection of large areas using relatively simple probes. It is to note that this technique involves relatively large digital data sets, that after several processing stages could lead to a good pattern recognition. The subject is obviously worth of much detailed investigation and a great deal of work remains to be done to provide a reliable indication of defect location in solid wood as well as in wood composites.

6 - REFERENCES

(1) DERESIEWCZ H., MINDLIN R.D., 1957 : Waves on the surface of a crystal. J. Appl. Phys. 28 (6), 669-671. (2) FARNELL G.W., 1970 : Properties of elastic surface waves. Physical acoustics, vol. 6, 109-166, Masson W.P. ed., Academic Press, New York. (3) WHITE R.N., 1981 : Surface elastic waves. Methods of Experimental Physics, vol. 19, 495-532. Edmonds P. ed., Academic Press, New York. (4) ROYER D., DIEULESAINT E., 1984 : Rayleigh waves velocity and displacement in orthorombic, tetragonal, hexagonal and cubic crystals. J.A.S.A., 76, 1438-1444. (5) BUCUR V., ROCABOY F., 1988 : Surface wave propagation in wood. Prospective method for the determination of wood off-diagonal terms of stiffness matrix. Ultrasonics, in press. (6) KOLLMANN F.F., COTE W.A., 1968 : Principles of wood science and technology. Springer Verlag, Berlin. (7) McINTYRE M.E., WOODHOUSE J., 1986 : On measuring wood properties. Part. 3. J.Catgut Acoust. S., 45, 14-23. (8) BUCUR V. : Wood structural anisotropy estimated by acoustic invariants. IAWA Bulletin, 9 (1), 67-74. (9) BUCUR V., PERRIN J.R., 1988 : Ultrasonic waves-wood structure interaction. Communication "Acoustics 88" Spring Conference, University of Cambridge, U.K., 5-8 April 1988. Proceedings Institute of Acoustics, vol. 10, part 2, 199-206, Edinburgh,U.K. (10) EFRANSJAH, BUCUR V., KILBERTUS G., 1988 : Impact of water stored treatment on spruce mechanical properties detected by ultrasonics. W.Sci.and Techno. (submitted for publication). (11) WOODHOUSE J., 1985 : Personal communication. (12) VAUTRIN A., HARRIS B., 1987 : Acoustic emission characterization of flexural loading damage in wood. J.M. Sci. 22 (3), 707-717. (13) BOND L.J., SAFFARI M., 1984 : Mode conversion ultrasonic testing. Nondestructive Testing. Vol. 7. Ed.

R.S. Sharpe, Academic Press, London. (14) WEDGEWOOD A., 1987 : Data processing in ultrasonic. NDT Conf. Proc. Ultrasonic International 87, 381-386. (15) McDONALD K.A., 1978 : Lumber defect detection by ultrasonics. Res. Paper F.P.L. 311, USDA,Forest Service, U.S.A. (16) McCRACKEN F.I., WANN S.R., 1983 : Sound can detect decay in standing hardwood trees. Res. Paper SO-195, USDA, Forest Service, U.S.A. (17) OKYERE J.G., COUSIN A.J., 1980 : On flaw detection in live wood. Materials evaluation, March, 43-47. (18) CHAZELAS J.L., VERGNE A., BUCUR V., 1988 : Analyse de la variation des propriétés physiques locales autour des noeuds. Actes du Colloque "Comportement mécanique du bois", Bordeaux, 8-9 Juin, 376-386. (19) SZABO T., 1978 : Use of ultrasonics to evaluate or characterize wood composites. Proc. 4-th Nondestructive testing of wood., 239-260, Washington State Univ. Pullman.

Ultrasonic Selection of Spruce for Factory Manufacture of Harp Soundboards

I M Firth
Department of Physics and Astronomy
University of St Andrews
North Haugh
St Andrews
Fife KY16 9SS
Scotland

ABSTRACT

A digital ultrasonic velocity meter was constructed to enable factory-floor measurements to be made on spruce tone-wood planks. The meter accepted keyed-in dimensions of the planks, and calculated velocity (c), density (ρ), Radiation Ratio (c/ρ) and Young's modulus ($c^2\rho$). Processing of time of flight measurements of an ultrasonic pulse, and the other calculations was based on a Z8 single chip micro-computer. Radiation Ratio and Young's modulus were used to select planks so that soundboards (made of about twenty five planks edge glued) were made with planks of similar or graded parameter values. Results for different types of spruce from Europe are presented.

INTRODUCTION

Musical instrument makers have always attached great importance to the selection of wood, not only species but also characteristics of the particular piece to be used. Selection becomes more important as high quality tone-woods become more scarce. Acousticians have measured elastic properties, density, and damping coefficients but to take scientific advantage of such measurements procedures to assess the properties of wood have to be as simple as possible. Simplicity, and speed of measurement are particularly required in the case of the large area spruce soundboards of the harp and piano which are made by edge jointing many quarter sawn spruce strips on the factory floor.

This paper describes a microcomputer controlled ultra-sonic meter which has been constructed to measure Young's modulus ($Y = c^2\rho$) and Radiation Ratio ($R = c/\rho$), two parameters which enable spruce to be evaluated for static strength, and acoustical output. The meter has been used in selection of wood in the construction of concert harp soundboards. Long spruce strips, machined to standard dimensions, are tested in a jig which measures the time of flight of an ultrasonic pulse along their lengths. Standard dimensions of the strips are keyed into the microcomputer and as each strip is measured its mass is inputted.

The microprocessor calculates, displays and records longitudinal velocity (c, ms^{-1}), density (ρ, kq m^{-3}), Young's modulus ($Y = c^2\rho$, Nm^{-2}), and Radiation Ratio ($R = c/\rho$, $kg^{-1}m^4s^{-1}$).

Measurements take place simply on the factory floor, and enable the soundboards of the harp, which are constructed from about 25 edge jointed strips, to be matched or graded along their lengths from treble strings to bass strings for static strength (Y) to withstand the static tension of the strings, or acoustic output (R).

WOOD PARAMETER

Static Parameters

The soundboard of the concert harp has to withstand the static tension of its strings, numbering 52, which rise at about 30° to the soundboard. Overall tension of the harp strings is nearly 10,000 N (50 N in the treble rising to 500 N in the bass). Young's modulus (Y) is the parameter which is important here for a given soundboard shape, and can be evaluated from the velocity (c) of an ultrasonic pulse and the density (ρ) of the wood as $c^2\rho$.

The microcomputer controlled ultrasonic meter described here evaluates c from pulse time of flight and the length of a spruce strip, density from the mass of a strip and its dimensions, and Young's modulus from the quotient c^2/ρ.

Acoustic Parameters

There are three parameters which have been suggested as indicators of acoustic quality of tone wood. These are

1. Radiation Ratio (R) where

$$R = c/\rho$$

introduced by Schelleng[1,2]and discussed by Ghelmezin and Beldie[3].

2. Loudness Index (LI) where

$$LI = R_L\, R_T\, Q_L\, Q_T/\pi^2$$

is based on the idea of loudness $L = R\, Q/\pi$ where Q is the quality factor of[4] a particular resonance, and the subscripts L and T represent measurements longitudinal and transverse to the grain of the wood in the quarter-sawn samples. Haines[4] discusses the significance of this term which can only be evaluated using a resonance method usually at the fundamnental resonant frequencys of a wood strip.

3. Utility (U) where

$$U = \pi LI\frac{1}{2}$$

has been introduced by Dettloff[5] as a single parameter which would characterise the usefulness of a sample of wood as "tone wood".

The statistical correlation between these parameters have been investigated by several workers. Ghelmezin and Beldie[3] and Dettloff[5] report the following

significant correlation between	Y_L and ρ
	R and ρ
	L_L and ρ
negative correlation between	L_L and Y
	R_L and Y

This introduces a difficulty in the selection of tone wood spruce for the harp because the acoustically optimum wood will be comparatively weak.

Both LI and U are measured at the lowest resonant frequencies of strips cut longitudinally and transversely to the grain using the xylophone method. This method is slow to execute as resonant frequencies and bandwidths have to be measured for each sample of wood.

In harp or piano making the strips of spruce used for making the soundboards are long, so that resonant frequencies are low and bandwidths particuarly narrow so that the xylophone method is impracticable on the factory floor. The only realistic parameter which can be measured for assessing the acoustic quality of spruce soundboard wood is therefore the Radiation Ratio (R) which can be automatically calculated from ultrasonic (longitudinal) velocity and density.

Radiation Ratio was introduced by Schelleng[1] in a discussion on the flexural similarity of tone words. When it is required to duplicate the acoustical behaviour of a plate with a different material the flexural behaviour will be similar if the stiffness per unit length and the densities per unit area of the plates are equal. Stiffness per unit length is proportional to Y and the cube of the thickness (t). Hence

$$Y_1 t_1^3 = Y_2 t_2^3.$$

Equal densities per unit area require that $t_1\rho_1 = t_2\rho_2$, and with $Y = c^2\rho$ it follows that $c_1/\rho_1 = c_2/\rho_2$.

This means that materials can be duplicated if this ratio, the Radiation Ratio (R), remains the same.

The Radiation Ration enters into other acoustical circuit relations. The acoustic radiation from a plate is largely determined by impedance which can be expressed at a resonant frequency (f_b) of a soundboard as

$$Z \propto (\rho/c)\, f_b^2 l$$

where f_b and l are not determined by the species or sample of tone wood.

This expression indicates that for the same instrument made of different woods the one having the lower (ρ/c) (higher Radiation Ratio $R = c/\rho$) will present the lower impedance. Hence for a given force from the strings, the instrument having the higher Radiation Ratio will produce larger soundboard motion and therefore the larger sound radiation. An increase of Radiation Ratio of 12% will increase sound output by 1 db.

APPARATUS

The apparatus is designed around a central processing unit (CPU), type MCS65C02. On demand from the CPU the transmitting (TX) piezoelectric transducer is excited (approx 800V applied across the TX) and the timing sequence started. On receiving a signal form the receiving (RX) piezoelectric transducer the timer is stopped, read by the CPU and the calculation sequences for velocity, Young's modulus and density of sample and Radiation Ratio initiated.

TIME OF FLIGHT CIRCUITRY

The CPU has control only over starting the sequence above: the stop sequence for the timer is controlled by logic due to the slow processor response. This is shown diagramatically in Figure 1.

DESIGN OF TIME-OF-FLIGHT LOGIC CIRCUITRY

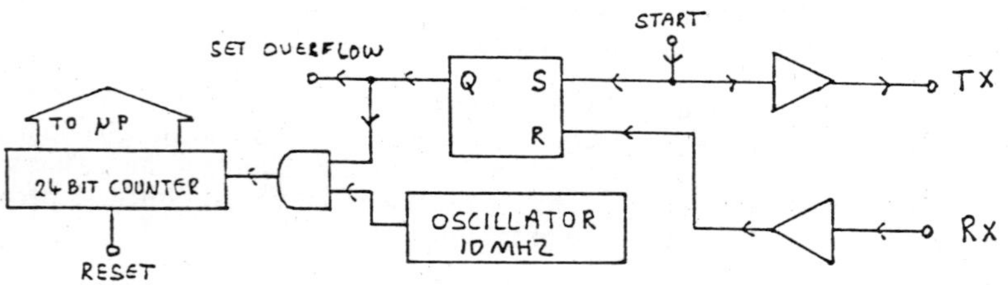

Block schematic for time-of-flight circuitry

Figure 1

As the timing resolution was specified as 100ns the timing clock was required to run at 10 MHz. A simple but highly stable crystal oscillator was designed around a schmitt trigger input inverted logic gate. The crystal is connected across the invertor as positive feedback for oscillation to occur and the crystal was trimmed with a variable capacitor to exactly 10 MHz.

On the starting command from the CPU logic feeds the 10 MHz clock to the timer. The logic is a latch and a gate. The latch is reset by the RX signal at the end of the timing period. The counter used in the apparatus was the 65C22VIA which can toggle up to 10 MHz and be connected directly to the CPU.

TX Stage

For adequate operation a TX pulse of 800V is required. This pulse is produced here by a high voltage power supply whose output is in series with the TX piezoelectric crystal and an electronic switch controlled by the CPU. Athough more complex, this method produces a single voltage edge to the TX crystal which maintains initial phase coherence in the output ultrasonic pulse.

The power supply is constructed from an invertor to multiply the main supply of 12V to 800V. A Hartley style oscillator at 50kHz was used with ferrite core transformers. At this high frequency the ferrite transformers are operating at high efficiency and the size of the smoothing capacitors in the secondary circuit is reduced.

To prevent the chance of voltage breakdown in the secondary, a voltage doubler is employed which requires the secondary to operate at 400V but produces the required 800V at the TX crystal.

The 800V pulse to the TX crystal is produced as follows, as shown in Figure 2. The TX crystal can be thought of as a capacitor with inherent high impedance. A SM6 Ω resistor is connected in series with a 10nF capacitor and the TX crystal across the 800V supply. The electronic switch is placed between the resistor and capacitor and ground, and when switched on has the result of producing a -800V pulse across the transducer. The descending edge time of the pulse is about 20ns. A 'Gate Turn Off' device (BTW58- 100R) was found ideal from its small physical size, its low gate turn on current, and that it could be driven straight from TTL logic gates via a suitable current limiting resistor .

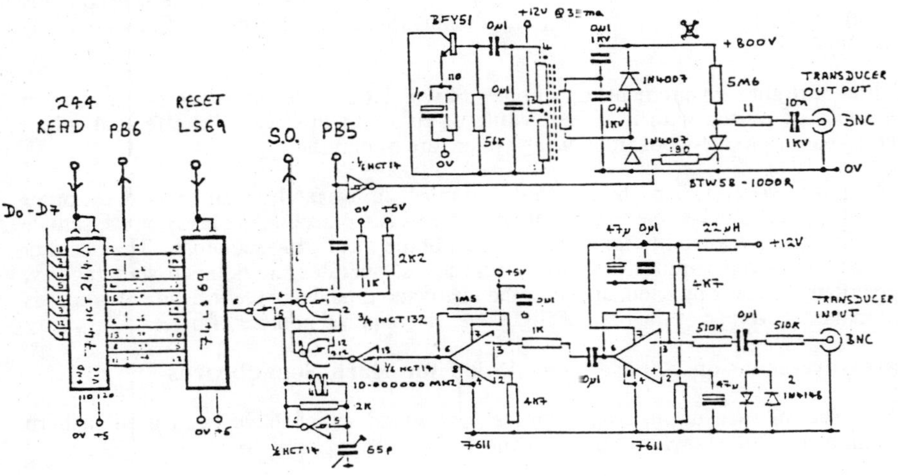

Figure 2 TIME OF FLIGHT CIRCUITRY AND WAVEFORMS.

RX Stage

The RX crystal minimum signal was specified as 5mV and this had to be amplified to 5V (logic level) using a high input impedance amplifier (1MΩ was adequate) to match the inherent high impedance of the RX crystal. Amplifier biasing was used to ensure that only the positive half cycle (1st rise of RX signal) was amplified so that only the positive half cycle was used to trigger the latch. This signal was fed to a schmitt trigger invertor device to give a clean signal synchronised with the leading edge of the RX crystal signal. This logic signal reset the timing control latch and stopped the timing counter.

The arithmetic calculations on the CPU are programmed to take effect after the time of flight of the ultrasonic pulse has been measureed. The operator keys in the length of the wood sample and its mass. On key demand the velocity, Young's modulus and Radiation Ratio are calculated and displayed.

There is a key facility to demand a hard copy print-out on a small portable printer.

MEASUREMENT FACILITIES

The transmitting and receiving piezoelectric heads are connected to the instrument via BNC coaxial cables. The heads have a thin soft rubber covering as a means of making good ultrasonic contact between head and the wood sample. The heads are mounted on a jig into which the long wood samples are placed, so that a constant pressure can be made between heads and wood. In measuring spruce tone-wood for the harp the heads are placed against the end grain of the strips.

The instrument is calibrated before use because the inherent time delays in the system and transducers must be zeroed. Placing the piezoelectric heads together and using the Calibration Mode the instrument makes a five pulse average of the delay and then automatically uses this value to zero the system: the zero set delay is subsequently subtracted from all timer values, and displays and uses for calculations only this calibrated value.

Time of flight measurements on samples of wood are made as averages over a preset number of pulses with a default value of five. In this way the effect of minor perturbations to the whole measuring system are minimised.

When the system is hand-held-operated another timing facility can be used. In order for the operator to be able to position the TX and RX piezoelectric leads on the sample by hand there is a facility for a delayed start of the measuring sequence. The operator keys in the time required to position the leads, operates the delayed start facility, positions the leads by hand, and after the instructed time delay the instrument proceeds to make an average of the time of flight over the required number of pulses.

Measurements on Spruce Tone Wood for Harp Soundboards

Two sets of measurements on spruce tone wood which have been used in harp soundboards are shown in Figures 3 and 4.

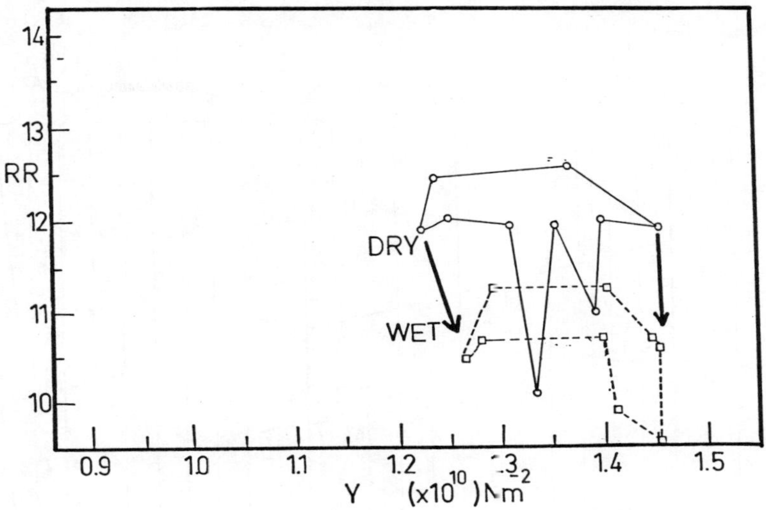

Figure 3

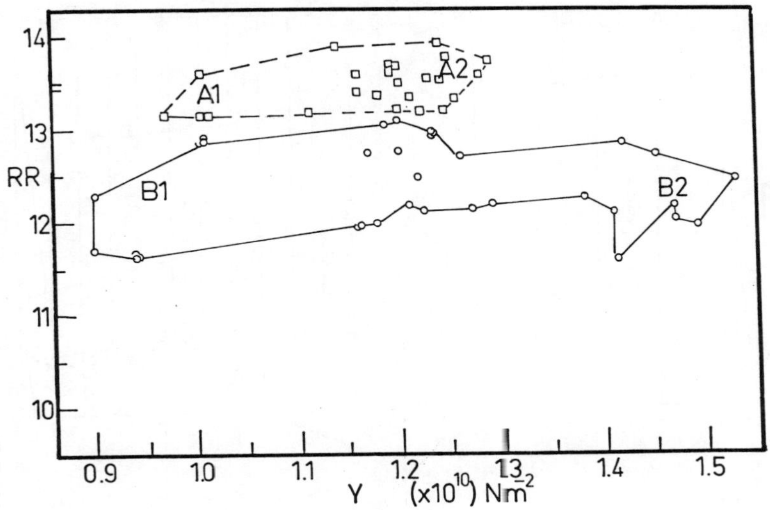

Figure 4

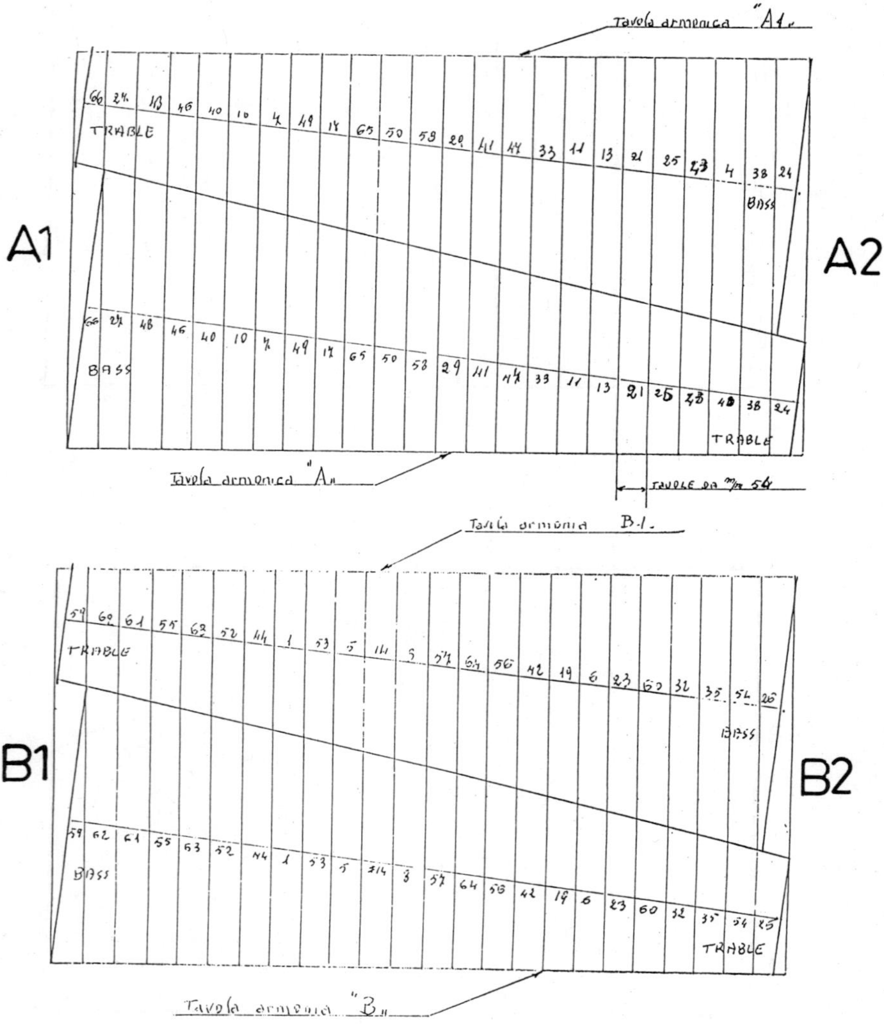

Figure 5

The spruce samples in Figure 3 are *Picea Abies* from Sweden which have been selected and microwave dried for use in harp and piano soundboards: Polygon A for dry wood as supplied, Polygon B after this wood had been exposed to air at 60% relative humidity, 18°C, for 5 days. Radiation Ratio ($R = c/\rho$) is plotted against Young's modulus ($Y = c^2\rho$). Dry samples are defined by the upper polygon of the figure which moves to the lower when the samples have absorbed moisture from the air. Moisture absorption displaces the samples to higher Young's modulus but lower Radiation Ratio bearing out two of the statistical correlations given by Ghelmezin and Beldie(3) (+ve between Y and ρ, -ve between R and Y).

Acoustical performance on the harp would be determined by the 'wet' samples (Polygon B), and if a comparison is made with the data presented in Figures 3 and 4 it is seemnthat Swedish tone wood is strong but has a lower Radiation Ratio than the French spruce described in Figure 4.

Figure 4 shows a factory selection of spruce from a supply in France and is again a plot of Radiation Ratio versus Young's modulus for 66 samples.

The measurements of Figure 4 show that the spruce from the same supply can be considered to fall into two categories: the polygon A of high Radiation Ratio, and polygon B with lower Radiation Ratio. Samples A do not have the same variation of strength as sample B, and in general are weaker.

On the basis of the measurements of Figure 4 four soundboards for the harp were made for testing. The sample A soundboards are made by edge jointing strips in order from A1 to A2 and cutting the triangular soundboards as shown in Figure 5. In this way one soundboard has high Young's modulus wood at the bass strings and low Y at the treble, and the Second Soundboard has positions reversed. Likewise Samples B have been made into the soundboards of similarly positioned Young's modulus. Soundboards A have an average Radiation Ratio of 13.7 and a range of Young's modulus from 0.97 $10^{10}Nm^{-2}$ to 1.3 $10^{10}Nm^{-2}$, and Soundboards B an average Radiation Ratio of 12.4 and range of Young's modulus 0.9 $10^{10}Nm^{-2}$ to 1.53 $10^{10}Nm^{-2}$.

These four soundboards have been built into concert harps and will be evaluated by listening tests.

ACKNOWLEDGEMENTS

The support of Salvi Harps (Halia)SpA is gratefully acknowledged.

REFERENCES

1. Schelleng, J.C., "The Violin as a Circuit", J. Acoust. Soc. Am. 35, 326 (1963).

2. Schelleng, J.C., "Wood for Violins", Catgut Acoust. Soc. Newsletter 37, 8 (1982).

3. Ghelmeziu, N. and Bedie, I.P., "On the Characteristics of Resonance Spruce Wood", Catgut Acoust. Soc. Newsletter 17, 10 (1972).

4. Haines, D., "On Musical Instrument Wood", Catgut Acoust. Soc. Newsletter 31, 23 (1979).

5. Dettloff, J., "Statistical Relationships between Acoustical Parameters of Violin Tonewoods", J. Catgut Acoust. Soc. 43, 13 (1985).

DIGITAL SIGNAL PROCESSING IN ULTRASONIC NON DESTRUCTIVE TESTING OF COMPOSITE METAL BONDED MATERIALS.

R. Zilber

N.D.T. Department, Soreq Nuclear Research Center,

Yavne 70600, Israel.

INTRODUCTION.

The use of industrial adhesives in critical structural components is becoming crucial, in particular for joining metals to composite materials - especially in the aircraft industry. Different adhesive and glues are being developed to be used advantageously in the solving of joining problems where other classical methods cannot be used. One of the major limitations for the use of adhesives for structural joints is associated with the difficulty encountered in making an accurate determination of bond quality.

We have developed a non destructive technique to evaluate the debonding that may be present in a metal-to-composite bonded sample. Ultrasonic signal properties have already been considered in problems of this kind[1,2,3,]. The approach used here is characterized by two features, namely - the use of a simple digital processing of the ultrasonic signal and a dry coupled transducer. We have succeeded in reaching two targets: 1) locating and characterizing unbonded areas and 2) we have found the basis for a criterion to pinpoint areas of poor quality of bonding suspiscious of being neither de-bonded nor perfectly bonded.

SAMPLE DESCRIPTION.

The sample is manufactured as a 2mm thick metal sheet bonded to a 4mm thick carbon-carbon composite plate. The thickness of the adhesive material is about 2 - 3mm.

The sample was prepared with deliberatedly included defects, in order to imitate debonding (Fig. 1). Two kinds of such debonding were simulated:

1) debonding between the metal plate and the adhesive;

2) debonding between the adhesive and the composite plate.

The defects were simulated by a foil of polyethylene which acts as air for the ultrasound, i.e. it stops the propagation of an ultrasonic beam and reflects it back.

The sample is intended neither to be wet nor immersed in water.

EXPERIMENTAL.

A 2.2 MHz, 13mm diameter dry coupled transducer (WD50-2 manufactured by Ultran) was used in this study. The equipment used includes an ultrasonic pulser/receiver analyser, a sophisticated programmable digital oscilloscope, an IBM PC/XT compatible, a six-color plotter (Fig. 2). A software was written to handle the data acquisition and an advanced software package was used for the processing of the data and the graphical display of the results (Fig. 3).

Because of the dry contact probe, no continuous scanning was considered. Instead, the probe was evenly pressed at 10 points over the sample: 5 at the simulated defects, 5 at randomly chosen locations were flaws are not expected (Fig.1).

At every location the waveform displayed on the oscilloscope was digitized and memorized after 10 averagings. The result (512 points representing the ordinates of the signal) were then transferred to the computer. The sampling frequency was 25MHz.

DATA PROCESSING.

It should be emphasized in the first place that classical analysis through A-scan examination is out of question. The reason is the overlapping of the different peaks from the various interfaces added to their harmonics, due to the low thickness of the sample. In such a case no useful information can be extracted from the experiment. Consequently a digital signal processing of the signal was considered. Two operations were performed upon the waveform y(t):

1) A 64 points Hamming window was applied to y(t):

$$W(t) = 0.54 - 0.46 \cos(2\pi(t-224)/64) \quad (1)$$

The result is a function x(t):

$$x(t) = [W(t)][y(t)] \quad (2)$$

2) An FFT algorithm was then applied to x(t).

A typical result of the operations performed and of the resulting power spectrum (Fourier Transform) is shown in Fig. 3.

The window, set by trial-and-error, is wide enough to include a useful part of the signal (about two cycles) and far enough from its starting point in order not to include the broadcast pulse. The resulting power spectrum is characteristic of the returning ultrasonic wave from the sample, in other words it gives an image of the interaction between the ultrasonic wave and the material in which it propagates.

DISCUSSION.

It is known that the frequency of an ultrasonic wave decreases as the attenuation increases. Therefore it is expected that the frequency components in the analysis performed will be relatively low if the material is well bonded. In this case the penetration of the wave in the sample is maximum and so will be the attenuation. Instead, when the ultrasonic beam impinges a debonded aerea, i.e. strikes an air interface or equivalent (polyethylene foil), the wave path in the material is smaller and so is the attenuation. As a consequence, the frequency spectrum is expected to have higher frequency components.

The results are summarized in Figs 4 and 5 about which the following observations can be made:

- at well bonded areas (measurements No. 6, 7, 8) a strong component at 1.15 Mhz is obtained;

- at areas unbonded between the metal sheet and the adhesive (measurements No. 2, 5) a strong well defined component appears at 2.55 MHz;
- at areas supposed to be de-bonded because of lack of adhesive during manufacture (measurement No. 1), we have found no anomaly; such a result is obviously due to the filling of the gap by the adhesive although it was intentionally left clear to create a defect;
- at one position where the flaw is supposed to be located between the adhesive and the composite (measurement No. 4), there is very little difference with the result obtained at a well bonded area;
- at another such area instead (measurement No. 3), we find two strong components, at 1.15 MHz and 2.38 MHz, indicating that there is some debonding under the metal sheet (2.38 MHz) but there is also an indication of interaction between the ultrasonic beam and a deepest zone in the sample (1.15 MHz);
- at the areas of the measurements No. 9 and 10, we discover that these supposed to be flawless areas do not satisfy the expectations. The power spectrum shows components of various frequencies which means reflections from several interfaces indicating several flaws, i.e. unperfect bonding.

Going back to the spectra of well bonded areas, the small component at 2.5 MHz corresponds to the small amount of the signal coming back from the interface metal-adhesive.

The set of experiments performed for the study presented here is obviously incomplete since we have been unable yet to obtain any information about the different layers composing the sample. This kind of information should be now lookerd after. Moreover, mechanical destructive tests ought to be performed on the sample in order to establish a correlation between the ultrasonic speculations and reality.

CONCLUSION.

We have in this study found a way to determine the unbonding between the metal sheet and the adhesive using a dry contact transducer. The way to determine the unbonding between the adhesive and the composite is not yet clearly established. However we have found the basis to determine the quality of the bonding in the case in which the adhesive has not perfectly adhered to the other components of the material. More work is necessary to correlate the power spectrum and the strengh of the bond.

REFERENCES.

1. Meyer, P. A. and Rose, J. L., in J. Appl. Phys. 48, 3705 (1977).
2. Rose J. L. and Thomas G.H., in Br. J. Non-Destr. Test. 21(3), 135 (1979).
3. Rose J. L., Avioli M. J. and Bilgram R., in Br. J. Non-Destr. Test. 25(2), 67 (1983).

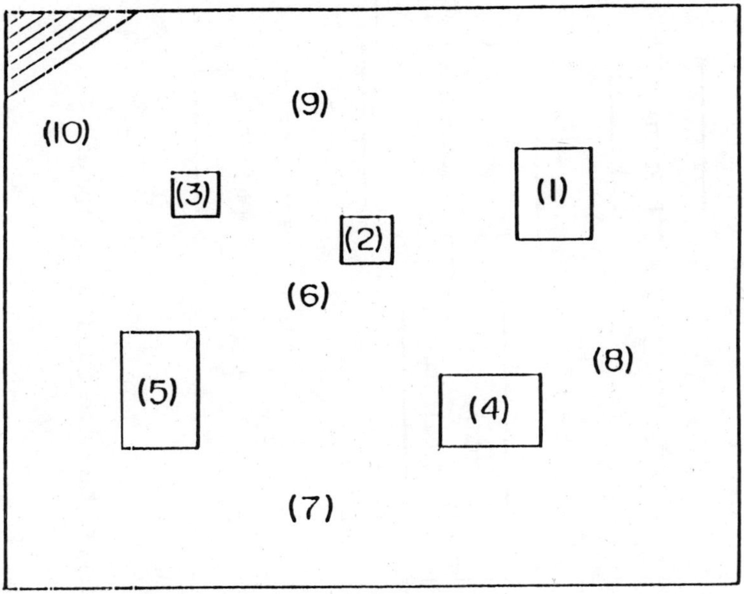

Figure 1. Sample with simulated defects.
Size of sample: 100x125 mm;
size of defects: squares - 10x10 mm, rectangles - 20x15 mm
(1) lack of adhesive
(2),(5) polyethylene foil between metal and adhesive
(3),(4) -- " -- -"- -- " -- adhesive and composite
(6),(7),(8),(9),(10) no defects implanted

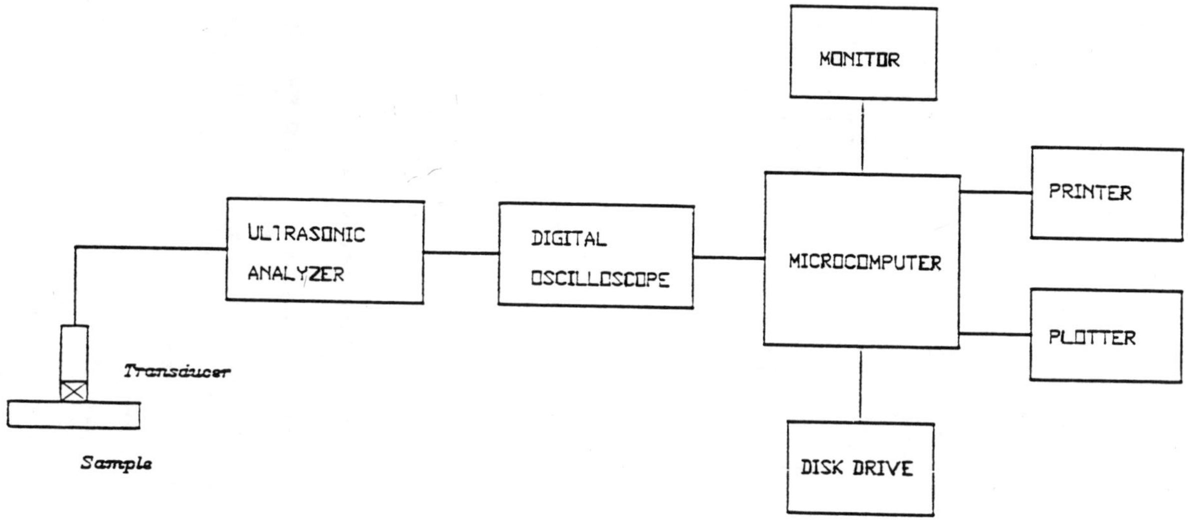

Figure 2. Bolck diagram of the experimental equipment.

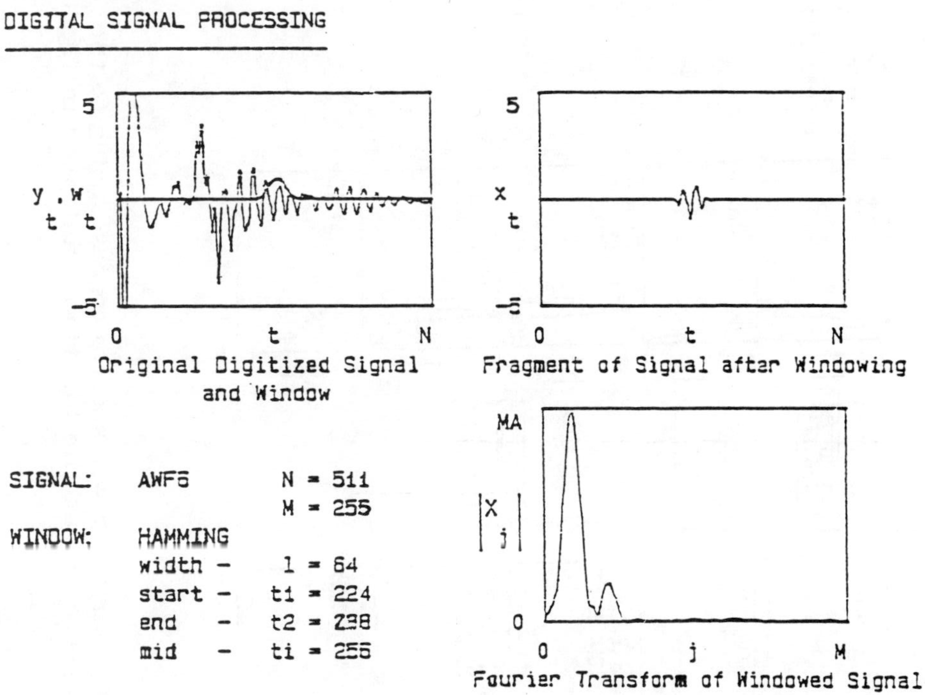

Figure 3. Graphic display of waveform with window, windowed signal and power spectrum (FFT) of windowed signal.

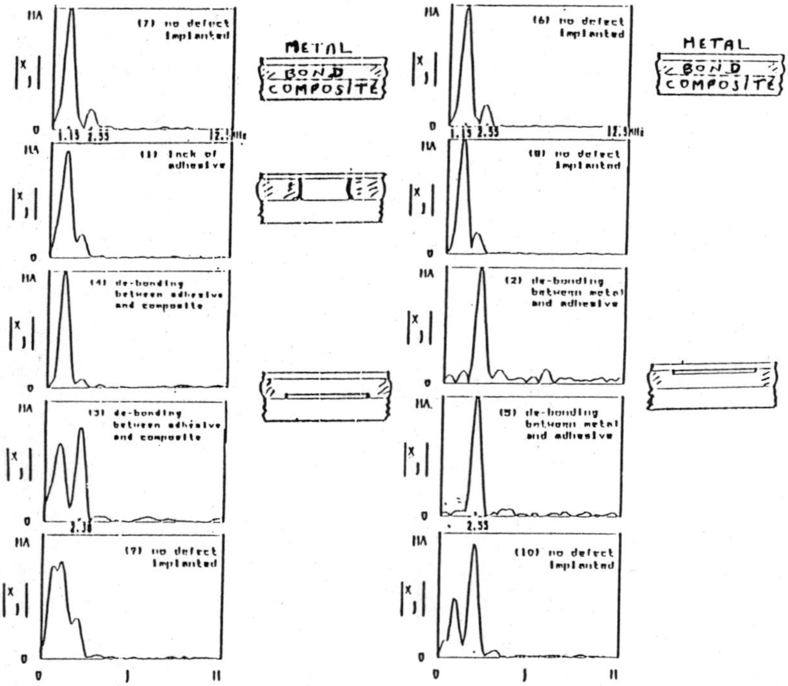

Figure 4. Summary of power spectra corresponding to the ten tests performed at the locations indicated in fig. 1.

The Application of SSBW and Acoustic Modes of Layered Structures to Planar Acoustic Devices

V.I. Anisimkin, I.M. Kotelyansky, M.A. Magomedov,
P.N. Mavlenkov, V.I. Fedosov

Institute of Radioengineering and Electronics
USSR Academy of Sciences
18 Marx Ave., Moscow 103907, USSR

Introduction

Interdigital transducers (IDTs) and piezoelectric single crystals are the main components of most modern microwave acoustic devices used for signal processing. In general, these devices are now based on surface acoustic wave of Rayleigh type. To improve the device performance one can, for example, develop new types of IDTs and gratings, or improve the technology of their preparation, by using more perfect substrates or new crystal cuts with unique acoustic properties. An alternative way to improve the performance is to use use another type of acoustic waves instead of Rayleigh waves which are commonly used now.

Here we overview our last results on application of the Gulyaev-Bleustein wave (GBW) to wrapped around delay lines, of acoustic modes of layered structures to delay lines and filters and of surface skimming bulk waves or surface transverse waves to planar acoustic resonators.

Wrapped around delay line on GBW

The main question here is whether GBW can or cannot propagate through the curved interface between two crystals. The question is not so obvious because the penetration depth of GBW is much larger than that of Rayleigh wave; therefore, the GBW propagation through the curvature has to be more complicated as compared with Rayleigh wave propagation. Fig.1 shows the frequency dependence of GBW propagation loss at cylindrical curvature of CdS crystal. Note, that due to GBW transformation to a number of modes [1], the transmission loss is a complex function of the frequency with narrow maxima and minima at selected frequencies. Fortunately, it is these minima which make it possible for GBW to propagate in a wrapped around manner, although only in a narrow frequency range of about 1 MHz. On CdS we obtained 4 round trip transits with total time delay of about 200μs [2]. A pulse acoustoelectronic amplification enabled partial compensation of the propagation and transmission losses as well as delayed pulse selection.

Acoustic modes of layered structures.

These modes can allow one to obtain unique acoustic properties which are impossible in the case of single crystals. For example, the velocity of Sezawa mode in a ZnO/Si structure can be as large as V=7000 m/s, i.e. 1.5 times larger than in Si (Fig.2). The coupling constant $K^2 = \frac{2\Delta V}{V}$ of Rayleigh mode in ZnO/LiNbO structure achieves 9%, which is 2 times as large as in YZ-LiNbO itself (Fig.3). Moreover, choosing materials with opposite signs of the temperature coefficients of delay (TCD) it is possible to obtain TCD = 0 for the layered structure as a whole. Unfortunately, however, the layered structures also possess some disadvantages as compared with single crystals. One is the insufficient reproducibility of the film parameters (thickness, density and grain size). Another one is the high acoustic attenuation of the films which restricts the application of the layered structures for practical devices. We try to estimate the minimal value α_f^R of SAW attenuation in an ideal film, i.e. in a film which consists of close packed fiber-like crystallines of hexagonal symmetry with C-axis perpendicular to the free surface. The expression is [1] :

$$\alpha_f^R = A\alpha_f^L + (1 - A) \cdot \alpha_f^T \tag{1}$$

where α_f^L and α_f^T are the attenuation coefficients of longitudinal and transverse bulk waves in the film, and A is a function of the $\frac{V_L}{V_T}$ ratio [1]. The values of α_f^L and α_f^T for different mechanisms were calculated from well known theories (for details see [3]). Viscosity and thermoelasticity, as well as SAW scattering due to rough surface and crystalline misorientation were all taken into account. ZnO, CdS, AlN textured and SiO_2 fused films were considered. The calculations were carried out at room temperature, for frequencies from 10 to 1000 MHz and for films with grain sizes of 300 and 1000 Å. All SAW energy was assumed to be confined inside the film.

Even in the ideal case, when the film does not have any voids or other imperfections besides the grain boundaries, we obtained that SAW attenuation in the textured film is 2 - 3 times higher than in the single crystal of the same material. It is these boundaries which are responsible for higher viscosity loss in films, viscosity being the dominant effect up to 500 MHz while at frequencies higher than 1 GHz the SAW scattering by roughnesses of the surface surface become more relevant. Also, the presence of the crystalline boundaries increases the thermoelastic loss and changes its frequency dependence from square to square root. But the value of the thermoelastic loss is still much less than the viscosity or the scattering loss. The scattering due to grain misorientation is negligible at all the considered frequencies. So, the value of α_f^R in ideal textured ZnO, CdS, AlN films is mainly due to two reasons: increased viscosity and roughness of the surface. The frequency dependence of α_f^R is also unusual. It changes from $\alpha_f^R \sim F^2$ at $F < 500$ MHz to $\alpha_f^R \sim F^5$ at $F > 1$ GHz (Fig.4). Note, that for nonideal films which have some more imperfections, the total SAW attenuation is even more increased.

If only one part r of the SAW energy propagates inside the film, as in layered structures, the value of SAW attenuation and TCD can be calculated as follows (Fig.5, Fig.6)

$$\begin{aligned} \alpha_{ls}^{R,S} &= r\alpha_f^{R,S} + (1-r)\cdot\alpha_s^{R,S} \\ TCD_{ls}^{R,S} &= rTCD_f^{R,S} + (1-r)\cdot TCD_s^{R,S} \end{aligned} \tag{2}$$

where (ls), (f) and (s) stand for layered structure, film and substrate, respectively, (R) for Rayleigh and (S) for Sezawa wave.

Planar Acoustic Transverse Wave Resonator

Over the past years a new class of crystal filters has emerged. These filters make use of surface acoustic multiple reflection between grating mirrors formed on the crystal surface. Higher frequency surface skimming bulk waves (SSBW) devices are the most promising. To make use of SSBW a periodic loading of the cavity region is necessary so as to convert a bulk transverse wave - SSBW - into a surface transverse wave (STW) and then achieve a surface wave behaviour.

We present experimental results on STW resonant filters. It is shown that the periodic grooves of the mirrors as well as the inlaid interdigital transducer in the cavity region make it possible to obtain high quality Q STW resonant filters. The advantages of such filter over their SAW counterpart include higher velocity hence higher frequency of operation, for the same period of interdigital transducers, reduced sensitivity to the surface contamination, and low spurious response.

To compare STW and SAW properties in resonant structures we have prepared test devices which were typical one-port planar resonators (Fig.7). Both devices were fabricated on the same substrate (ST-SiO_2) by the same photomask. Quartz substrates were aligned to launch STW orthogonal to and SAW parallel to x-direction. The depth h_m of the grooves, the thickness h_{Al} of aluminum interdigital electrodes and the length L_m of the cavity (Fig.7) were varied from sample to sample. For STW-resonators $h_m = 0.20$, 0.34, 0.39, 0.61, 0.83, and 1.32 μm. For SAW resonators $h_m = 0.38$ μm. The thickness h_{Al} was changed from 0.2 to 0.4 μm, thus allowing to vary the effective groove depth in the cavity ($h_c = h_m - h_{Al}$). The length L_m was changed from 42.5 periods Λ of the grating to 44.3 Λ through 0.2 $\Lambda = 2$ μm, allowing the existence of an effective resonance in one of the samples even though the exact value of the phase of the wave reflection coefficient ϕ_r is unknown. The interdigital transducer was placed symmetrically at equal distances from both mirrors. The used one-port resonators permitted to record only one type of acoustic wave excited in the cavity. SAW devices did confirm the version of monomode regime (Fig.8, column A). The experiments clearly revealed only one high quality Q resonance, which was observed for sample with L_m approximately equal to the even number of Λ (Fig.8, column A, line 9). In accordance with [6],[7], it gives zero phase ($\phi_r = 0$) of the reflected wave at the reference plane (4-4' plane on Fig.7). For other SAW samples (other L_m and the same h_m and h_{Al}) the quality factor Q became smaller, and for L_m approximately equal to the odd number of Λ SAW resonance completely disappeared (Fig.8, column A, line 4).

The characteristics of STW resonators were totally different (Fig.8, column 5,B). There were two types of resonances which were observed for the same one-port resonant configuration. The first resonance (C_1) had $\phi_r = (-0.7 \pm 0.2)\pi$, while the phase ϕ_r of the second resonance (C_2) was $\phi_r = (0.5 \pm 0.2)\pi$ (Fig.9).

Furthermore, C_2 resonance was observed for all h_m, h_{Al} used, while C_1 resonance existed only when the cavity grooves were almost totally filled by Al. With the decrease of $\frac{h_m}{h_{Al}}$ ratio C_1 resonance tends to disappear. Finally, the C_1 resonance was not noticeably sensitive to grooves metalization, while C_2 resonance revealed higher quality factor for mirrors full of Al (compare columns 5 and B, lines 6-10, Fig.8). On the basis of the above results we identified C_1 and C_2 resonances as two types of transverse waves existing in rotated Y-cuts of SiO_2 [8]. The waves are SH-bulk wave nearly parallel to the crystal surface (C_2 resonance) and SH-surface wave or Gulyaev-Bleustein type wave (C_1 resonance). The resonant frequencies, hence the values of propagation velocities, are very close, while the phase of reflection of the waves, hence the displacement behaviour in the vicinity of the surface, are remarkably different. By varying the construction parameters (h_m, h_{Al}, L_m) high quality Q STW resonators were actually obtained (Q $=$ 9000 at aproximately 249 MHz). Note that the SAW resonator with approximately the same maximum Q-value operates at about 1,5 times lower frequency (F $=$ 157 MHz).

References

[1] Viktorov I.A., *Sound Surface Waves in Solids*, Moscow, Nayka, 1981.

[2] Anisimkin V.I., Morozov A.I., Jap.J.Appl.Phys., 1977, v.16, Supplement 16-1, p.527.

[3] Anisimkin V.I., Kotelyansky I.M., Soviet Phys.: Solid State Phys. 1988, v.30, No.3, p.853.

[4] Auld B.A., Gagnepain J.J., Tan M., Electr.Lett., 1976, v.12, No.24, p.650.

[5] Gulyaev Yu.V., Plessky V.P., Soviet Phys.: JTP Letters, 1977, v.3, No.5,

p.220.

[6] Gulyaev Yu.V., Plessky V.P., Soviet Phys.: Radioengineering & Electronics, 1984, v.29, No.7, p.1301.

[7] Anisimkin V.I., Gulyaev Yu.V., Magomedov M.A., Soviet Phys.: JTP Letters, 1983, v. 9, No.9., p. 555.

[8] Wilcox J.Z., Yen K.-H., IEEE Trans., 1981, v.SU-28, No.6, p.449.

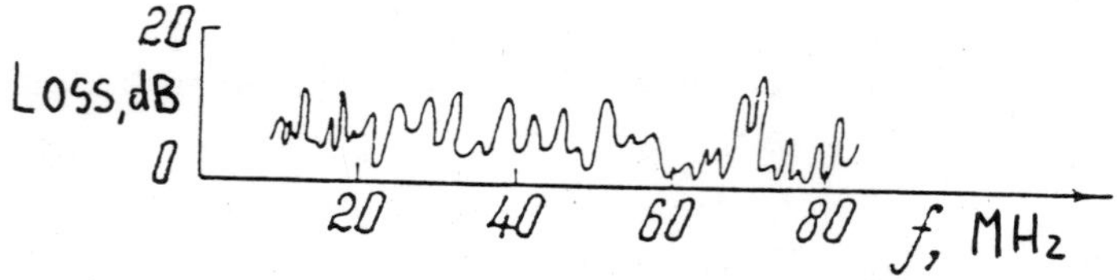

Fig.1: Frequency dependence of GBW losses.

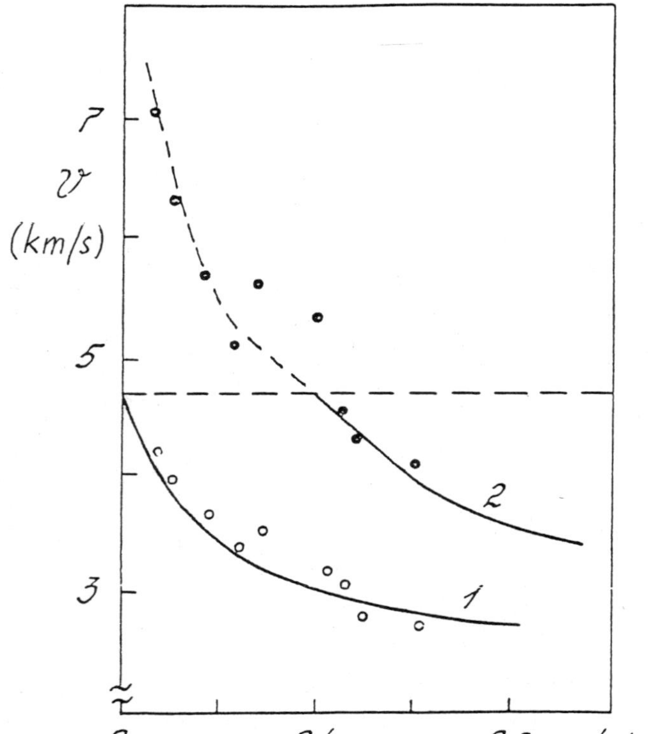

Fig.2: Velocity of Sezawa wave in ZnO/Si structure (curve 2), compared with SAW velocity in SI (curve 1).

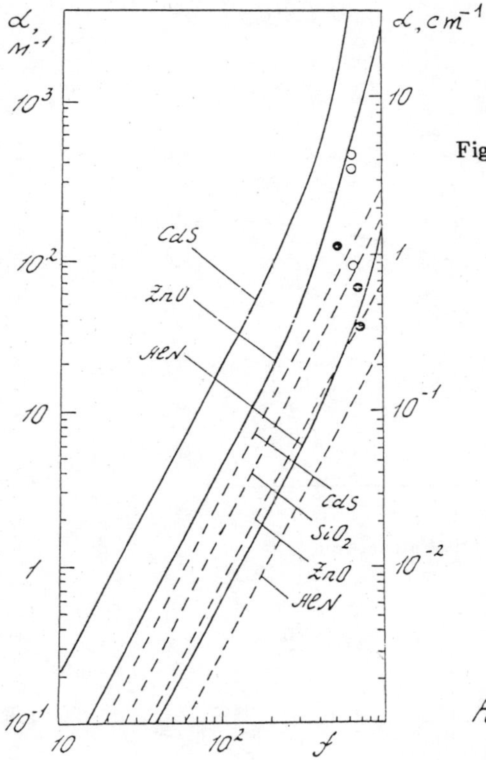

Fig.3: Coupling constant K^2 of Rayleigh mode.

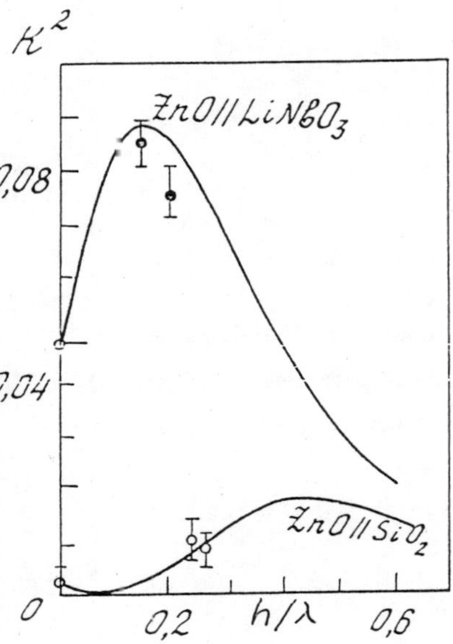

Fig.4: Frequency dependence of attenuation constants.

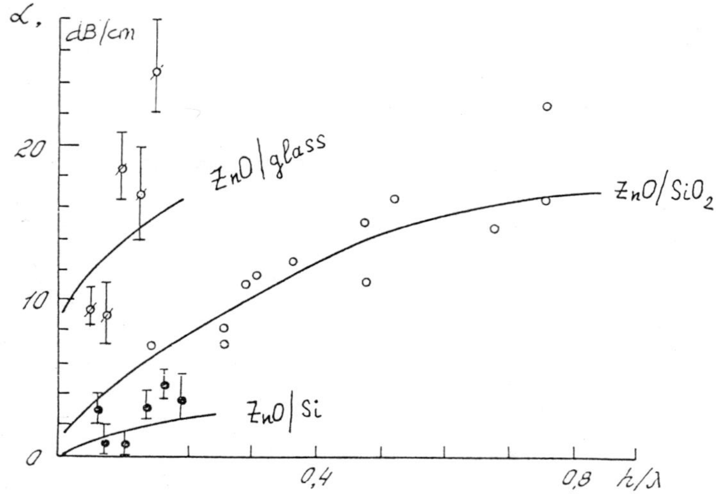

Fig.5: SAW attenuation in different structures.

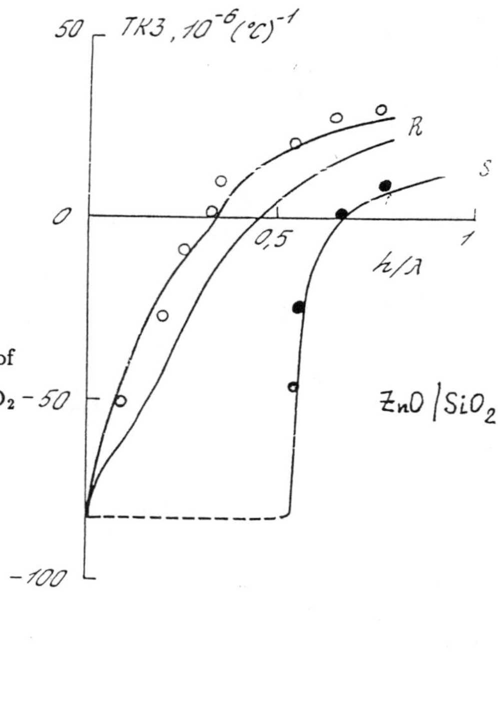

Fig.6: Thermal dependence of SAW parameters in ZnO/SiO_2 structure.

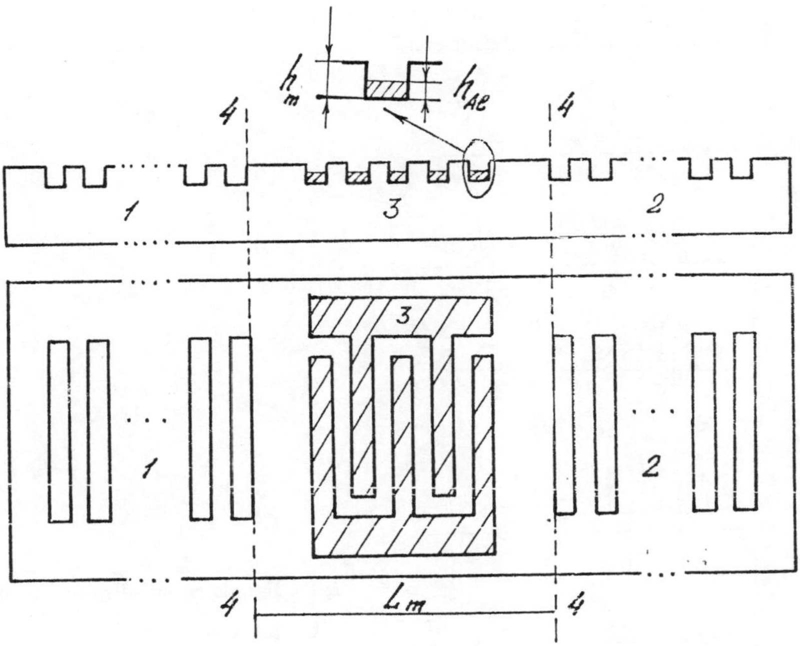

Fig.7: Model of one-port planar resonator

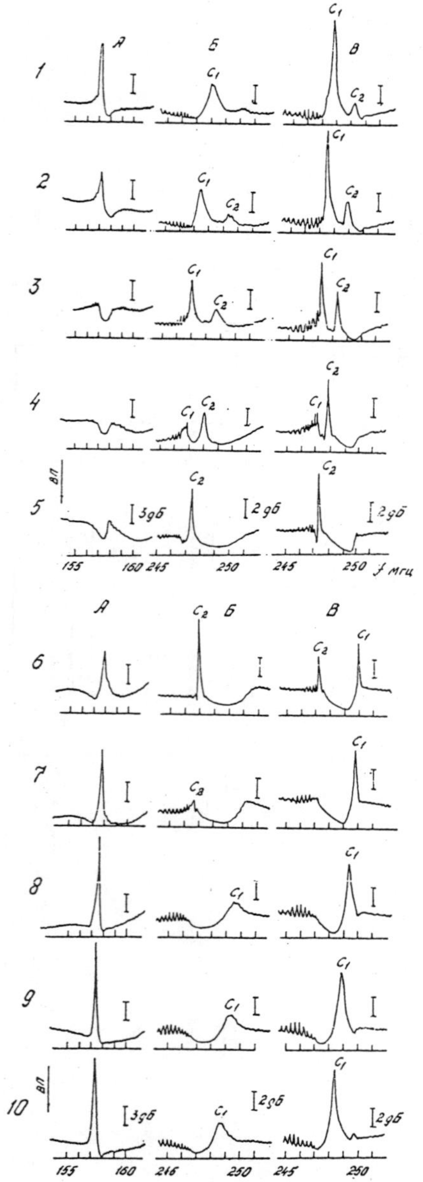

Fig.8: Signals received in one-port SAW resonator

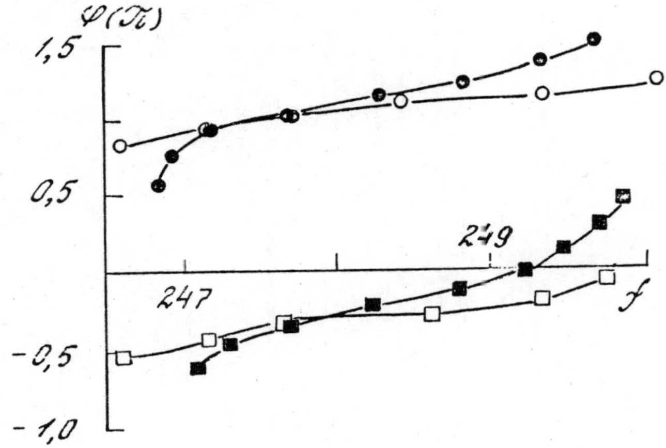

Fig.9: Phase dependence of signals received in STW and GBW resonators.

Resonant and nonlinear acoustooptic interaction in solids and possible applications

Yu.V. Gulyaev and G.N. Shkerdin

Institute of Radio Engineering and Electronics of the Academy of Sciences of the USSR, Moscow

1 Introduction

Acoustic waves cause modulation of the dielectric permittivity of the crystal when they propagate. Thus acoustic wave in crystals are accompanied by waves of dielectric permittivity. When propagating in a crystal an electromagnetic wave (EMW) diffracts on this permittivity wave produced by acoustic wave (AW), and acousto-optic (AO) interaction arises. This effect is the basis of a whole class of AO control systems of electromagnetic radiation, of information processing and others. AO methods have also found wide applications in physical investigations: with their help information can be obtained on the characteristics of AW (propagation velocity, attenuation and so on), as well as on crystal properties in which the interaction occurs.

One of the most important parameters defining AO interaction is sound modulation amplitude of crystal dielectric permittivity $\Delta\varepsilon_{ik}$. The value $\Delta\varepsilon_{ik}$ is a function of EMW frequency ω, and can experience resonant increase at certain frequencies. In the optical range of EMW frequencies the main contribution in crystal dielectric permittivity is made by crystal electrons; therefore, the frequency behavior of $\Delta\varepsilon_{ik}$ in this range reflects electronic spectrum peculiarities depends on electron crystal properties.

In the present paper, we consider some typical resonances of function $\epsilon_{ik}(\omega)$, in the presence of external magnetic and electric fields: we show the relation between resonances and electron properties of crystals, and what information on this properties can be obtained from the experimental investigations of AO resonance interaction.

2 Results and discussion

When a harmonic AW propagates in a crystal, the dielectric permittivity ϵ_{ik} can be written in the following way:

$$\epsilon_{ik} = \epsilon_{ik}^{o} + \Delta\varepsilon_{ik} \ , \quad where \quad \Delta\varepsilon_{ik} = \sum_{n} \Delta\varepsilon_{ik}^{n} exp[in(\vec{q}\vec{r} - \Omega t)] \tag{1}$$

with $\vec{q}$, $\vec{r}$ are the wave vector of AW and the displacement, respectively.

In the effective mass approximation, expressions for resonant part of $\Delta\varepsilon_{ik}^{rn}$ can be written in the following way:

$$\Delta\varepsilon_{ik}^{rn} = \frac{4\pi e^2}{m^2\omega^2 V} \sum_{\mu,\mu_1} \frac{(f_\mu - f_{\mu_1}) L_{ik} \left\langle \mu | e^{i\vec{k}\vec{r}} | \mu_1 \right\rangle \cdot \left\langle \mu_1 | e^{i\vec{k}_n\vec{r}} | \mu \right\rangle}{\hbar\omega + E_{\mu_1} - E_\mu + \frac{i\hbar}{T_{\mu\mu_1}}} \tag{2}$$

where e, m are the charge and the mass of free electron, V is the crystal volume, μ_i are a full set of quantum numbers of electron states in external fields in the absence of EMW, E_μ, f_μ are energy of electron states and function of electron distribution, $T_{\mu\mu_i}$ is relaxation time of induced current produced by EMW due to a transition $\mu \rightarrow \mu_1$, $\vec{k}$ is the wave vector of EMW in a crystal, $\vec{k_n} = \vec{k} + n\vec{q}$, $L_{ik} = \langle \mu | \hat{p}_k | \mu_1 \rangle \langle \mu_1 | \hat{p}_i | \mu \rangle$, $|\mu\rangle$ are smooth envolopes of wave functions for the states μ, $\vec{p} = \frac{\hbar}{i} \frac{\partial}{\partial \vec{r}}$.

The expression (2) describes electron contribution to $\Delta\varepsilon_{ik}$ and consequently its principal resonant peculiarities connected with crystal electron properties. On the basis of expression (2) various resonances have been studied in the behaviour of $\Delta\varepsilon_{ik}$ such as resonances at the edge of fundamental absorbtion region where $\hbar\omega \simeq E_g$ (E_g is the width of forbidden gap) [1], resonances on exciton levels [2], resonances on local impurity levels [3], [4] and others. Information has been obtained from experiments on resonant AO interaction on the spectrum of resonant electron states as well as on parameters of the states such as deformation potentials of states, relaxation times of resonant transitions T_r and else. Further, we will consider resonant properties of photoelasticity in external magnetic and electric fields. This problem is also of interest as there is a possibility of controlling resonant properties of photoelastic tensor.

2.1 Crystals in external magnetic field

We are going to consider the case of strong magnetic fields when quantization of electron spectrum with magnetic field is clearly revealed. It occurs under the conditions: $\omega_H^v T_v > 1$, $\omega_H^c T_c > 1$ (here $\omega_H^{c,v} = \frac{eH}{m_{c,v}c}$, $m_{c,v}$ are the effective masses of electrons of conduction band (c) and of holes of valent band (v), c is velocity of light, H is the value of magnetic field, $T_{c,v}$ the relaxation times of electron states in bands). In this case the presence of magnetic field can strongly affect the value $\Delta\varepsilon_{ik}$ as well as the efficiency of AO interaction on the whole. Here resonances take place close to the absorption lines on Landau levels. The strongest resonances exist near regions of cyclotron resonances in the bands, when $\omega \simeq \omega_H^{c,v}$, and near regions of interband optical transistions on Landau levels.

2.1.1 The region of cyclotron resonance

This case is the simpliest one for collinear AO interaction when EMW and AW are propagating along the direction of magnetic field. In the isotropic case resonant AO interaction can be realized only by using circular polarization of EMW which experiences resonant cyclotron absorption. In the region of resonance the values of $|\Delta\varepsilon_{ik}^r|$ increase by a factor $\xi = \omega_H^{c,v} T_{c,v}$ that is supposed to be greater than one. It means that under corresponding conditions it leads to an increase of diffraction efficiency by a factor ξ^2.

2.1.2 The region of interband resonances on Landau levels

In the linear approximation of AW amplitude the expression for $|\Delta\varepsilon^r_{ik}|$ can be written in the following way:

$$|\Delta\varepsilon^r_{ik}| \simeq \frac{e^2|L_{ik}|}{4m^2\omega^2\pi}\left(\frac{2\mu}{\hbar^2}\right)^{\frac{3}{2}} \frac{\hbar\omega^*_H|\Delta E_g|}{\left|\hbar\Delta\omega + \frac{i\hbar}{T_{cv}}\right|^{\frac{3}{2}}} \tag{3}$$

where ΔE_g is the modulation amplitude of E_g due to AW , $\Delta\omega$ the frequency tuning of strict resonance, $\omega^*_H = \frac{eH}{\mu c}$, $\mu^{-1} = m_c^{-1} + m_v^{-1}$. When $\omega^*_H T_{cv} > 1$ the magnetic field (as in the previous case) increases the maximum resonant values of $\Delta\varepsilon_{ik}$ by a factor $\xi_1 = \omega^*_H T_{cv}$. Thus, the efficiency of AO interaction in strong magnetic fields is defined by values of E_g, T_{cv}, as well as by effective electron masses. In the region of strict resonance diffraction efficiency v is proportional to H^2 (when $v \ll 1$). In the region of cyclotron resonance the efficiency of AO interaction is defined by the modulation amplitude of free electron concentration by AW instead of ΔE_g. Thus the study of resonant AO interaction can give information about all these values. It should be noted that in the region of interband resonances the effect can be considerable only in case that $\omega^*_H T_{cv} > 1$. In fact it can occur only at low temperatures and for $H \geq 10^4 Oe$.

Let's consider briefly the experiment on resonant collinear back propagating AO interaction in the vicinity of cyclotron resonance in n-InSb specimens. Because of the high mobility of free electrons in this materials it is easy to reach the condition $\omega^*_H T_{cv} > 1$ even for moderate values of magnetic field. The electron concentration and electrons mobility of the sample are about $2.6 \cdot 10^{12}\text{cm}^{-3}$ and $3.5 \cdot 10^5\text{cm}^2/\text{V sec}$. The polarization of incident radiation is circular and corresponds to necessary polarization for cyclotron resonance. The reflection of EMW from back facet of this sample is nearly complete.

As a result of calculation we obtained the following expression for the relative amplitude of EMW reflected from the sample for the described situation:

$$\xi = \frac{E_r}{E_o} = \frac{\exp(2\lambda d)\,[\beta_- + \exp(iqd)]\,(1 + R\beta_+) - [1 + \beta_+ \exp(iqd)]\,(R + \beta_-)}{\exp(2\lambda d)\,[\beta_- + \exp(iqd)]\,(R + \beta_+) - [1 + \beta_+ \exp(iqd)]\,(1 + R\beta_-)} \quad (4)$$

here: R is the reflection coefficient of EMW from input facet of sample, d is the sample width along wave propagation, $\lambda = \frac{\sqrt{\gamma_{+1}\gamma_{-1} - \delta^2}}{q}$, $\beta_\pm = \frac{\gamma_{\pm 1}}{\delta - i\lambda q}$, $\gamma_\pm = -\frac{\omega^2}{c^2}\,[\varepsilon_{zz\pm} + i\varepsilon_{yz\pm}]$, $\delta = q\left(\frac{\omega}{c}\sqrt{\varepsilon^o_{zz} + i\varepsilon^o_{yz}} - \frac{q}{2}\right)$, q is the wave vector of acoustic wave. It's considered that the dielectric constant of the sample under the pertubation of acoustic waves is the following:

$$\varepsilon_{ik} = \varepsilon^o_{ik} + \varepsilon_{ik+}\exp(iqx) + \varepsilon_{ik-}\exp(-iqx) \quad (5)$$

and circular polarization of EMW is $E_+ = E_z + iE_y$. The absorption of EMW is resonant when $\omega \simeq \omega^c_H$, namely, for this EMW polarization, if $\omega^c_H T^c > 1$ and $\vec{H}$ is directed in -x direction. It's supposed that wave vectors of EMW and acoustic wave are directed along x axis.

In the presence of external magnetic field we have the following expressions for ε_{zz} and ε_{yz} in n-InSb semiconductor:

$$\begin{aligned} \varepsilon_{zz} &= \varepsilon_o\left\{1 - \frac{\omega_p^2 T_c}{\omega}\frac{\omega T_c + i}{(\omega T_c + i)^2 - \omega_H^2 T_c^2}\right\} \\ \varepsilon_{yz} &= i\varepsilon_o\frac{\omega_p^2 T_c}{\omega}\frac{\omega_H T_c}{(\omega T_c + i)^2 - \omega_H^2 T_c^2} \end{aligned} \quad (6)$$

here: $\omega_p^2 = \frac{4\pi n l^2}{\varepsilon_o m_c}$, n, m_c are concentration and effective mass of conduction electrons $\omega_H = \omega^c_H$. From (6) we have the following expression for the value of $\varepsilon_{zz} + i\varepsilon_{yz}$ that determines resonant properties of expression (4):

$$\varepsilon_{zz} + i\varepsilon_{yz} = \varepsilon_o\left\{1 - \frac{\omega_p^2 T_c}{\omega}\frac{1}{(\omega - \omega_H)T_c + i}\left[1 + \frac{n_+ e^{iqx}}{n_o} + \frac{n_- e^{-iqx}}{n_o}\right]\right\} \quad (7)$$

It is seen that values $\varepsilon_{zz} + i\varepsilon_{yz}$, $\gamma_{\pm 1}$ and ξ are increased resonantly for $\omega \simeq \omega_H$ and $\omega_H T_c >$. Thus we have a resonance increase for AO interaction efficiency for $\omega \simeq \omega_H$.

In the experiments, EMW radiation with wavelength $\lambda = 0,645\mu$ and power $P = 15$mW was generated by a backward wave lamp. This beam fell on a dividing element after passing through a polarizer and lens system. The dividing element was oriented to pass through the incident radiation completely. Then the radiation was directed on a $\frac{\lambda}{4}$ wave plate that was oriented to produce the circular polarization of EMW necessary for resonance . This radiation with an aperture of 5mm fell normally on the input facet of sample. The sample of size $10x10x15$mm^3 was placed inside a solenoid that was cooled at temperature $T = 77^o$K. The shear piezoactive acoustic wave was generated by piezotransducer made of $LiNbO_3$ attached to the back facet of the sample. Pulses of acoustic waves with time durations - $(0.1 \div 10)\mu$sec, intensities -$(0 \div 20)$W/cm^2 and frequencies -$(25 \div 32)$MHz were generated in this sample by pulsed generator. Values of magnetic field were changed up to 2,7 KOe

The polarization of the beam reflected from the sample after $\frac{\lambda}{4}$ wave plate was orthogonal to initial one. For this reason, the backward radiation was reflected completely from the beam splitter. Then, it was to a radiation detector constructed on the basis of Matt barrier. Finally, the signal was registered after amplification.

Calculated dependencies of the relative intensity changes of the reflected radiation due to $|\Delta\varsigma|^2$ AO interaction are shown on Fig.1. It is interesting to note that the position of reflected signal maximum weakly depends on the acoustic wave intensity and it is defined by condition $x \sim d$ where x is absorption length of EMW in the crystal in the presence of magnetic field.

Analogous dependencies for different concentrations of electrons in a sample are given on Fig.2. Experimental results for $n = 2.6 \cdot 10^{12}$cm^{-3} are shown. Rather good agreement between theory and experiments is seen: the decrease the of effect with the increase of concentration is connected mainly with an increase of absorption. For the same reason maxima of reflected signal are moved to the region of smaller magnetic fields with an increase of n.

The calculation of the harmonic amplitudes in reflected radiation as a function of the acoustic wave amplitude is shown in Fig.3. Experimental results for harmonic amplitude are also given. Almost all the energy of the reflected radiation is converted into harmonic of frequency $\omega + \Omega$ when the amplitude of acoustic wave is increased. This is easy to understand since in this situation the harmonic is generated during AO interaction near the input facet of the crystal. The harmonic of reflected radiation at frequency $\omega - \Omega$ can be generated only with the help of EMW reflection from back facet of crystal, when the intensity of EMW is small for strong AO interaction. Finally analogous results versus the value of magnetic field are given in Fig.4. It's seen that maximum of harmonic amplitude of frequency $\omega+\Omega$ corresponds again to the condition $x \sim d$. The value of this harmonic is dominant for strong magnetic fields. This effect is connected also with the decrease of radiation energy near the back facet because of strong absorption in crystal.

So AO interaction in semiconductors in millimiter and submillimiter regions turns out to be rather strong and sensitive both to internal parameters of materials (concentration, mobility of electrons) and to external magnetic field. These circumstances give us the hope that resonant AO interaction can be used not only for measurements of internal parameters of semiconductors, but for registration and control of EMW.

One of the most interesting effects from the viewpoint of possible applications is the effect of frequency shift of EMW. This effect can be helpful for detection of weak signals.

2.1.3 The case of strong and high frequency acoustic wave.

An interesting situation for high frequency AW in magnetic field appears when $|\Delta V_{c,v}| > \frac{\hbar^2 c^2}{2m_{c,v}}, \frac{\hbar}{T_c}, \frac{\hbar}{T_v}$ (here $\Delta V_{v,c}$ are amplitudes of electron potential energy in the field of AW). Here the electron states bound by AW are in fact a set of narrow acoustic minibands . In case when $\vec{q} \parallel \vec{H}$ transverse electron movements with respect to magnetic field are quantized by the magnetic field. Here in the dependence of $\Delta\epsilon^n_{ik}(\omega)$ there are a number of narrow peaks corresponding to allowed transition between both acoustic minibands of valent band and conduction band.

Moreover maximum values of $|\Delta\epsilon^n_{ik}|$ can exceed proper values of $|\Im\epsilon_{ik}|$ in the region of interband resonances on Landau levels in the absence of AW by a factor of $\varsigma = \sqrt{\frac{\hbar^2 q^2 T_{c,v}}{2\mu\hbar}}$ (if $\varsigma > 1$).

This is the only case known to us when such considerable (gigantic) resonsnces may originate. The present effect is connected with considerable reconstruction of electron spectrum under AW effect.

In reality, the effect can occur at about helium temperatures. Only then quantization phenomena of electron spectrum by AW can take place. In order to observe such effect at $\frac{\hbar}{T_{c,v}} \sim 10^{-4}\ eV$ AW frequency of the order of 1GHz and AW intensity $\geq 1\ W/cm^2$ are required.

Considerable resonances in $\Delta\epsilon^{rn}_{ik}$ at $\vec{q} \parallel \vec{H}$ predicted near optical interband transitions on acoustic minibands can be rather useful for experimental discovery and investigation by AO technique of peculiarities of quantization of electron spectrum by AW.

2.1.4 Crystal in cross electric and magnetic fields.

In the previous case resonances took place in the region of optical transition between Landau levels of valent band and conduction band with equal Landau level indices $N_c = N_v$. It was found that if to impose an addition cross electric field, new resonances appear in photoelasticity in the region of optical transition between Landau levels of valent band and conduction band with $N_c \neq N_v$. At $\beta = \dfrac{(m_c + m_v)c\lambda E}{2\hbar H} \sim 1$ ($\lambda = \sqrt{\dfrac{\hbar c}{|eH|}}$ is a megnetic length, E- the value of electric field) the given resonant increase of $|\Delta\epsilon_{ik}|$ is nearly equal to this resonant value for the previous case at $N_c = N_v$ taking place in the absence of transverse electric field.

At $H \simeq 10^4\ Oe$, $m_c \simeq 0.2 \cdot 10^{-27} g$, $m_c < m_v$ it is possible to obtain strong resonances in $|\Delta\epsilon_{ik}|$ on Landau levels with $N_c \neq N_v$ if the value of electric field $E \simeq (200 \div 300) V/cm$. Thus electric field can efficiently control resonances at interband transitions on Landau levels and create new tipe of resonances at transitions where Landau level indices are different. As a result information on corresponding resonant electron transitions can be obtained.

3 Conclusions.

1. The presence of sufficiently strong external electric and magnetic fields leads to the appearence of strong resonances in photoelastic tensors in the region of optical transitions on Landau levels of conduction and valent bands. The position and value of resonant coefficients $\Delta\epsilon_{ik}^{rn}$ depend on the characteristics of resonant electron levels, such as deformation potentials, time of relaxation and else. They are also defined by values of external fields of E and H and can be strongly changed by the fields.

2. The existence of strong cyclotron resonance in the presence of external magnetic field in AO interaction can be used for realization of effective AO interaction in millimeter or submillimeter regions of EMW wavelengths.

3. The influence of spatial dispersion effects on $\Delta\epsilon_{ik}^{rn}$ is becoming substantial at rather high frequences of both EMW and AW. The appearing of strong resonances

in $\Delta\epsilon_{ik}^{rn}$ at $\vec{q} \parallel \vec{H}$ due to acoustic quantization of electron spectrum in bands is one of the most interesting phenomena. Experimental investigations of peculiarities in resonant AO interaction can be fruitful to discover and study the effects of quantization of electron spectrum by acoustic waves.

4. The effects of the influence of external fields on resonance photoelasticity which have been considered here show a possibility of control ling resonant photoelasticity constants by means of external fields: information can be obtained from experimental results on properties of electrons in crystals, such as the values of components of deformation potentials, relaxation times of electron states and other values.

References

[1] R. Loudon, Proc. Royal Soc., 275, 218 (1963).

[2] S. Adachi, C. Hamaguchi, Phys. Review, 19B, 938 (1979).

[3] G.N. Shkerdin, Yu.V. Gulyaev, Fiz. Tverd. Tela, 20, 945 (1978).

[4] Yu.V. Gulyaev, S.M. Movsisyan, G.N. Shkerdin, Fiz. i Techn. Poluprov., 14, 638 (1980).

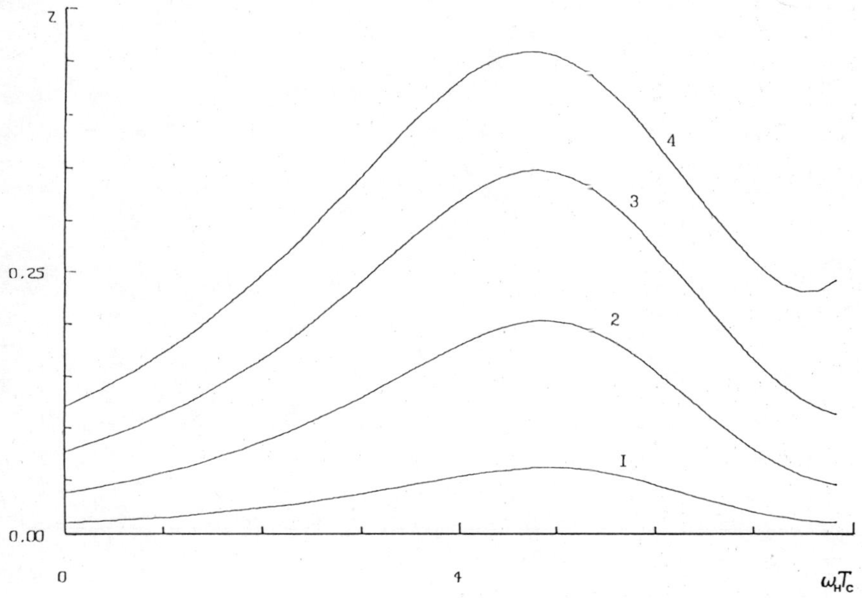

Fig.1: Relative intensity of reflected radiation $|\Delta\xi|^2$ as a function of magnetic field $\Omega_H T_C$ for different depth of electron concentration as modulated by acoustic waves $\left(\frac{\Delta n}{n_0}\right)$.

1 - $\frac{\Delta n}{n_0} = 0.2$

2 - $\frac{\Delta n}{n_0} = 0.4$

3 - $\frac{\Delta n}{n_0} = 0.6$

4 - $\frac{\Delta n}{n_0} = 0.8$

Here: $n_0 = 2.6 \cdot 10^{12} cm^{-3}$, R=0.

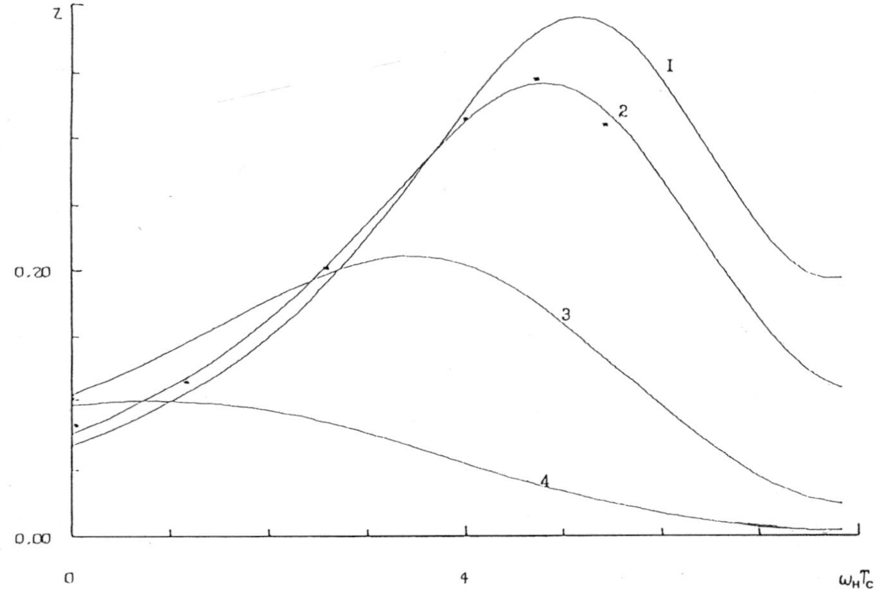

Fig.2: Dependencies of $|\Delta\xi|^2$ as a function of $W_H T_C$ for different electron concentration n_0

1 - $n_0 = 2 \cdot 10^{12} cm^{-3}$

2 - $n_0 = 2 \cdot 6 \cdot 10^{12} cm^{-3}$

3 - $n_0 = 5.2 \cdot 10^{12} cm^{-3}$

4 - $n_0 = 1.3 \cdot 10^{13} cm^{-3}$

Here: $\frac{\Delta n}{n_0} = 0.6$, R=0

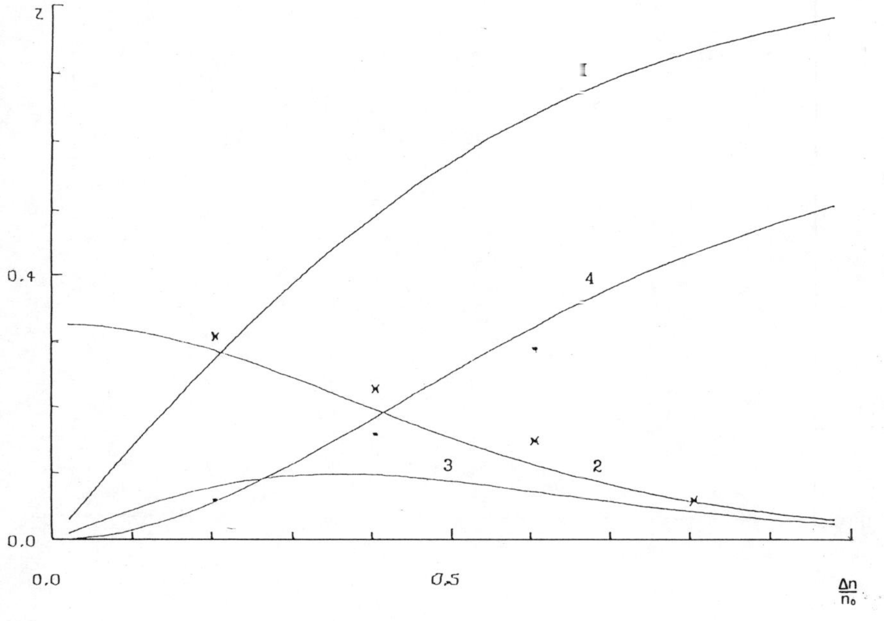

Fig.3: Dependencies of spectral harmonic amplitudes as a function of depth of electron concentration as modulated by acoustic waves.

1 - relative amplitude of harmonic with frequency $\omega - \Omega$

2 - relative amplitude of harmonic with frequency ω

3 - relative amplitude of harmonic with frequency $\omega - \Omega$

4 - value of $|\Delta\xi|^2$

Here: $n_0 = 2.6 \cdot 10^{12} cm^{-3}$, R=0

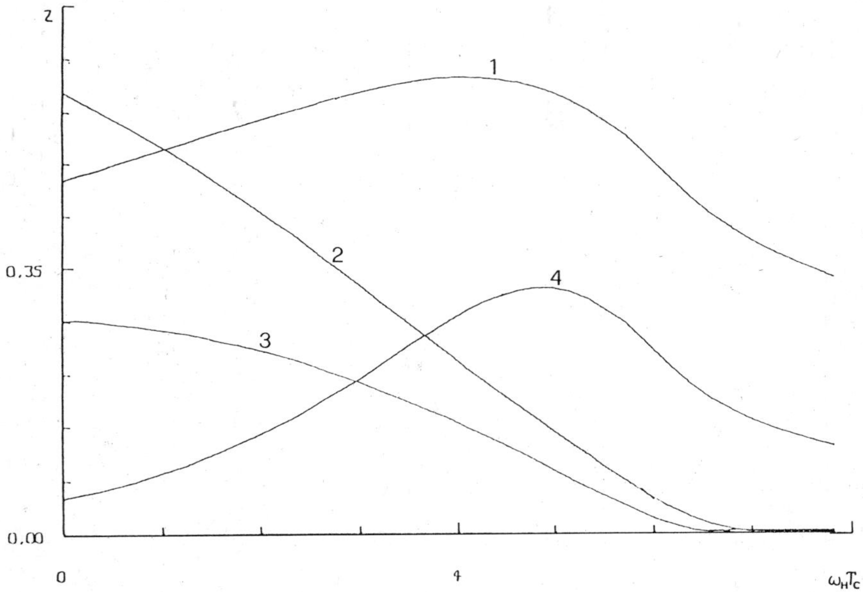

Fig.4: Dependencies of spectral harmonic amplitudes as a function of $\Omega_H T_C$

1 - relative amplitude of harmonic with frequency $\omega + \Omega$

2 - relative amplitude of harmonic with frequency ω

3 - relative amplitude of harmonic with frequency $\omega - \Omega$

4 - value of $|\Delta\xi|^2$

Here: $n_0 = 2.6 \cdot 10^{12} cm^{-3}$, R=0, $\frac{\Delta n}{n_0} = 0.6$

Attendants in the Erice S.Francesco's courtyard

International School of Physical Acoustics
Third course on
ULTRASONIC SIGNAL PROCESSING
Erice, Italy, October 20-30, 1988

LIST OF PARTICIPANTS

L. Adler
The Ohio State University
Nondestructive Evaluation Program
190 West 19th Avenue
Columbus OH 43210-1182
USA

V. Anisimkin
Institute of Radioengineering
and Electronics
Academy of Sciences of the USSR
18, Marx Ave.
103907 Moscow
USSR

J. Attal
USTL
Lab. d'Automatique et de
Microelectronique
Place Eugène Bataillon
34060 Montpellier Cedex
FRANCE

B.A. Auld
Stanford University
Edward L. Ginzton Laboratory
Stanford CA 94305
USA

A. Bettucci
Dip. Fisica
Università di Perugia
Via Elce di Sotto
06100 Perugia
ITALIA

J.R. Blakey
Physics Department
University of Surrey
Guildford, Surrey GU2 5XH
U.K.

V. Bucur
Centre de Recherches Forestières de Nancy
Champenoux B.P.35
54280 Seichamps
FRANCE

A. Cand
Université Paris VII
Group de Physique des Solides
Tour 23 2,Place Jussieu
75251 Paris Cedex 05
FRANCE

F. Craciun
Istituto di Acustica, CNR
Via Cassia 1216
00189 Roma
ITALIA

E. Danicki
Inst. Fundamental Technol. Research
Polish Academy of Sciences
21 Swietokrzyska Str.
00-049 Warsaw
POLAND

N. Denisenko
Inst. Ciencias des Materiales
Serrano 144
28006 Madrid
ESPANA

I. Eick
Inst. für Angewandte Physik
Schlossgarten Str. 7
D-61 Darmstadt
FRG

I.M. Firth
Dept. Physics and Astronomy
University of St. Andrew
North Haugh, St. Andrew
Fife KY16 9SS
SCOTLAND

D. Fitting
National Inst. Standards & Technology
Fractural & Deformation Div.
325 Broadway
Bolder CO 80303
USA

G. Floridi
Dip. di Fisica
Università di Perugia
Via Elce di Sotto
06100 Perugia
ITALIA

I. Gabrielli
Dip. di Fisica
Università di Trieste
Via Valerio 2
34127 Trieste
ITALIA

J.J. Gagnepain
CNRS
Lab. Physique et Metrologie
des Oscillateurs
32, Avenue de l'Observatoire
FRANCE

A. Galbato
Dip. di Energetica
Università di Roma I
Via A. Scarpa 14
00161 Roma
ITALIA

L. Giovannini
Dip. di Fisica
Università di Perugia
Via Elce di Sotto
06100 Perugia
ITALIA

G. Guidarelli
Istituto di Acustica, CNR
Via Cassia 1216
00189 Roma
ITALIA

Y.V. Gulyaev
Institute of Radioengineering
and Electronics
18 Marx Ave.
Moscow 103907
USSR

D. Hauden
LPMO-CNRS
32 Avenue de l'Observatoire
25000 Besançon
FRANCE

H.W. Jones
Technical University of Nova Scotia
P.O. Box 1000
Halifax, Nova Scotia B3J 2X4
CANADA

E.R. Jorzik
CCR EURATOM
21020 Ispra VA
ITALIA

A. Jungman
Univ. Paris VII
Group de Physique des Solides
Tour 23 2, Place Jussieu
75251 Paris Cedex 05
FRANCE

K. Kaufmann
Werkstoffe der Elektrotechnik
Universität Duisburg
Kommandantenstr. 60
D-4100 Duisburg 1
FRG

C.V. Kimball
Schlumberger Doll Research
Old Quarry Rd.
Ridgefield CT 06896
USA

A. Korpel
University of Iowa
Dept. Electrical & Computing Eng.
Iowa City IA
USA

A. Kulik
EPFL
Institut de Génie Atomique
Ecublens
CH-1015 Lausanne
SWITZERLAND

N. Lamberti
Ist. Ingegneria Elettronica
Università di Salerno
84081 Baronissi SA
ITALIA

M. Laschena
Dip. Energetica
Università Roma I
Via A. Scarpa 14
00161 Roma
ITALIA

M. Locatelli
Centre d'Etudes Nucleaires SBT/L.C.P.
Bp 85 X
38041 Grenoble Cedex
FRANCE

E. Molinari
Istituto di Acustica, CNR
Via Cassia 1216
00189 Roma
ITALIA

A. Morsch
Fraunhofer IzfP
Universität, Gebaude 37
D-6600 Saarbrücken
FRG

L. Palmieri
Dip. di Fisica
Università di Perugia
Via Elce di Sotto
06100 Perugia
ITALIA

S. Pangraz
Fraunhofer IzfP
Universität, Gebaude 37
D-6600 Saarbrücken
FRG

G. Quentin
Univ. Paris VII
Group de Physique des Solides
Tour 23 2, Place Jussieu
75251 Paris Cedex 05
FRANCE

Longsheng Qian
Changshung Institute of Optics
and Fine Mechanics
P.O. Box 1024 Changshung
P.R. CHINA

D. Raeside
2121 Grand View Avenue
Norman OK 73072
USA

R. Reibold
Physikalish-Technische Bundesanstalt
Bundesallee 100
Postfach 3345
D-3300 Braunschweig
FRG

S. Rocchi
Dip. Ingegneria Elettronica
Via S. Marta 3
50139 Firenze
ITALIA

G. Scarano
Istituto di Acustica, CNR
Via Cassia 1216
00189 Roma
ITALIA

F. Soares Schlindwein
Leicester Royal Infirmary
Dept. Medical Physics & Clinical Engin.
Leicester 1 5WW
U.K.

S. Shatish
Institut de Genie Atomique
Dept. de Physique
EPFL
CH-1015 Lausanne
SWITZERLAND

G.N. Shkerdin
Institute of Radioengineering
and Electronics
Academy of Sciences of the USSR
18 Marx Ave.
103907 Moscow
USSR

E. Soczkiewicz
Institute of Physics
Silesian Technical University
ul. Krzywoustego
44-100 Gliwice
POLAND

P. Verardi
Istituto di Acustica, CNR
Via Cassia 1216
00189 Roma
ITALIA

E. Verona
Istituto di Acustica, CNR
Via Cassia 1216
00189 Roma
ITALIA

J.W. Wolf
Georgetown University
Physics Dept.
Washington DC 20057
USA

R. Zilber
Soreq Nuclear Research Center
NDT Dept
Yavne 70600
ISRAEL